THE MESSIANIC REDUCTION

MERIDIAN

Crossing Aesthetics

Werner Hamacher

Editor

Stanford
University
Press

―――――――

Stanford
California

THE MESSIANIC REDUCTION

Walter Benjamin and the Shape of Time

Peter Fenves

Stanford University Press
Stanford, California

©2011 by the Board of Trustees of the Leland Stanford Junior University. All rights reserved.

No part of this book may be reproduced or transmitted in any form or by any means, electronic or mechanical, including photocopying and recording, or in any information storage or retrieval system without the prior written permission of Stanford University Press.

Printed in the United States of America on acid-free, archival-quality paper

Library of Congress Cataloging-in-Publication Data

Fenves, Peter D. (Peter David), 1960- author
  The messianic reduction : Walter Benjamin and the shape of time / Peter Fenves.
      pages cm.--(Meridian, crossing aesthetics)
  Includes bibliographical references and index.
  ISBN 978-0-8047-5787-4 (cloth : alk. paper)--ISBN 978-0-8047-5788-1 (pbk. : alk. paper)
  1. Benjamin, Walter, 1892-1940. 2. Phenomenology. 3. Time. I. Title. II. Series: Meridian (Stanford, Calif.)
  B3209.B584F46 2011
  193--dc22   2010043505

Typeset by Bruce Lundquist in 10.9/13 Adobe Garamond

For Zoli

Contents

Acknowledgments — xi

Note on Abbreviations and Translations — xiii

Introduction: The Course of the Argument — 1

1. Substance Poem Versus Function Poem: "Two Poems of Friedrich Hölderlin" — 18

2. Entering the Phenomenological School and Discovering the Color of Shame — 44

3. "Existence Toward Space": Two "Rainbows" from Around 1916 — 79

4. "The Problem of Historical Time": Conversing with Scholem, Criticizing Heidegger in 1916 — 103

5. Meaning in the Proper Sense of the Word: "On Language as Such and on Human Language" and Related Logico-Linguistic Studies — 125

6. Pure Knowledge and the Continuity of Experience: "On the Program of the Coming Philosophy" and Its Supplements — 152

7. The Political Counterpart to Pure Practical Reason: From Kant's *Doctrine of Right* to Benjamin's Category of Justice — 187

Conclusion: The Shape of Time — 227

Appendix: Translations

    The Rainbow: Dialogue on Fantasy   247

    The Rainbow; or, The Art of Paradise   254

    Notes on an Afternoon Conversation   256

    From a Notebook Walter Benjamin Lent to Me [Gershom Scholem]: "Notes Toward a Work on the Category of Justice"   257

Notes   259

Bibliography   291

Index   307

Acknowledgments

The origin of this study was a conversation with Jacques Derrida in October 1990 concerning the relation among his own early inquiries into Husserl's genetic phenomenology, the early writings of Walter Benjamin, and the early years of the phenomenological movement. The argument I develop here has profited immensely from many conversations—real and virtual—with Werner Hamacher, Giorgio Agamben, Carol Jacobs, Avital Ronnell, Samuel Weber, and Géza von Molnár. I have also had the privilege of discussing my work with many other outstanding scholars, including Stéphane Mosès, Thomas Schestag, Rainer Nägele, Marc Shell, Susan Meld Shell, Bonnie Honig, Timothy Bahti, Henry Sussman, Kevin McLaughlin, Susan Bernstein, Robert Gooding-Williams, Daniel Heller-Roazen, Burkhardt Lindner, David Ferris, Richard Block, Paul North, Paul Reitter, Michael Jennings, and Stanley Corngold. I am especially thankful to Julia Ng for sharing with me some results of her research at the Benjamin archive in Berlin and the Scholem archive in Jerusalem. Suhrkamp Verlag kindly granted me permission to translate the two texts entitled "The Rainbow" that Giorgio Agamben recovered. Werner Hamacher and Jörg Kreiernbrock reviewed my translations, which appear in the appendix to this volume. Two earlier attempts to capture the argument proposed here—both of which are hereby superseded—can be found in my contribution to *Theoretical Questions*, edited by David Ferris (Stanford, Calif.: Stanford University Press, 1996), and my own *Arresting Language* (Stanford, Calif.: Stanford University Press, 2001).

Above all, I am grateful for the loving support of Susannah Gottlieb and the indispensable insights of Inbo Gottlieb Fenves and Zoli Gottlieb Fenves, who have given me an impression of what it's like to see colors.

Abbreviations and Translations

Except where otherwise indicated, citations in the body of the text refer to the following volume:

Walter Benjamin, *Gesammelte Schriften*. Ed. Rolf Tiedemann and Hermann Schweppenhäuser. 7 vols. Frankfurt am Main: Suhrkamp, 1972–91.

The following abbreviations are used throughout the text:

C   Hermann Cohen, *Logik der reinen Erkenntnis*. 2nd ed. Berlin: Verlag Cassirer, 1914.

GB   Walter Benjamin, *Gesammelte Briefe*. Ed. Christoph Gödde and Henri Lonitz. 6 vols. Frankfurt am Main: Suhrkamp, 1995– .

He   Martin Heidegger, *Frühe Schriften*. Ed. Friedrich-Wilhelm von Hermann. Frankfurt am Main: Klostermann, 1978.

Hu   Edmund Husserl. *Husserliana: Gesammelte Werke*. Ed. Husserl archive in Leuven, under the directorship of H. L. van Breda. 40 vols. to date. The Hague: Nijhoff, 1950– .

K   Immanuel Kant, *Gesammelte Schriften*. Ed. Königlich Preußische [later, Deutsche] Akademie der Wissenschaften. 29 vols. to date. Berlin: Reimer; later, Walter de Gruyter, 1900– . In the case of the *Critique of Pure Reason*, references are to the first edition (= A) and the second edition (= B).

S   Gershom Scholem, *Tagebücher, nebst Aufsätzen und Entwürfen bis 1923*. Ed. Karlfried Gründer, Herbert Kopp-Oberstebrink, and Friedrich Niewöhner, in association Karl Grözinger. 2 vols. Frankfurt am Main: Jüdischer Verlag, 1995–2000.

WBA   Walter Benjamin Archive, Akademie der Künste, Berlin (according to the manuscript number).

All translations are my own, unless otherwise indicated. Many of Benjamin's writings under discussion in this volume are translated in the first volume of *Selected Writings*, ed. Michael Jennings (Cambridge, Mass.: Harvard University Press, 1996–2000). Most of the letters to which I refer can be found in *Correspondence of Walter Benjamin*, trans. Manfred Jacobsen and Evelyn Jacobsen (Chicago: University of Chicago Press, 1994). A partial translation of Gershom Scholem's youthful diaries has been published as *Lamentations of Youth*, ed. and trans. Anthony Skinner (Cambridge, Mass.: Harvard University Press, 2007). Translations of Kant's writings almost always include references to the Akademie edition of his works, and translations of Husserl's writings similarly rely on *Husserliana*. Many of the texts collected in Heidegger's *Frühe Schriften* have been translated in *Becoming Heidegger: On the Trail of His Early Occasional Writings*, ed. Theodore Kiesel and Thomas Sheehan (Evanston, Ill.: Northwestern University Press, 2007). Unfortunately, little of Hermann Cohen's systematic work has yet been translated into English.

THE MESSIANIC REDUCTION

Introduction

The Course of the Argument

In one of the curriculum vitae that Walter Benjamin wrote in 1928, shortly after his hopes of securing an academic position had collapsed, he represents his early philosophical inquiries in terms of four names—Plato, Kant, Husserl, and Marburg: "In particular and in ever-repeated reading, during my time as a student, I concerned myself with Plato and Kant, in connection with Husserl's philosophy and the Marburg school" (6: 218). In similar documents of the period he says much the same thing—minus the reference to "Husserl's philosophy." And the same subtraction is palpable in the reception of Benjamin's work from its very beginning. Gershom Scholem, who was the first reader of numerous texts under consideration in this volume, and who also co-edited the first collection of Benjamin's writings, considered his friend something of a phenomenological neophyte: "[Benjamin] gained an indistinct notion of [Husserl's] *Logische Untersuchungen* [Logical investigations] during his time in Munich."[1] The other co-editor of the first collection of his writings, Theodor Adorno, was similarly dismissive—not so much of Benjamin's training in phenomenology as of the phenomenological program as a whole. The collapse of Benjamin's academic ambitions followed upon the rejection of his *Habilitationsschrift* or second dissertation on the *Ursprung des deutschen Trauerspiels* (Origin of the German mourning play), the preface to which makes no mention of the founder of phenomenology but at a crucial point quotes a long passage from one of Husserl's students, namely Jean Héring, from whom Benjamin takes the idea of "essentiality" (*Wesenheit*). Adorno, after borrowing extensively from Benjamin's preface for the successful completion of his *Habilitationsschrift*, conceived of a plan to earn a Ph.D. at Oxford by writing a polemic against the phenomenological movement, which would be more or less guaranteed to find a friendly reception among Oxford professors of philosophy.[2] In response to a report of Adorno's plans for a British doctorate, Benjamin posed a conciliatory question, which suggests that Husserl should not be condemned for the aberrations of his students: "I am eager some day to know more about your annihilation of 'the intuition of essence.' Wouldn't Husserl reconcile himself to such an annihilation, after he could take into account what purpose this instrument could serve in the hands of a Heidegger?" (*GB*, 5: 110).

In the curriculum vitae where Benjamin briefly discusses his early interest in phenomenology, he summarizes the methodological program underlying the *Origin of the German Mourning Play* with reference to the very "instrument" Adorno would later set out to "annihilate." Because, as Benjamin explains, his "mode of investigation" acknowledges that every work of art is "incomparable and one-time [einmalig], it stands closer to an eidetic way of taking appearances into consideration than to an historical one" (6: 219)—a notable claim, since it implies that the historical mode of inquiry, however much its proponents may protest to the contrary, tends to deny the incomparability and timeliness of the works under study. The claim is notable for another reason as well: Benjamin emphasizes the proximity of his "mode of investigation" to that of Husserl, but he also refrains from specifying what ultimately separates his work from Husserl and his followers. The study undertaken here seeks to make up for this lacuna by determining the point where Benjamin's philosophical investigations, which culminate in the "Epistemo-Critical Preface" to the *Origin of the German Mourning Play*, part ways with "Husserl's philosophy."

This study is guided by the following thesis: in response to the debates that Husserl unleashes among his students with the introduction of the idea of the phenomenological reduction, Benjamin begins to work out his own version of the reduction, in which the so-called "natural" attitude gets "turned off" (in German, *ausgeschaltet*) and is thus brought to a "halt" (in Greek, *epochē*). The supposedly "natural" attitude—which Benjamin will associate with mythology—consists in the general premises that there is a world of substantial things that lie outside of our consciousness and that our experience is the result of the manner in which these things affect us. The initial aim of the reduction that Husserl first publicly proposes in *Ideen zu einer reinen Phänomenologie und phänomenologischen Philosophie* (Ideas pertaining to a pure phenomenology and to a phenomenological philosophy) lies in gaining a stance of pure receptivity. Once the phenomenologist has achieved this stance, phenomena give themselves as they are, without distortions that result from theoretical presuppositions, including the "natural" theory that experience derives from causal interaction between the mind and the world. What ultimately separates Benjamin's mode of thought from Husserl's, then, is this: from its title onward, *Ideas* proceeds as though the philosopher is fully capable of "turning off" the attitude that bars access to phenomena and can thus enter into the sphere of "pure phenomenology" on the strength of will; Benjamin, by contrast, makes no such concession to the profession of philosophy.

Yet Benjamin does not consequently look for an alternative subject—a nonphilosopher or mystic, for example—who could successfully "turn off" the attitude in question. A thoroughgoing reduction of the "mythological" attitude cannot be accomplished by anyone, including any communal "one," who would seek to do so. The "oneness" of whatever or whomever accomplishes the "turning off" of the attitude that sees itself as natural is of a higher "power"—in mathematical terms—than any unity of consciousness or community. Thinking in light of this accomplishment, which can be neither anticipated nor foreseen, thus acquires a paradoxically open-ended character. What Benjamin says about the structure of messianicity in the opening sentence of his so-called "Theologisch-Politisches Fragment" (Theological-political fragment), which was probably written in the early 1920s, goes for the structure of the reduction as well: just as the one who "turns off" the "natural" attitude is alone capable of establishing what has thus been accomplished, so, according to Benjamin, "the messiah alone . . . first redeems, completes, creates the relation [of every historical event] to the messianic" (2: 203). The reduction is messianic for this reason: only the unity of a higher "power" than that of consciousness or community can accomplish it. A particular phenomenon will be identified in the course of this study that nevertheless guarantees the existence of a fully "reduced" sphere in the absence of its accomplishment: the coloration of shame. And a name will emerge for this sphere: *time*. The term *time* in this case refers neither to the time of "inner-time consciousness" (Husserl) nor to time as the "possible horizon for any understanding of being" (Heidegger), but, rather, to a "plastic" time, which is shaped in such a way that its course is wholly without direction, hence without past, present, or future, as they are generally understood.[3] Time is thus released from what Benjamin identifies as the "highest category" of "world history," namely "guilt" (*Schuld*), which stamps every "world-historical moment" with its "unidirectional" character (6: 93)—toward ever-deeper guilt. The task around which Benjamin's work comes to revolve does not consist in accomplishing the reduction of the natural-mythological attitude through a heroic exertion of philosophical will but, rather, in discovering the tension between the nondirectionality of time and the unidirectionality of history. This tension itself has a direction, which can be discerned in certain works of art and stretches of time: "toward the messianic" (6: 124).

～

Husserl is only one of the four names that Benjamin mentions in the aforementioned curriculum vitae, and of the four, it is the name he mentions the least often. The absence of Benjamin's name from accounts of the early years of the phenomenological movement or, conversely, the absence of phenomenology in accounts of Benjamin's early philosophical writings is hardly an accident, and it cannot be ascribed simply to Scholem's low opinion of Benjamin's phenomenological acumen or Adorno's low estimation of phenomenology in general. Rarely in the writings under discussion in this study does the term *phenomenology* emerge. One reason for Benjamin's reticence, beyond the fact that his early writings tend to remain silent about the work of his contemporaries, can be discerned from the thesis sketched above: with very few exceptions, Benjamin declines to position himself as the "I" to whom phenomena appear and who is also in command of a language that could describe them as they are given. The other three names Benjamin mentions in connection with his philosophical studies—Plato, Kant, and Marburg—are associated with similar forms of reticence. Plato does not describe what he once saw on "the plane of truth," to cite a famous phrase from *The Phaedrus* (248b); rather, the task of describing what was once seen is generally left to a character named "Socrates," who must himself be forced to speak of such things, and even when he does, he does not speak directly of what he has seen. For Kant, phenomena are not only never purely given; they are always only a product of a synthesis, whose synthesizer, under the name of "the transcendental unity of apperception" (K, B 133) or "consciousness in general" (K, 4: 300), is an empty function, which lies at the basis of possible experience but is not itself a possible object of experience.

And as for the members of the Marburg school, especially Hermann Cohen, who first established its program of research, and Ernst Cassirer, who developed the program in his own distinct manner, there is simply no question that the "I" is in no position to intuit essences or receive phenomena, even if consciousness is qualified as "transcendental." Instead of seeking the given within the limits of its givenness, Cohen works out the process of object-generation in the construction of empirical science, and in place of the interpretation of concepts as substances, which are derived from a process of abstraction, Cassirer draws on contemporary mathematics and proposes that they be understood as functions, which are concrete universals, since they describe the law or principle through which all of their values can be determined. The Marburg school thus replaces Kant's critique of the "faculties" of reason and judgment with "epistemo-critique," which not only eliminates all talk of mental "faculties" but also does away with any reference to consciousness in general: objectivity is a matter of categorial coherence, not correspondence with things in the world. As their subtitles indicate, both Cohen's *Princip der Infinitesmal-Methode und seine Geschichte* (Principle of the infinitesimal method and its history) and Cassirer's *Substanzbegriff und Funktionsbegriff* (Concept of substance and concept of function) present themselves as "epistemo-critical prefaces."[4] And whereas the slogan Husserl announces in *Logical Investigations*, "to the things themselves" (Hu, 19: 10), captures much of what he is after, the one for which Cohen is doubtless best known, "the fact of science" (C, 57), says very little, unless it is recognized that the "fact" in question does not consist in established bodies of knowledge but, rather, serves as a methodological replacement for transcendental subjectivity. The first chapter of this study shows how Benjamin adopts certain methodological principles from the Marburg school and seeks to discover through an analysis of Hölderlin's late poetry a point of departure for philosophy, beginning with "pure aesthetics" (2: 105), that is even less amenable than the Marburg school to the concepts of substance and subject alike. The name of this counterpoint to the "fact of science" is *Lehre* (doctrine, teaching, theory), which "teaches" only the transient moment of its transmission.

∽

Just as the antagonism between various schools of neo-Kantianism and various versions of phenomenology frames many of the philosophical discussions conducted in German academic philosophy from the publication of *Logical Investigations* in 1900 through the tumultuous decade of the 1920s, so does the tension between the Marburg school and "Husserl's philosophy" traverse many of the essays, dialogues, sketches, and fragments Benjamin wrote since he matriculated at the University of Freiberg in 1913 until the completion of the *Origin of the German Mourning Play*, whose preface is itself an exponent of this tension: the principal term of its title, *epistemo-critique*, is drawn from the Marburg school, but it avoids any engagement with neo-Kantianism and makes no mention of Kant at all. An inverse tension can be discerned in the relation between Benjamin's writings during his period as a student and the seminars he attended. A significant number of these writings contribute to a broadly conceived "critical altercation [Auseinandersetzung] with Kant and Cohen" (*GB*, 1: 441), and yet nowhere was he similarly engaged with any of the neo-Kantian professors whom he encountered. Cohen retired from teaching in 1912, and although Benjamin briefly attended the lectures of Cassirer in Berlin, he showed little interest in them.[5] As a student in Freiburg, he came into contact with Heinrich Rickert, whose version of neo-Kantianism differed from the Marburg school, especially in its emphasis on the idea of value, but was similarly concerned with the methods by which the object of knowledge is constructed.[6] In the winter semester of 1913–14 Benjamin attended Rickert's lectures on the "logic as the foundation of theoretical philosophy" as well as his seminar on Henri Bergson's theory of time. As it happens, Martin Heidegger was also present in both courses. In a letter to Rickert from the following summer, Heidegger offers his teacher "warmest thank for the strongly philosophical stimulation and instruction that I was able to take away from your lecture course and, above all, from your seminar."[7] Benjamin was less impressed with the seminar—or at least less obsequious with regard to its instructor: "I sit there," he writes to a friend, "and nibble on a sausage" (*GB*, 1: 112).[8]

After Benjamin moved to the University of Munich for the winter semester of 1915–16, he encountered for the first time—and perhaps for the last time as well—a professor of philosophy whose seminar sustained his interest. In a long letter to Fritz Radt in December 1915, which begins with a lively description of his usual "disappointment" (*GB*, 1: 296) with the quality of his classes, especially those of the art historian Heinrich Wölfflin, Benjamin registers a different kind of dissatisfaction in the case of Moritz Geiger, complaining that his "seminar has too few hours"

(*GB*, 1: 301).⁹ In a curriculum vitae written in the year of his death—and three years after Geiger's—he refers to this experience: "The classes of the Munich philosopher Moritz Geiger left me with a lasting impression" (6: 225). A descendent of an illustrious German-Jewish family—his grandfather was Abraham Geiger, founder of modern liberal Judaism, and one of his more distant relatives, Ludwig Geiger, was a remarkable philologist who produced innovative studies of linguistic and perceptual history, while his uncle became a major Goethe scholar—Geiger contributed to a wide range of philosophical topics, ranging from the theory of quantity in psychology (where he argued in favor of intensive magnitudes), through the theory of the unconscious, to the philosophy of mathematics (especially the axiomatic foundations of Euclidean geometry).¹⁰ The text under discussion in Geiger's seminar was Kant's *Kritik der Urteilskraft* (Critique of judgment), which is itself concerned with a broad range of topics, beginning with its analysis of the feeling of the beautiful and concluding with a reflection on the final purpose of creation as a whole. Solely in terms of breadth, the primary text for the seminar that left a lasting impression on Benjamin was particularly well suited to the inclinations of its instructor.

Despite the European war—about which Benjamin remained almost entirely silent, as if even words of disgust were somehow implicated in the celebrations of the war that he unequivocally deplored—the winter of 1915–16 was an auspicious time, and the University of Munich a propitious place, to discuss the foundations of aesthetics: "If anything, the Munich circle [of phenomenology, founded by Alexander Pfänder] was even more gregarious than the Göttingen group, meeting frequently for regular discussions and informal study groups."¹¹ In 1914, under the title *Ästhetik des reinen Gefühls* (Aesthetics of pure feeling), Hermann Cohen had published a major revision of Kant's critique of taste. And in the previous year Geiger had singled himself out among Husserl's students with the publication of *Beiträge zur Phänomenologie des ästhetischen Genusses* (Contributions to the phenomenology of aesthetic enjoyment) in the first number of the *Jahrbuch für Philosophie und phänomenologische Forschung* (Yearbook for philosophy and phenomenological research), which he co-edited with Husserl, Pfänder, Max Scheler, and Adolph Reinach, all of whom contributed similarly pathbreaking studies to its first volume, beginning with Husserl's *Ideas*: Scheler published a treatise on "material ethics," Pfänder an extensive paper on the psychology of dispositions, and Reinach a remarkable reflection on the a priori foundation of civil law in the speech act of promising.[12] The degree to which Benjamin schooled himself in "Munich phenomenology" can be discerned from one of the major theses he develops in his *Habilitationsschrift*, which aptly summarizes the research program pursued by Pfänder and Geiger: "Every feeling is bound up with an a priori object, the exposition of which is its phenomenology" (1: 318). And at least three of the four names Benjamin would associate with his student years—Kant, Husserl, and Marburg—converge in Geiger's seminar.

Of course, nothing can be determined with certainty about what came under discussion during the seminar; but in his *Contributions to the Phenomenology of Aesthetic Enjoyment* Geiger briefly outlines a critique of Kant's aesthetics that accords with the program that guides *Logical Investigations*: Kant, according to Geiger, is far more concerned with completing the system of transcendental idealism than with the primary phenomenon under investigation, namely the delight in beautiful appearances.[13] In a review from 1928 Benjamin succinctly expresses a generalized version of this critique: "Husserl replaces the idealistic system with discontinuous phenomenology" (4: 536). The same line of criticism would presumably apply to Cohen's *Aesthetics of Pure Feeling*, which serves as the third and final part of his "system of philosophy." Nevertheless, there is reason to suppose that around 1916 Geiger would not have simply reiterated his earlier critique of Kantian aesthetics—or at least would have done so with some hesitation. With the introduction of the idea of a phenomenological reduction, Husserl so altered the character of his philosophical program, particularly in view of its starting point, that his students and disciples had no choice but to reevaluate what they took to be its direction, even if they ultimately decided that they would continue along the lines sketched out in *Logical Investigations*. And something akin to the phenomenological reduction can be discerned in the opening paragraphs of the *Critique of Judgment*. As Geiger indicates, the "disinterestedness" that Kant attributes to anyone who undertakes a specifically aesthetic judgment requires a major modification of the "ordinary attitude."[14] It is not, however, the philosopher *cum* phenomenologist who prompts this modification; rather, the "natural" or "ordinary" attitude is "turned off" by nature itself—or more exactly, by "free beauties" of nature (K, 5: 229), which are "there" only as correlates of aesthetic delight.

However the discussion may have gone in Geiger's seminar, it drew Benjamin into the orbit of phenomenology, which he had previously encountered only in the programmatic form of Husserl's "Philosophie als strenge Wissenschaft" (Philosophy as rigorous science): "Geiger's seminar has too few hours. He is assaulted by [military] service, and the problems are too difficult to yield much during a seminar. I am going through a phenomenological work of his [*Contributions to the Phenomenology of Asthetic Enjoyment*]. I am also reading Husserl's difficult, principal groundwork [either *Logical Investigations* or *Ideas*], so as to gain entrance into his school" (*GB*, 1: 301–2).[15] It is in this context that, as Chapter 2 of this study shows, Benjamin goes so far as to experiment with the "instrument" of "intuition of essences" in response to a paper published in *Kant-Studien* that sought to demonstrate that *Ideas* carries out the Copernican revolution in philosophy that Kant left incomplete.[16] And it is in the same context, as the volume as a whole proposes, that Benjamin begins to develop his own version of the phenomenological reduction under the paradoxical premise that it cannot be accomplished by anyone who would seek to do so. It is not as though, for Benjamin, it is simply impossible for things to give themselves as they are; they can appear—but not to Benjamin "himself." Instead, they appear to others—for example, to children. Or they appear to a certain Margarethe, whose description of a dream sets into motion the "dialogue of fantasy" Benjamin probably wrote in late 1915 or early 1916, under the title of "The Rainbow." And phenomena also appear to artists with an ambiguity that derives from their artistic intention. Chapter 2 of this study discusses Benjamin's "entrance" into Husserl's "school," while Chapter 3 concentrates on the two texts he wrote under the title of "The Rainbow," the first of which accords with the *Critique of Judgment* in its exposition of natural beauty—starting with the rainbow—as the "reduction" or "leading back" of phenomena to their origin in the innocent sphere of fantasy. By contrast, the second "Rainbow," which makes no mention of any natural rainbows, departing from the first, proposes in a highly concentrated manner that painting be seen as the "art of paradise." Because a painting cannot purely and simply disclose its painterly character—by inscribing, for instance, "this is a painting" into the painting without making the inscription part of the painting—Benjamin poses for himself the following question: how does a painting acquire a name that is precisely *its* name and not an arbitrary designation? The distinction between name and designation stands, in turn, at the basis of his contemporaneous inquiries into the foundations of logic and the theory of language.

∼

The first of the "logical investigations" that Husserl undertakes in the eponymous treatise begins with a discussion of a distinction that subtends not only the following five investigations but the unfolding of *Ideas* as well, namely the distinction between expression and indication, the rudiments of which can be briefly sketched as follows: whereas an expression embodies "meaning" (*Bedeutung*), an indication does not. In a now famous paragraph of the first investigation Husserl identifies the "solitary life of the soul" (Hu, 19, 1: 35) as the sphere in which discourse firmly detaches itself from all indicative entanglements. As long as I speak only with myself, my speech is purely expressive: what is meant is given in the very act of meaning to say something. As soon as I seek to communicate what I mean, however, my speech falls into the sphere of indication, beginning with the word *I*. In a series of logico-linguistic studies from around 1916, many of which are concerned with Russell's set-theoretical paradox, Benjamin adopts and transforms the distinction with which *Logical Investigations* begins. In place of the opposition between expression and indication, Benjamin introduces a distinction between "judgments of predication" and "judgments of designation" (6: 9): only in the case of the former can one speak of meaning, properly speaking, whereas in cases of designation, there is merely "inauthentic meaning [uneigentliche Bedeutung]" (6: 10). Whenever a term is said to meaning something, its meaning is categorically different from the meaning of a term that means something in the absence of any stipulation as to what it is supposed to mean. The "something" that a properly meaningful term means is, however, first and foremost meaning pure and simple. For this reason, the term in question cannot properly be called a "term" but is in a certain sense indeterminate, which is to say, infinite in its own peculiar way. In still other words, it is a name, the meaning of which derives from the thing named instead of from the speaking subject who would presume to give names to things.

The point of retreating into inner monologue, for Husserl, lies in reducing the scope of indication to zero and thereby allowing the phenomenon under investigation—especially pure logical meaning—to appear as such. It anticipates in this way the theme of the phenomenological reduction, as Jacques Derrida has persuasively argued.[17] In order to achieve a corresponding reduction of designation, for Benjamin, a very different movement is required: not a reversion to the ipseity of the self, which cuts itself off from all communication, but rather a restitution of limitless communicability. The meaning of things derives from the things meant in their names rather than from an act of meaning-bestowal by which the speaking subject relates a word to a thing. As Chapter 5 of this study argues, Benjamin thus replaces Husserl's theme of monologue with the idea of panlogue. And instead of seeing communication as a fall into the sphere of indication, he proposes a reading of the Book of Genesis in which the expulsion from paradise results from the "excitement" (2: 153) of the judging word, which designates things and condemns the speaking subject in the same stroke. In the course of working out his exegesis of Genesis, however, Benjamin gets caught up in a particularly difficult problem: is the original language of human beings constitutively singular or potentially plural? In order to solve this question, he turns away from the schemata developed in response to *Logical Investigations* and adopts a framework of thought that derives from the *Critique of Pure Reason*: for Kant, divine intuition—to the extent that one can speak of such a thing—is altogether spontaneous, whereas human intuition is only receptive; similarly, for Benjamin, the creative word is purely spontaneous, whereas the original language of human beings is spontaneous only to the extent that it is primarily receptive. Because of its residue of spontaneity, however, Adamic language crosses into the sphere of designation, which crystallizes in the form of the proper name. In his preface to his translation of Baudelaire's *Tableaux parisiens* (Parisian scenes), which he published in 1923 under the title "Die Aufgabe des Übersetzers" (The task of the translator), Benjamin does not so much abandon the idea of a lost panlogue as replace it with the concept of "pure language" (4: 13), which never manifests itself as such but which is nevertheless meant by every language as a whole, in complementary relation to every other. Just as the tension between phenomenology and Kantian critique traverses his studies of color and fantasy, so does it propel his contemporaneous reflections on logic and language.

∼

In addition to drawing Benjamin into the orbit of phenomenology, Geiger's seminar on the third *Critique* solidified his sense that there is something singular about Kant's work, in relation to which philosophy acquires a certain continuity. One of the presentations he prepared for the seminar apparently dealt with the second section of the third *Critique*, which circumscribes the applicability of teleological judgment to both organic beings and the phenomenon of nature as a whole. Although Kant makes little mention of nonnatural history in the context of the *Critique*, all of his occasional essays on human history are intimately bound up with the status of teleological judgment. Before having read the relevant writings, Benjamin conceived of a plan to write his dissertation on "Kant and history" (*GB*, 1: 390). Disappointed upon reading a few of the relevant essays, he proposed a dissertation on the Kantian and neo-Kantian idea of the "infinite task." Although this, too, never came to fruition, he did not thereby relinquish his conviction that Kant's work, especially its terminology, is comparable to no other, with the possible exception of Plato's. Nowhere does Benjamin express this conviction more forcefully than in the following passage in a letter to Gershom Scholem from July 1917: "As far as the question goes, which has been demanded for so long: how can I *live* with the position I've taken with regard to the Kantian system?—I am constantly at work on making this life possible through insight into the theory of knowledge and must have diligence and patience for the enormous task that, with all due respect, this life means for people of our attitude" (*GB*, 1: 402).

Benjamin's most extensive effort to gain "insight into the theory of knowledge" in this period can be found in an essay he wrote in late 1917 and early 1918, "Über das Programm der kommenden Philosophie" (On the program of the coming philosophy), the central thesis of which can be formulated as follows: the purification of the theory of knowledge provides the basis for a concept of "higher experience." Epistemology gains purity to the extent that it breaks free from all "epistemo-mythology" (2: 161), which ultimately consists in the conviction that experience results from causal interaction between subjects and objects. In this way, "epistemo-mythology" is equivalent to what Husserl calls the "natural" attitude, and Benjamin specifically mentions phenomenology in conjunction with the attempt to work out the structure of "pure epistemo-theoretical (transcendental) consciousness" (2: 162), without the presupposition that consciousness is another name for the subject-term of the subject-object relation. As for the Marburg school, it does not so much lack a sufficiently pure concept of knowledge—Cohen's major epistemo-critical study, after all, is called *Logik der reinen Erkenntnis* (Logic of pure knowledge)—as fail to understand the essentially continuous character of experience. Under the guidance of the "fact of science," which requires that philosophy abstain from giving direction to ongoing research, experience breaks up into the various fields of empirical sciences, which doubtless are all subject to the same "infinite task" but which misunderstand the infinitude of their task by representing it as an endless approximation. In the attempt to grasp the "infinite task" of science as something other than an asymptotic approach to the ideal of perfect knowledge, Benjamin borrows the term *power* (*Mächtigkeit*) from transfinite set theory: just as the size of the linear continuum is of a higher power than that of the set of integers, so the unity of science as a whole is of a higher power than any given unit of scientific inquiry. In later texts, Benjamin replaces "science as a whole" with "philosophy as a system of problem" and then again with "truth" *simpliciter*. Chapter 6 of this study analyzes the drift of these inquiries into the concepts of knowledge and experience. Because both Kantian and neo-Kantian versions of "epistemo-critique" understand the continuity of experience only as a regulative idea, not as the critical element of its constitution, Benjamin develops a Platonic counterpart to the "Kantian typic" (2: 160) in the "Epistemo-Critical Preface," where a "primordial listening to essentialities" replaces the "vision of essences" that finds expression in the phenomenological "instrument" of *Wesensschau*. It almost goes without saying that anyone whose intention lies in listening to essentialities cannot perceive them for this reason.

∼

Whereas Chapter 6 takes up Benjamin's transformation of Kantian theoretical philosophy, Chapter 7 is concerned with his corresponding transformation of Kant's practical philosophy. The primary text under consideration is a set of notes that Scholem transcribed into his diary in the fall of 1916. The opening claim of these "Notizen zu einer Arbeit über die Kategorie der Gerechtigkeit" (Notes toward a work on the category of justice) asserts that a "possession-character" (S, 1: 401) accrues to goods by virtue of their transience; but it is nevertheless unjust for anyone, including society at large, to make them into actual possessions. Benjamin thus responds to the opening paragraphs of Kant's *Doctrine of Right*, which seek to expand the concept of right beyond the sphere of its immediate applicability—the human body, roughly speaking, along with whatever anyone happens to be holding at any given moment—on the basis of the postulate that every nonpossessed thing must be available for my use under the condition that I be able to bring it "under my control" (K, 6: 246). The phrase *under my control* is a translation of "in meiner Gewalt." *Gewalt*, for Kant, is distinguished from *Macht* (power) insofar as it consists in an "act of the elective will" and so can be described as minimally rational. By virtue of its rationality, *Gewalt*—which Kant himself translates both as *potestas* (authoritative power) and as *violentia* (violence)—thus establishes the starting point of law. At the outset of the *Doctrine of Right*, then, is the following thesis: *Gewalt* in all its troubling ambiguity prepares the ground for right, which should eventually extend to every part of the globe under the sign of "eternal peace."

Declining to affirm the initial steps of the *Doctrine of Right*, Benjamin seeks to identify the category of justice, which requires, in turn, a supplementary critique: in this case, a critique of *Gewalt*. For Kant, law (*Recht*) is the political correlate to pure practical reason, for it lays out a stable structure of relations among rational beings within which they are able to use both themselves and things without degrading themselves or others in the process. For Benjamin, by contrast, the political—or perhaps more exactly, the real or effective—counterpart to pure practical reason consists in pure *Gewalt*, which destroys the structure of right and thus does away with the legally sanctioned illusion that any use of persons or things can be disentangled from the complex of guilt. In the event of pure *Gewalt*—which can never been recognized as such—the mechanism that stores up, monopolizes, and delivers the current of power as though it were a natural resource is suddenly "turned off" (*ausgeschaltet*). This is what Benjamin means by the term "presiding power [waltende Gewalt]" (2: 203), which closes his contribution to the critique of *Gewalt*.

∼

The central chapter of this volume revolves around a "very difficult remark" that Benjamin proposed at the start of an afternoon-long conversation with Scholem in August 1916: "The years are countable but, in contrast to most countables, not numerable" (S, 1: 390; 2: 601). As a student of mathematics, Scholem was in a good position to help Benjamin develop the "problem of historical time"—a phrase Benjamin borrows from Georg Simmel—in conjunction with this remark. In reflecting on the results of their conversations in August 1916, Benjamin began a long letter to Scholem on the theme of language and mathematics, which he ultimately abandoned because of its immense difficulty, instead writing a "little treatise" (*GB*, 1: 343) under the systematic title "Über Sprache überhaupt und über die Sprache des Menschen" (On language as such and on human language). In the letter where he announces his near completion of the "little treatise" he returns to the "problem of historical time" and briefly outlines a critique of a recent essay by a fellow participant in Rickert's seminar on Bergson—namely Heidegger, whose "Zeitbegriff in der Geschichtswissenschaft" (Concept of time in historical scholarship), according to Benjamin, "documents in an exact manner how *not* to go about this matter" (*GB*, 1: 344). Reversing the direction of Heidegger's methodology yields the following result: as long as years are countable, they cannot be numbered; once they are numbered, there are none left to count. The number of every historical year is always in a certain sense "one," regardless of the calendrical system that a regime has stipulated and made so conventional that it comes to appear natural.

In the course of the aforementioned conversation Scholem poses a question that runs congruent with, yet remains distinct from, the problem of historical time: what is the shape of time? In "Zwei Gedichte von Friedrich Hölderlin" (Two poems of Friedrich Hölderlin), Benjamin had earlier developed the enigmatic concept of "temporal plastics [zeitliche Plastik]" (2: 119) in order to address this very same problem. Associating the "plastic" shaping of time with Bergson's idea of duration, Benjamin draws on a striking phrase from the second of the two poems he analyzes: "turn of time [Wende der Zeit]." Far from being a particular point of time—a revolutionary moment, for instance—"turn of time," for Benjamin, and perhaps for Hölderlin as well, describes the structure of time in general, which is somehow always "turned." During their conversation Scholem in August 1916, Scholem and Benjamin dismiss the "metaphysical" notion that the course of time must be represented as an irreversible motion along a straight line; instead they associate time with a variety of more complicated curves, the last of which curiously accords with Benjamin's interpretation of the phrase from Hölderlin: a curve, namely, that is continuous yet nondifferentiable. No tangent lines can be drawn in relation to a curve of this kind; hence it has no direction: every point takes a sharp turn and cannot therefore be "touched" (4: 19) in accordance with a law given by the curve itself. Not only is the shape of time unimaginable; it is inviolate—and thus innocent.

Shortly before his death Scholem claimed to have in his possession a detailed document of the conversation in August 1916; but it has apparently been lost. The conclusion to this study proposes that Benjamin takes up the suggestion that fleetingly emerged in his conversation with Scholem in August 1916. The course of time is captured by a curve that is everywhere continuous yet nowhere differentiable: it is so sharply "turned" at every point that it proceeds without direction, neither progress nor regress, and every one of its stretches is not only like every other but also like the course of time as a whole. For the same reason, every time recapitulates—without ever exactly repeating—the whole of time. In this way, Benjamin responds to the Nietzschean idea of the eternal return of the same. Time, so construed, runs counter to history, and the tension between time and history, which repeats itself in the tension between history and myth, points toward a messianic resolution, in which history suddenly forms in conformity with the nondirectional, hence "senseless," course of time.

∽

The ordering of the chapters in this volume is not so much chronological as systematic. The first two chapters describe Benjamin's "critical altercation" with the two major programs of philosophical research undertaken at the universities where he studied: first, neo-Kantianism, then phenomenology. The middle three chapters are concerned with the structure of space, time, and meaning respectively. And the final two chapters describe Benjamin's transformation of Kantian theoretical philosophy, on the one hand, and Kantian practical philosophy, on the other. The order of chapters does not conform to the contours of a philosophical system in order to project the system of philosophy that Benjamin somehow failed to develop; rather, the systematic structure of the exposition is meant to serve as a grid through which the texts under consideration can assert their independence from the protocols and procedures of traditional philosophical discourse. The argument thus unfolds in conjunction with readings of specific items in the dossier of Benjamin's early writings. Because each chapter is a relatively self-contained analysis, certain themes are necessarily repeated in summary form. And in certain places, particularly the central chapter, the analysis of the text requires, to a certain extent, its reconstruction.

There is no presumption here of "full coverage." Many of the important documents for an understanding of Benjamin's "intellectual development" are missing, such as, for instance, his relation to Gustav Wyneken in particular and to the German youth movement in general. With the exception of Benjamin's dense essay on Hölderlin, little is said in this volume of his writings on German literature, or of his own private contribution to the German literary tradition in the form of a sonnet sequence written in memory of the poet Friedrich Heinle, which begins with an epigraph from Hölderlin's "Patmos." With regard to Benjamin's dissertation on Friedrich Schlegel and Novalis, with its afterword on Goethe, there are only brief discussions; and even less is said about Benjamin's major essay on Goethe's novel *Wahlverwandtschaften* (Elective affinities). The greatest lacuna, however, is the almost total absence of any reference to the texts Benjamin wrote after the rejection of his *Habilitationsschrift*. This absence is not generated by the expectation that there will be a second volume. What Fritz Heinle writes to one of Husserl's sons shortly before he committed suicide and they were sent into battle is probably true for Benjamin as well: "I read a lot, much philosophy, but it does not go very far."[18]

§1 Substance Poem Versus Function Poem

"Two Poems of Friedrich Hölderlin"

A Technical Term

Benjamin generally shies away from introducing technical terms. Under certain circumstances he uses unfamiliar terms, many of which are derived from the addition of the suffix *-barkeit* ("ability") to a common noun. Among terms of this type, which can be found throughout his writings, "translatability," "criticizability," and "knowability" are particularly prominent.[1] But none of these terms is technical in a strict sense: they do not owe their origin to authorial fiat. A typical formula for the creation of a technical term runs as follows: "Terminologically, I will call such-and-such *x*." The advantage of a sentence of this kind lies in its capacity to disambiguate the term in view of future use. The disadvantage lies in the prominence of the "I" who presumes that it has a right to dispose over language. In his *Berliner Chronik* (Berlin chronicle) Benjamin formulates a stylistic imperative his work has hitherto obeyed: "Never use the word 'I' except in letters" (6: 475). The imperative corresponds to the requirement that technical terms be avoided whenever possible. In the "Epistemo-Critical Preface" to his *Origin of the German Mourning Play* Benjamin explains the rationale for the imperative that gives shape to his work: "The introduction of new terminologies, as long as they are not intended solely for the domain of concepts, is therefore worrisome within the philosophical domain. Such terminologies—a misfortunate naming, in which intention [*Meinen*] takes a greater share than language—betrays the objectivity that history has given the principal coinages of philosophical reflection" (1: 217).

On perhaps no other point is Benjamin more consistent. In all his writing, from beginning to end, he generally declines to introduce technical terms, for wherever they occur they are evidence of a certain disorder, in which the intention of the subject takes precedence over the historically canonized objectivity of philosophical terminology. In some of his later works, Benjamin draws attention to certain terms that are unknown to Greek, Latin, French, and German philosophical traditions; but these terms, including *aura* and *flânerie*, are neither newly devised nor defined solely for the sake of the subsequent discussion. They retain their aura, so to speak, and—to the consternation of many commentators—meander in their meaning. The exercise in category construction that characterizes the kind of philosophical project Heidegger undertakes in *Sein und Zeit* (Being and time) is altogether foreign to Benjamin's work, both early and late. The sense that his work is unsystematic and should thus be classified as "unphilosophical" stems in no small part from his refusal to construct a table of categories, even if only in the mode of its negation. For this reason, however, it comes as something of a surprise that an essay from late 1914 and early 1915, entitled "Two Poems of Friedrich Hölderlin"—which he would later characterize as his "first major work" (*GB*, 3: 157)—introduces a technical term in its opening paragraph, *das Gedichtete*, which will henceforth be translated as "the poetized."

In *Einbahnstraße* (One-way street), Benjamin proposes a series of recommendations for the production of "thick books." One of these tips corresponds to the disavowal of technical terms that finds expression in the previously quoted passage from "Epistemo-Critical Preface": "Terms for concepts are to be introduced that never appear in the entire book except in those places where they are defined" (4: 104). If "Two Poems of Friedrich Hölderlin" can indeed be called Benjamin's "first major work," then the proposal for producing "thick books" applies to his literary corpus as a whole. A newly devised term, *the poetized*, is laboriously defined, only to be abandoned in everything that follows. Stranger still is the fact that, around the time of Benjamin's death at the beginning of World War II, the term he invented at the start of World War I returns in its original context—as a way to capture the singular character of Hölderlin's late poetry. Benjamin is *not* responsible for this term when it returns, however; in his *last* major work, he explicitly sets himself against the one who takes over the idea of "the poetized," namely Martin Heidegger.[2]

In retrospect, then, it is apparent that *the poetized* is like very few others terms in Benjamin's carefully crafted lexicon. And yet, the point of introducing the term is entirely in line with the stylistic imperative it violates: to the extent that commentary is concerned solely with the poetized, it disregards the empirical circumstances of an artistic production. By introducing *the poetized*, Benjamin seeks to solve the problem Nietzsche posed in the fifth section of *Die Geburt der Tragödie* (The birth of tragedy), which radicalizes and transforms Schopenhauer's aesthetics by demonstrating that there is nothing subjective about *any* work of art, including lyric productivity: "How is the 'lyric poet' possible as an artist," Nietzsche asks, appropriating a Kantian formulation, and immediately adds: "[the lyric poet] who, according to the experience of all times, always says 'I' and sings before us the entire chromatic scale of his passions and desires."[3] In order to establish the context in which a solution can be found, Nietzsche draws on the authority of Schiller, who identifies "musical mood" as the poetic equivalent to the Kantian a priori. When Benjamin concludes the penultimate paragraph of "Two Poems" with a quotation from Schiller (2: 125), he may be drawing an implicit distinction between his solution to the problem and that of Nietzsche, which makes every lyric poem into the objectification of a certain mood. With the introduction of term *poetized* into aesthetics, Benjamin in any case seeks to solve the problem Nietzsche posed in his first major work. And in so doing, he likewise advances a cause espoused by a wide variety of philosophical movements of the period: the critique of psychologism.[4]

Just as the idea of "pure knowledge," for Hermann Cohen, and "pure logic," for Edmund Husserl, have nothing in common with the empirical inquiries into the minds of scientists or logicians, so, for Benjamin, does "pure aesthetics" (2: 105) remain unconcerned with the moods of artists and audiences alike. In the case of lyric poetry, the aesthetic counterpart to pure knowledge and pure logic identifies the artistic task that any given poem is meant to solve, regardless of what its author wished to say or how its audiences may have felt. Once the spheres of logic and mathematics are liberated from psychologism, the same can be done with aesthetics, beginning with the form of fine art that is most susceptible to psychological speculation by virtue of, in Nietzsche's words, the "little word 'I'" for which it provides an expressive vehicle. Unlike the advocates of the New Criticism, who will also dispense with psychological and biographical considerations, the program Benjamin sketches in "Two Poems" is less concerned with grasping the verbal artifact by means of close reading than with approaching the methodological ideal of the "pure poetized" (2: 108), which, according to Benjamin, is either itself a poem or identical to life itself.

Antipsychologism

The sheer density of "Two Poems," which rigorously follows the stylistic imperative formulated in *Berlin Chronicle* and makes no reference to an "I," invites its readers to speculate about the psychological or biographical conditions of its author, as if it were a second-order lyric poem that opens up its secrets as soon as the interpreter gains insight into what the author was experiencing when it was written. And if one wanted to discern Benjamin's state of mind at the time, one need look no further than the paragraph of *Berliner Chronicle* that begins with a reflection on his stylistic imperative. After noting that, for a long time, he refrained from using "I" outside of letters, Benjamin enigmatically discloses his relation to a poet, namely Fritz Heinle, who killed himself in the early days of World War I (6: 475–80). The fact that Friedrich Heinle and Friedrich Hölderlin share the same initials lends credence to the following line of interpretation: "Two Poems of F . . . H . . ." is a prose elegy in memory of a poet whom Benjamin encountered "not in 'life' but, rather, only in his poetry" (6: 477).[5] An interpretation of this kind betrays the methodological stipulations of Benjamin's essay and at the same time ironically fulfills them: the betrayal consists in disregarding Benjamin's effort to develop a mode of "aesthetic commentary" (2: 105) that makes no mention of the psychological state or empirical condition of the author; the fulfillment consists in respecting the premise on which this effort rests—that commentary not be based on speculation into what the author wants to say. But this fulfillment by way of betrayal moves in precisely the opposite direction from the one Benjamin sketches in "Two Poems." Instead of aiming for the truth of the work, it concerns itself with the psychological state of its author. And this concern can be amplified by means of further psychological speculation: the reason commentators on Benjamin's essay want to disregard its guiding intention lies in their own discomfort with its density, which can be interpreted as sign of psychological distress.

Instead of further discussing the psychological state of commentators who seek to understand commentators on Benjamin's commentary as an expression of his psychological state—and thereby risking a perpetuation of this potentially endless series of remarks by inducing other commentators to discuss the motivations for my own attempt to do away with this line of interpretation—I want to interrupt this process and pose a question: what is truth? Or, more in keeping with the problem at hand, in what way can a lyric poem be true? The second question can be formulated in another manner as well: how can a lyric poem transcend the poetic subjectivity it apparently expresses and thus achieve its own specific mode of objectivity? The methodological remarks with which Benjamin prefaces his analysis of Hölderlin's poetry are directed at questions of this kind:

> An aesthetic commentary on two lyrical poems is going to be attempted, and this intention demands a few preliminary remarks on method. The inner form, which Goethe designates as content [Gehalt], should appear among these poems. The poetic task, as a presupposition of an evaluation of the poem, is to be identified. . . . Nothing about the process of poetic creation will be discussed, nothing about the person or *Weltanschauung* of the creator; on the contrary, only the particular and unique sphere in which the task and presupposition of the poem reside will emerge. This sphere is at once the product and object of the investigation. It can no longer be compared with the poem but is, rather, the only thing that is ascertainable [das einzig Feststellbare] in the investigation. This sphere, which for every poem has a particular shape [Gestalt], will be designated as the poetized [das Gedichtete]. In this sphere a peculiar region should be disclosed—the region that contains the truth of poetry. This "truth," which is precisely what the most serious artists so urgently claim for their creations, is to be understood as the objectivity [Gegenständlichkeit] of their creative activity, as the fulfillment of any given artistic task. "Every artwork has an ideal a priori, an inner necessity to exist" (Novalis). (2: 105–6)

The quotations from Goethe and Novalis with which Benjamin punctuates the opening paragraph of "Two Poems" demonstrate that the goal of his study is not altogether unprecedented. Certain poets have demanded of poetry that it be understood in terms of an "inner" structure that precedes rather than expresses their psychological states. But neither "content" (Goethe) nor "inner necessity" (Novalis) suggests a particular method through which the structure in question can be identified. Benjamin seeks to develop such a method without relying on the sterile distinctions between inner and outer or content and form. As a first approximation, he defines "the poetized" as both the "product and object" of the very investigation he is about to conduct—which means at the very least that the existence of the poetized is predicated on there being an investigation of its existence. This is not to say that the poetized is a fantasy of the investigator's making; but Benjamin is obviously aware of this danger, so much so that his further exposition of the poetized is meant to thwart it. Insofar as the object of the investigation is finite yet unbounded, it lends itself to the image of a sphere. The sphere under construction is not, however, homogeneous; the point of generating a sphere lies in describing the region that, while lying on the sphere, transcends it as well. This region, which the investigation does not so much generate as disclose, is the "truth" of poetry, where "truth" is not understood as the correspondence between thinking and being or between subject and object but, rather, in terms of "fulfillment" (*Erfüllung*), on the one hand, and "objectivity" (*Gegenständlichkeit*), on the other. In 1901 Husserl had launched the phenomenological movement by explicating the phenomenon of truth as the intuitive "fulfillment" of a certain mode of signifying intention (Hu, 19, 2: 590), and in the following passage from the manifesto he published under the title "Philosophy as Rigorous Science," which Benjamin read in the summer of 1913 (*GB*, 1: 144), Husserl presents the task of phenomenology as the disclosure of those "objectivities" whose "being" is immanent to the structure of consciousness and therefore owes nothing to transcendent objects:

> If the theory of knowledge wants to investigate the problems of the relation between consciousness and being, it can have being only as a correlative of consciousness before its eyes—as an "intended" [Gemeintes] on the order of consciousness: as a perceived, remembered, expected, represented through an image, supposed, valued, and so forth. . . . For the essence-analysis of consciousness, the clarification of all fundamental kinds of objectivities [Gegenständlichkeiten] is unavoidable and is indeed contained in this analysis; still, however, only in an epistemological analysis that sees its task in the investigation of correlations. For this reason, we conceive of all such studies, even if they are relatively separate from one another, under the title of *phenomenology*. (Hu, 25: 15–16)

Benjamin adds a term to the list of "accomplishments" Husserl compiles: in addition to perceiving, remembering, expecting, representing through an image, supposing, and valuing, there is also the act of poetizing, whose objective intention is categorically different from any other act. Regardless of what particular poets may be thinking when they create their poems, to the extent that they are poets, they poetize, and the "intended" of their poetizing can be called, without further qualification, "the poetized." The goal of "aesthetic commentary" lies in so isolating the poetized that nothing of the corresponding act of poetizing remains. In this way, the "sphere" generated by the investigation is reduced to the "region" that represents the "truth of poetry," regardless of what shape the poetized takes in any given instance. Described in these terms, however, "aesthetic commentary" both accords with and parts ways with what Husserl calls "phenomenology" in the preceding passage. The "intended" of the accomplishments he catalogs—perceiving, remembering, expecting, and so forth—can be brought to consciousness. There doubtless remains a question as to the specific character of the consciousness to which "objectivities" thus appear; but there is no question, for Husserl, that the appearance of such things is bound up with the coming philosophy. For Benjamin, by contrast, the isolation of the poetized has nothing to do with psychic acts. Least of all can the poetized, as an eidos of sorts, be "seen" in accordance with the governing idea of *Wesensschau* (intuition of essence). "Truth," for Benjamin, as for Husserl, lies in the fulfillment of a certain mode of signifying intention; but the fulfillment is precisely not "intuitive," for Benjamin. For this reason, he makes no mention of the act-side of the accomplishment; that is, he says nothing about poetizing in his exposition of what is meant by "the poetized." The operative opposition in "Two Poems" does not reside in the distinction between the act- and object-side of the total accomplishment but, rather, in the tension internal to the object-side: the poetized in its particular shape, understood as the particular poetic intention that the poem under consideration fulfills, contrasts with the "pure poetized," which is only a methodological ideal.

Limit Concept and Function

By making the poetized radically unavailable to intuition, the methodological program for "pure aesthetics" decisively distances itself from the phenomenological program laid out in both *Logical Investigations* and "Philosophy as Rigorous Science," while simultaneously entering into close proximity with the principles of the Marburg school. Nowhere is the "Marburgian" character of Benjamin's undertaking more readily apparent than in the exposition of the poetized as a "limit concept [Grenzbegriff]" (2: 107). In a sense, Benjamin has no choice but to describe the poetized in these terms, given the exclusion of intuition. To say of the poetized that it is a limit concept means, in Kantian terms, that no intuition corresponds to it—but it is not therefore empty. Rather, it functions as a methodological principle that guides the investigation in which it is generated. Only once in the *Kritik der reinen Vernunft* (Critique of pure reason) does Kant use the term *limit concept*: "the concept of the noumenon is merely a limit concept, used for the purpose of limiting the pretensions of sensibility, and is therefore only of negative use" (K, A 255; B 310–11). The program of the Marburg school consists, to a certain extent, in a massive expansion of this remark, which provides a clue for resolving the basic problem of Kantian epistemology: the "thing-in-itself" is supposed to "affect" our cognitive faculties, and yet it cannot be an object of knowledge. To the extent that the thing-in-itself can be interpreted as the limit concept of a scientific enterprise, the problem is resolvable in the following way: every science generates its own limit concept, through which its object is progressively constructed as an immanent thought-object. What Benjamin says of the "pure poetized" then applies verbatim to the object of the sciences in general: it is the "absolute task" and, for this reason, it "must . . . remain the purely methodological, ideal goal" (2: 108). The principle under which the object of knowledge is generated, for Kant, is the "transcendental unity of apperception" (K, A 108, B 139), whereas, for Cohen—who removes consciousness from epistemology and plants it unambiguously in the science of psychology—the object is generated on the basis of its inner "lawfulness" (C, 55). And for Benjamin, who closely follows Cohen in this regard, the principle of the poetized lies in "the law of identity," which assumes the central function that Kant accords to the identity of self-consciousness without representing identity in terms of consciousness: "The law, according to which all apparent elements of sensibility and ideas show themselves as a complex of essentially, in principle, infinite functions, is called the law of identity. The synthetic unity of the functions is thereby designated. It is known in its particular shape [Gestalt] as an a priori of the poem" (2: 108).

The "law of identity" opposes a basic thesis of Kantian critique and accords with the founding principle of the Marburg school. The *Critique of Pure Reason* begins with a "Doctrine of Elements" that identifies two irreducible sources of knowledge: sensibility, which is itself divided into the "pure intuitions" of space and time, and the understanding, which consists in the discursive faculty of concept-construction. By the time Cohen writes *Logic of Pure Knowledge*, he had identified only a single source of knowledge: "pure thinking" (C, 90). By formulating the poetized in terms of a "law of identity," Benjamin follows Cohen to the letter: not only are spatial, temporal, and spiritual elements supposed to interpenetrate each other; the result of this "spatio-temporal permeation of all shapes" is a purely "intellectual complex" (2: 112) that can be understood as the aesthetic counterpart to what Cohen calls "pure knowledge." The poetized is a synthetic a priori, but the synthesis is not that of sensibility and conceptuality; rather, it is the synthesis of those "infinite functions" that represent the spatio-temporal-intellectual complex of the poem under investigation. By formulating "the law of identity" in these terms, however, Benjamin turns away from Cohen, who detects a certain relativism in the concept of function, and associates his program with the version of "epistemo-critique" that Ernst Cassirer had recently begun to elaborate in conjunction with his own "cognitive commentary" on the development of the modern mathematical sciences.[6]

Of particular importance for the program Benjamin develops in "Two Poems" is the epistemo-critical program Cassirer wrote in 1910 under the slightly misleading title of *The Concept of Substance and the Concept of Function*. The title is misleading to the extent that it fails to account for the antagonistic relation not so much between the two concepts as between two interpretations of conceptuality in general. Although Cassirer avoids the legal language with which Kant first introduced the idea of critique, his treatise can be understood as a renewed "tribunal of reason" (K, A xi), in which the epistemo-critical interpretation of concepts of function argues its case against the metaphysical interpretation of concepts as substances. For the proponents of the metaphysical tradition, the process of abstraction gives rise to concepts by reducing the range of particulars, until there is only a generic concept, which corresponds to the substance that acquire specificity by way of predication; according to the proponents of epistemo-criticism, concepts, interpreted as functions, are concrete universals, which describe the law or principle that determines every value within its range. With only slight changes in terminology, "Two Poems" restages the same tribunal in the field of aesthetics. The choice of precisely *two* poems is evident from this perspective: the poetized of the first poem under discussion, "Dichtermuth"

(Poetic courage), consists of particular beings that acquire unity only through a process of abstraction that disregards what makes each of them different from one another. By disregarding all differences, the poet, who is unlike everything in the cosmos, discovers his so-called courage. By contrast, the poetized of the second poem, which Hölderlin published under the odd title of "Blödigkeit" (Infirmity, Timidity, with a hint of Stupidity) resolves itself into a complex of infinite functions, the elements of which are all determined by the universal law of relationality that governs the whole.[7] When concepts are interpreted as substances, special cases are ignored or dismissed; when concepts are interpreted as functions, they themselves function as rules that expose the reason why special cases are indeed special. And the poet who emerges in the poetized of "Blödigkeit" is nothing if not a special—or, as Benjamin would later say, an "extreme"—case. Because "Blödigkeit" can be understood as a revision of "Dichtermuth," the sequence of poems corresponds to the sequence described in the title of Cassirer's treatise: just as the interpretation of concepts as functions overcomes the metaphysical illusions generated by the interpretation of generic concepts as substance, so does the function poem resolve the mythological illusions contained in the substance poem and thus create its own nonderivable "mythos."

In response to the question "Upon what do the limit concepts of particular sciences border?" Cassirer's version of epistemo-critique answers: reality, understood not as a metaphysical substance but only as the methodological goal of any given scientific activity, which expresses itself in a thoroughgoing connectivity among its concepts, each of which is itself "objective" to the extent that it contributes to this connectivity. In response to a similar question about the poetized, Benjamin provides a similar answer—with one major difference. Unlike the limit concept, which gives a purely methodological definition to the object of the other sciences, the poetized is defined by *two* poles, each of which is a functional unity in its own right: the unity of life, on the one hand, and the unity of the poem, on the other. Whereas the unity of the poem is the *terminus a quo* of "aesthetic commentary," its corresponding *terminus ad quem* is that of life: "This idea of the task is, for the creator, always life" (2: 107). The category of life thus corresponds to that of reality in epistemo-critique: the unreachable goal of the undertaking in question. And by specifying the two poles that define the poetized, Benjamin can characterize its geometrical character as a "transitional sphere" (2: 107), which proceeds from the unity of the poem to that of life. In this way, life—again only as a functional unity, not as biological substance or biographical subject—is the telos of the poetized. In the same way, the methodology accords with the second of the two poems, which assumes the critical role that Cassirer assigns the interpretation of the concept as pure function and accordingly directs poets to go "straight into life [bar ins Leben]."

Space and Time

So thoroughly does Cassirer's *Concept of Substance and the Concept of Function* enter into the methodological program of "Two Poems" that the distinction between function and substance determines even its minor claims, such as the following one, which concerns a rhyme sequence in "Blödigkeit": "The identity as law is not given substantially but, rather, functionally" (2: 117). Seen from the perspective of Cassirer's version of epistemo-critique, the direction of Benjamin's analysis, far from being hopelessly opaque, is readily apparent. In the poetized of "Blödigkeit," as opposed to that of "Dichtermut," space and time are neither immediate data of consciousness nor independent elements of knowledge; rather, each so thoroughly permeates the other that there is neither space nor time but, rather, only a "spatio-temporal order" (2: 124). With respect to the temporalization of space, Benjamin singles out a word in "Blödigkeit" to describe its character: *gelegen* (opportune), a temporal term that derives from a spatial one (*legen*, "to lay"). The word appears in the fifth line of the poem: "Was geschiehet, es sei alles gelegen dir! [Whatever happens, everything for you should be opportune]." The fifth line of "Dichtermut," by contrast, conjures up a mythological motif: "Whatever happens, everything for you should be blessed [geseegnet]." The difference between *seegnen* and *gelegen* indicates, according to Benjamin, the degree to which the poetized of "Blödigkeit," unlike that of its predecessor, fulfills the "law of identity." The concept of blessedness presupposes a divided cosmos in which a higher sphere presides over a lower one; this division is then reproduced in the lower sphere, which separates sacred places from profane ones. The poetized of "Blödigkeit," by contrast, is undivided: no place is superior to any other; none inferior; every place is, in Benjamin's term, both "determining" and "determined," which means that the essence of each place, without distinction, lies in its "opportune" character: thus the poet can everywhere be a poet.

In analyzing the spatiality of "Blödigkeit," Benjamin draws on a rich complex of words related to *gelegen*, including *legen* and *Lage* (situation). The second line of the poem, which departs from the corresponding line in the earlier poem, is of particular significance in this context: "Geht auf Wahren dein Fuß nicht, wie auf Teppichen? [Does your foot not stride on the true as on a carpet]." In response to this line, Benjamin presents the spatial character of the poetized as the "truth of the situation [Wahrheit der Lage]" (2: 114). To the extent that the concept of truth accords with "the law of identity," it cannot be understood as the correspondence between two unrelated terms, such as thinking and being or subject and object; rather, the "truth of the situation" is simply the situation pure and simple: the *Lage* as it is *gelegen*; or the "situation" as the "making opportune" of what is otherwise only an empty medium that can be fulfilled by ascribing an intention and a corresponding telos. When the space in which everything takes place is sheer "opportunity," there can be no difference between a point of departure and one of completion; every action is already complete at any given point, and *completion* henceforth means precisely "starting out." Space does not, then, contract into a single point; rather, it infinitely expands—Benjamin emphasizes "extension" (2: 113) in this context—such that every place, delimited as it may be, is space in its entirety, without inside or outside. The "truth of the situation" can thus be understood as infinite openness, in which every step that the poet takes is a matter of "fate" or "destiny" (*Schicksal*) rather than a movement toward a destination defined independently of the step taken.

The law of identity requires that time be as thoroughly spatialized as space is temporalized. In the dimension of temporalized space, Benjamin draws on the language of "Blödigket"; in the corresponding dimension of spatialized time, he inserts a technical term of his own: "plastics" (*Plastik*), which is drawn from the Greek word *plattein* (to mold, to shape) and generally applies to forms of "plastic art" such as sculpture, mosaic, and architecture. Benjamin first introduces the term in conjunction with a discussion of the following phrase from "Blödigket," which again departs from the corresponding line in "Dichtermuth": "den denkenden Tag [the thinking day]." Within the context of the philosophical controversies over the limits of Kantian critique to which Hölderlin himself contributed, the phrase lends itself to the following explication: *day* serves as a trope for appearances in general, *thinking* as a synonym for *intellect*, and the synthetic unity of the two yields *intellectual intuition*. Benjamin, however, declines to follow this line of interpretation. Just as the methodological remarks with which "Two Poems" begins grant no space for eidetic intuition, so does its analysis of the spatio-temporal character of the poetized avoid any mention of intellectual intuition, even when it can be unambiguously supported with reference to Hölderlin's own poetological reflections.[8] In order to identify the shape that time takes in the poem, he turns to a particularly striking phrase in the second poem—and interprets it in such a manner that it, too, is taken out of the context of Hölderlin's own times, so that it captures the structure of time in general: "Wende der Zeit [turn of time]." Instead of discussing this phrase in terms of a particular revolution, such as the recent French one or a potential German counterpart, Benjamin goes in the very opposite direction and presents it in relation to what Henri Bergson calls *durée* (duration), understood as "real" or "authentic time" in contrast to its spatialized and mathematicized deformation: "Entirely in opposition to 'fleeting time,' to the 'transient ones,'" Benjamin writes, quoting phrases from earlier poems of Hölderlin, "that which persists, duration in the form of time and human beings [das Beharrende, die Dauer in der Gestalt der Zeit und der Menschen], has been developed in the new version of these lines. The phrase 'turn of time' plainly captures the instant [Augenblick] of persistence, the moment [Moment] of inner plastics in time" (2: 119).

At no point in its uniformly dense argumentation is "Two Poems" more densely argued. A shorthand version of what Benjamin is doing in his discussion of the "turn of time" runs as follows: he discovers "duration" in a passage that would otherwise be reserved from "intellectual intuition." Both the concepts of *intellectual intuition* and *duration* are developed in explicit opposition to the *Critique of Pure Reason*. According to Kant, "intellectual intuition" is unavailable to human beings and can be attributed only problematically to a divine understanding (K, B 147)—to which Fichte, Schelling, and Hölderlin respond, each in his own way, by showing that philosophy or poetry has no greater task than showing how and when the intellect is itself intuitive. According to Kant, time is exactly like space in terms of its divisibility (K, A 25; B 41)—to which Bergson responds in his dissertation *Essai sur les données immédiates de la conscience* (Essay on the immediate data of consciousness) by denying that the qualities of time are in any way comparable to those of space. Under certain circumstances, for Bergson, time appears to be as divisible as space; but this appearance is an illusion that derives from the act of measurement, which takes canonical form in the modern mathematical sciences. Not only is time, for Bergson, indivisible; it is also malleable—elastic, if not exactly plastic—for depending on a range of factors, duration can be longer or shorter. But, of course, duration cannot be either longer or shorter. The use of spatial vocabulary in relation to time shows the degree to which time as it is really lived recedes from everyday language: "supposedly homogeneous time . . . is an idol of language."[9] An artistic task can thus be established in the context of Bergson's version of vitalism: the task of bringing time into speech without conforming to the idolatry of language.

For Benjamin, this task is an indispensable dimension of the poetized generated from "Blödigket." As it happens, Georg Lukács identifies the same task at the same time—the winter of 1914 to 1915, to be exact. Just as the theory of poetry that Benjamin works out in "Two Poems" culminates in the exposition of the phrase "turn of time" as the "moment of inner plastics in time," so does the theory of the novel Lukács proposes in a treatise published under this title culminate in an analysis of *L'Éducation sentimentale* in "durational" terms: quantifiable time appears to flow in Flaubert's novel, as one empty incident follows another; but the nullity of each incident allows time as it is actually lived—without incident—to manifest itself in its indivisible wholeness.[10] In comparison with Lukács's appropriation of Bergson's principal thesis, Benjamin's is exceedingly daring: when it comes to capturing the quality of duration, a four-hundred-page novel appears to be in a far better position than a poem of some twenty-four lines. But this kind of calculation is based on a misconception, for, despite what the word suggests, duration is not a "long time." To draw on the phrase in "Blödigkeit" through which Benjamin explicates the "plastic-intensive" character of time: duration is time "turned" away from its spatialization. And this is where the real daring of Benjamin's endeavor lies, for in the concept of temporal plastics he combines—without synthesizing—the antithetical philosophical programs undertaken by Bergson on the one hand and by the Marburg school on the other.

In "On the Program of the Coming Philosophy," as almost an aside, Benjamin suggests that the concept of a nonsynthetic relation between two concepts allows the "coming philosophy" to build on the Kantian trichotomy of relational categories without thereby supporting a dialectical progress of categorical thesis, hypothetical antithesis, and disjunctive synthesis: "Beyond the concept of synthesis, that of a certain nonsynthesis of two concepts will gain the highest systematic importance, since a relation other than synthesis is possible between thesis and antithesis" (2: 166). The idea of temporal plastics derives from a similarly nonsynthetic combination of antithetical concepts. From the perspective of epistemo-critique, temporal plastics represents a thoroughgoing permeation of space and time, which leaves no room for "independent data of consciousness." From the perspective of Bergson's *Essay on the Immediate Data of Consciousness*, temporal plastics represents time freed from the mistaken image of its measurable flow. The absence of a synthesis prompts a question: to whom does time appear "plastic"? Or, otherwise spoken, for whom is it "turned." In the context of epistemo-critique there is a clear answer: the "plastic" character of time does not appear to anyone but is, rather, a dimension of science. An antithetical answer can likewise be found in Bergson: the individual thinker, removed from prejudices enforced by science, can gain an intuition of the durational character of true time. By combining the two concepts without synthesis, Benjamin changes the field in which the question is posed: science is neither the solution to the problem nor the impediment that prevents a gifted individual from experiencing time as it truly is. The law of identity is once again the guide to Benjamin's exposition. Time appears plastic only to those to who are themselves plastic. Mortals are excluded from this category simply because their form is, by definition, always changing. Only the gods—or, more exactly, only the gods of form, whom Nietzsche dubs "Apollonian"—can experience the plastic character of time. And in the absence of a "rejuvenating" counterpart, who would function as a pure synthesis of the opposition between mortal and divine, the gods fall prey to their experience of the inner structure of time.[11]

God as Object

Just as Benjamin does not mention Husserl when he defines the poetized as a certain mode of objectivity, Cohen when he presents the poetized as limit concept, Cassirer when he distinguishes substantial from functional unity, and Bergson when the idea of duration enters into his analysis, so the name of Nietzsche does not make its way into "Two Poems" when the gods of Hölderlin's cosmos are described in terms of their "plastic" character. Benjamin's silence in the case of Nietzsche is particularly significant, for the absence of the term *Apollonian* corresponds to the absence of its mythic counterpart—despite the obvious fact that Dionysus plays no small role in Hölderlin's late poetry, beginning with "Wie wenn am Feiertage" (As when on a holiday). Among the earliest attempts to rescue Hölderlin from oblivion and place him in the company of Goethe and Schiller, the small treatise Friedrich Gundolf wrote in 1912 under the title "Hölderlin's Archipelagus" is of decisive importance, for Gundolf—a major critique of the period who participated in Stefan George's "circle"—seeks to show that Hölderlin's Hellenism overcomes the limits of German classicism and prepares the way for the dynamic Hellenism of Nietzsche and George. The Apollonian character of Hölderlin's Hellas, as Gundolf explains at the opening of his essay, emerges from an insight into the fundamentally Dionysian character of the natural world as "truly *natura*, as becoming."[12] In the middle of his essay on Goethe's *Elective Affinities* Benjamin subjects Gundolf's image of Goethe as hero to withering criticism (1: 159–67). In "Two Poems," by contrast, the criticism of the kind of commentary Gundolf epitomizes takes the form of silence: for Benjamin, there are no gods other than those of pure form, which is to say, there is no Dionysian counterpart to the Apollonian principle. But this denial does not represent a return to German classicism, which fails to appreciate the depth of Dionysian insight into the abyssal character of all culture. Far from affirming the stability of the gods, Benjamin affirms the very opposite: the gods "fall prey to their own plasticity" (2: 119).

The absence of the Dionysian in "Two Poems" expresses itself in a number of ways: as the absence of musical motifs, even when Benjamin draws attention to certain phonetic features of the poems; as the absence of any reference to ecstasis; as the absence of nature understood as primordial chaos; as the paradoxical absence of night, in contrast to which the day, including "the thinking day," appears; and especially as the absence of "the coming god," about whom Hölderlin famously writes in "Brot und Wein" (Bread and wine), referring at once to Dionysus and Jesus. The absence of Dionysus in the poetized of Hölderlin's late poetry doubtless means that Benjamin gives no credence to the sort of heroic Hellenism Gundolf, among others, seeks to reclaim; but he begins, as it were, with the premise on which the heroic version of Hellenism rests: left on its own, without a Dionysian renewal, the Apollonian world of culture falls into ruin. The law of identity demands nothing less. Because the gods are as "plastic" as the time they experience, they become immobile, hence "dead form" (2: 121). Here is where a Dionysian renaissance would begin. But Benjamin proceeds in a different direction: the gods are not even dead; rather, they are made into an object that falls into the hands of the poet, who is not thereby divinized but is, instead, brought into a strange sphere where the crime of hybris is no longer punishable.

The claim that the gods turn into objects is the last and boldest claim that Benjamin makes in the course of his analysis. The following lines from "Blödigkeit" are again decisive: "Unser Vater, des Himmels Gott, // Der den denkenden Tag Armen und Reichen gönnt [Our father, God of the heavens, does not begrudge the thinking day to the poor and the rich]." The crux of this image, according to Benjamin, lies in the verb *gönnen*, which is generally understood as a mode of giving, granting, or allowance. For Benjamin, however, the conception of time as a godsend belongs to the mythological tradition from which the "mythos" of Hölderlin's late poetry departs. The note of giving inherent to *gönnen* consists not so much in a relation of giver-god to receiver-mortal as in the equanimity of the former with regard to the latter's reception of the datum—in this case, the "thinking day." The supreme god, who represents the Apollonian order, is no longer enraged when all mortals, "rich and poor," can see what he sees: "the true." Mortals are thus in a position to assert their equality with the gods but do not thereby solicit a fateful reprisal. All of this means that the "thinking day" is the time in which hybris does not simply go unpunished, as if it were only an exceptional day, a Saturnalia of sorts; rather, the supreme god "allows" there to be no supreme law on the basis of which everyone would be assigned a discrete place and time. If, however, the gods abstain from enforcing the line that divides themselves from the mortals, they are not only merely "dead form" but are only as good as objects, which have no will of their own—objects, therefore, that poets can "bring."

Rigorously followed, the law of identity does not issue into a human-divine synthesis of either the Dionysian or the Christian type; rather, it yields a "spatio-temporal order" from which the gods have not exactly departed but in which they abstain from using their punishing power. This strange situation, ambiguous in the extreme, is perhaps what drew Benjamin's attention to "Blödigkeit" in the first place. Instead of concluding his exposition of the poetized with a discussion of the "coming god," Benjamin goes in an entirely different direction, discovering a god who can come only insofar as the poet "brings" him. For Benjamin, the brought-character of the coming god makes him into something of a trophy that crowns the total interpenetration of space and time:

> The god ceases to determine the cosmos of the song, whose essence [namely the poet], rather, freely selects the object: he brings the god, since the gods have already become in thought the reified being of the world. Here the admirable articulation of the final strophe is recognizable, in which the immanent goal of all formations of this poem join together. The spatial stretching of the living determines itself in the temporally inner intervention [Eingreifen] of the poet. . . . "Good also are we and skilled [geschick] at something for someone"—once the god has turned into an object in its dead infinity, the poet takes hold [eingreift] of him. (2: 121–22)

Benjamin's discussion of the final lines of "Blödigkeit" is based on an implicit thesis: the poet who grabs hold of the Apollonian god and intervenes into a world that would otherwise be dead form is not a placeholder for Dionysus, much less a figuration of "the Crucified One." In Hölderlin's terms, the poet is not even exactly a "half-god." Above all, though, the poet does not do anything to the world in which he intervenes, for, in colloquial terms, the world is perfect. In terms of the epistemo-critical lexicon that Benjamin borrows for the purpose of distinguishing the poetized of the later poem from that of the earlier one: the elements of sensibility so thoroughly interpenetrate one another that the resulting order is saturated. And the interruptive act of taking hold of a god takes place at a time-point of supersaturation, in which the thoroughgoing interpenetration of spatial-mortal and temporal-divine elements is confirmed.

The Meaning of Poetic Existence

The analysis of the "spatio-temporal order" in "Blödigkeit" shows that poets are so unlike the other element in the cosmos that they alone survive the total interpenetration of space and time; but the analysis fails to indicate what then can "bind" or "obligate"—*Verbindung* in German means both—poets to the world. "Binding" or "obligation" assumes the function that Kant cedes to transcendental synthesis, and the poet becomes the exponent of pure intellect, which must somehow be combined with the "spatio-temporal order" in order for the cosmos to be *one*. Without such unity, there is no *single* poetized. This, then, is the danger: that the elements of the world will be so tightly interpenetrated that they will repulse whatever remains—which is to say, once again, that they will repulse poetic existence. The courage of the poet, in turn, consists in being bound or obligated to the world despite its irreducible unfitness for a specifically poetic mode of existence. According to Benjamin, the courage of the poet in "Dichtermuth" is not of this kind; rather, the poem encourages the poet to be courageous by showing him that there is an affinity between his mode of existence and that of the world to which he is obligated. The affinity lies in the substantial character of both poet and world. For this reason, "the poetized of the first version knows courage only as a property" (2: 123). And for the same reason, the poetized of the first poem is not properly unified: "There can be no talk of a unity of the poetized in the first version" (2: 123). In the later poem, by contrast, poetic existence is essentially courageous and there can be therefore no place or time for encouragement.

The "peculiar paradox" of courage then defines the character of the poet: "whoever is courageous withstands the danger and yet does not heed [achtet] it" (2: 123). To the extent that the danger is not something in the world but, rather, the total interpenetration of its elements, the poet is a danger to the second power: a danger that makes the world dangerous once again. As the very danger in question, however, the poet cannot heed himself: a touch of "stupidity" (*Blödsinn*) informs his "infirmity" (*Blödigkeit*), which Benjamin associates with the perplexing image of the "lonely savage [einsam Wild]" (2: 125) around which "Blödigkeit" revolves. Something of Schiller's description of the "naïve" mode of poetizing poetry is discernable in Benjamin's discussion of poetic existence, and indeed he draws his analysis to a close by quoting from the conclusion to Schiller's *Briefe über die ästhetische Erziehung des Menschen* (Letters on the aesthetic education of the human being) (2: 125). But the nature of poetic existence cannot be determined on the basis of Schiller's distinction between naïve and sentimental: the poet is not naïve, for he is not united with nature; and he is not sentimental, for he does not reflect on his distance from the world. As soon as the poet pays attention to himself, his essence no longer converges with that of courage, and he therefore no longer exists as a poet. Not only, then, is the poet a danger to a fully saturated world, he is a danger to himself—a transcendental danger, one could say, who, by universalizing danger, overcomes it in the same stroke: "The greatness of danger emerges in the courageous: by first striking someone who is courageous, who gives himself entirely over to danger, danger strikes the world. . . . The world of the dead hero is a new mythic world, saturated with danger: this precisely is the world of the second version of the poem" (2: 123–24).

Attempts to represent "Two Poems" as Benjamin's indirect commemoration of Fritz Heinle, who killed himself at the beginning of the war, apparently so that he would not have to participate in it, can find justification in the brief discussion of the poet as "dead hero" (2: 122). And nowhere else is Benjamin's discussion of the poetized more distant from "Blödigkeit," which does not easily lend itself to an analysis in terms of courage. The opposite is true of "Dichtermut." In the earlier poem the situation of the poet is compared to that of the swimmer, whose watery world can be nicely captured by the phrase Benjamin emphasizes: "saturated with danger" (2: 124). The poets of "Blödigkeit," by contrast, are not only *not* compared to swimmers; they stand "upright" in open air, "drawn by golden strings." Far from succumbing to the relentless waves that overpower them; at the end of the poem the poets appear to be so very much alive that they bring a god. It is here, above all, that Benjamin must make a case for the later poem being "about" the same thing as the earlier: "the transformation of the duality of death and poet into the unity of a dead poetic world, 'saturated with danger'" (2: 124). As part of this effort, Benjamin suggests that the word *Einkehr* (stop) in the difficult third stanza of "Blödigkeit" can be seen as the "shape" of death; but his analysis is not based on this tentative suggestion, for death is no longer understood in relation to the category of "the living," much less in terms of biological substances or biographical subject; rather, death is the limit of life, which, when reached, does not revive, reanimate, or resurrect "the dead," but instead *means* a higher—hence purely "functional" rather than "substantial"—life. The poet thus acquires a function altogether different from that of a Dionysian hero, who rejuvenates a world that has fallen prey to sclerosis:

> Deposited [versetzt] into the middle of life, nothing remains to [the poet] but motionless being, complete passivity [das reglose Dasein, die völlige Passivität], which is the essence of courage; nothing but abandoning himself to relation. . . . He is nothing but the limit up against life, indifference, surrounded by enormous sensible powers and the idea that in itself protects his law. The last two verses [of "Blödigkeit"] contain most powerfully how very much he means the untouchable middle of all relation [wie sehr er die unberührbare Mitte aller Beziehung bedeutet]. (2: 125)

Some of Benjamin's earliest writings are concerned with a condition similar to that of the poet whom he thus describes: just as the courageous poet does not heed the danger, which is nothing other than himself, so does "youth." The absence of self-consciousness makes it possible for both poet and youth to "mean" something that otherwise escapes systems of signification. Thus, by distinguishing "confession" (*Bekenntnis*) from "recognition" (*Erkenntnis*), Benjamin arrives at the following formula in a contemporaneous essay on the relation between youth and religion: "Youth that confesses itself to itself means religion [bedeutet Religion], which does not yet exist" (2: 73). Youth "means" religion because the temporal structure of youthful signifier and religious signified are identical to each other. Similarly, the temporal-spatial-intellectual structure of the courageous poet is identical to what he means: both are, in a word, "untouchable" or "inviolate." To the extent that the term *touch* covers all of sensible existence, this means that the structure of the poet and the meaning of his existence consist in transcending "sensible existence." No wonder Benjamin refers to a "mystical" principle in this context. The sensible form of this transcendence is—and here Benjamin adopts and transforms a word from Hölderlin—"sobriety" (*Nüchternheit*). The more emotionless the poet, the less can be said about his psychological state. Apropos the wholly sober poet, there is nothing to say, for there is no psychological state at all. Unmoved in the extreme, poetic existence approximates nonexistence and becomes, as a result, the signifier of purely functional life.

In the Direction of "Doctrine"

At the end of "Two Poems," Benjamin, however, is at a loss for words. The technical term for which he searches in vain would name something that runs counter to myth without falling back into traditional mythology. The absence of this term is no small failure, moreover, for it is bound up with the meaning of "the poetized." A schematic version of the argument that leads Benjamin into a place where words fail him would proceed as follows. The poetized can be broadly defined as the artistic task that a particular poem fulfills, and the pure poetized is, in turn, the artistic task that poetry as such fulfills; in filling out the categories of the poetized as the "transitional sphere" between poem and life, however, "aesthetic commentary" provisionally relies on certain categories of myth, especially fate or destiny. "Pure aesthetics" can thus be characterized as the philosophy of myth—but once again, only to a certain extent. At a certain point the poetized not only departs from traditional mythology but gains freedom from its own "mythos" as well. The law of identity in this way gives rise to a counterlaw of difference: at some point in its transition from the unity of the poem to that of life, the sphere of the poetized can no longer be described in terms of the categories through which it is generated. Enigmatically—yet precisely—Benjamin identifies this murky point where divine and human spheres converge. In the case of Hölderlin's late poetry, this convergence takes the following form: "Myth is recognizable [erkennbar] in the inner unity of god and destiny" (2: 109). The poetized generated from "Blödigkeit," however, goes beyond the point of this unity. Here again, the absence of "the coming god" is decisive: the poets assert a difference from the heavenly sphere by the very act of bringing one of the gods. In making their songs, the poets establish this difference, which transcends the categories of myth: it is, as it were, not only beyond the Parcae but outside of destiny altogether.

In place of a mythic term such as *Dionysus*, Benjamin introduces into his conclusion of this discussion—"to be sure, without explicit justification" (2: 126), as he admits—a term drawn from the science of geography: "the oriental."[13] To a certain extent, the "orient" is to "Greece" in Benjamin's program for "pure aesthetics" as Dionysus is to Apollo in Nietzsche's *Birth of Tragedy*, for the oriental-Dionysian transcends the order established by the Greek-Apollonian: "This is the oriental, mystical principle, the one that overcomes limits, which in this poem again and again so clearly supersedes the Greek formative principle, a principle that creates a spiritual cosmos from pure relations of intuitions, of sensible existence, in which the spiritual is only the expression of the function that strives toward identity" (2: 124). "The oriental" is not a category of the poetized; it is, rather, its underlying directionality. For this very reason, however, a term such as *the oriental* goes only so far. Added to this, of course, is the fact that the term *oriental* is by no means neutral. Anti-Semitic discourse of the period often condemns the Jews for the "oriental" characteristics that make it impossible for them to be truly assimilated to European peoples and cultures; conversely, proponents of a radical renewal of Judaism, especially Martin Buber and his disciples, seek to strengthen its "oriental" character as an antidote to the paralysis to which, in their view, the West in general and Western Jews in particular have succumbed.[14] Benjamin may be alluding to Judaism and perhaps to Zionism when he speaks of "the oriental" at the end of "Two Poems." As soon as he had an opportunity to study Benjamin's essay, Gershom Scholem suggested as much: "Friedrich Hölderlin lived *the* Zionist life among the German people" (S, 2: 347). And Benjamin, for his part, may have gone even further in this direction. Scholem found the following remark of Benjamin disturbing enough to note at least twice: "There can be only *one* Zionist" (S, 2: 201; cf. 2: 44–45). In this context, however, its meaning becomes clear: *only* Hölderlin can be a Zionist, for only in his work does movement toward the "orient" find its proper form.

Regardless of what Benjamin's remark may have meant, *the oriental* is not the term his exposition of the poetized ultimately requires. It doubtless points away from both *the Greek* and *the fatherland*, but it gives no indication as to where another term could be found. Because the earth is spherical, it may be positively misleading, for the end of the orient lies in the occident, and the end of the occident lies in the orient—which gives rise to a doctrine of the eternal return. Yet "the oriental" in "Two Poems" has nothing to do with the cycle of return. Nor does the first term by which Benjamin elucidates the oriental go very far, namely *the mystical*. But at least the mystical gives a certain negative insight into the point at which the poetized of the late Hölderlin makes a transition to the pure poetized. In contrast to the "Greek" principle, according to which the spiritual or intellectual dimension is only an "expression" of "sensible existence" striving for identity, the mystical principle is altogether spiritual: "sensible existence" does not so much disappear as become totally transparent to itself. In the absence of mystical categories that are not at the same time mythic, however, the term *mystical* is likewise of little help in delineating what happens as the poetized begins to approach its methodological ideal.

All of this is to say, once again, that Benjamin does not yet find himself in possession of a word for what happens when the poetized finally departs from its corresponding "mythos." To be sure, there is a guide in this regard: the final stanza of "Blödigkeit," especially the "insistent haunting caesura" (2: 125) of the penultimate line, in which "Einen bringen [bring one (of the heavenly)]" is separated from "Doch selber [yet ourselves (the poets)]." Under the condition that the "brought" god is itself rigorously distinguished from "the coming god," whose name is either Dionysus or Jesus, the guiding line leads to an impasse. The stance of the poet does not consist in waiting for the "coming god." Such a stance is consistent:

the passive vigil of poets would be the property by which they would be knowable as poets. But the poets, according to the final lines of the poem, are decidedly active in their passivity, as they—"yet [them]selves"—bring their "skilled" hands. The perfection of passivity lies in a certain action, and the consummation of motionlessness requires a certain motion. This is, as it were, the impasse of passivity: there must be a supplementary or complementary action that completes passivity. Benjamin does not use the term *impasse* in this context; indeed, he scarcely seems to notice that there is something fundamentally incoherent in basing the idea of "motionless existence" on those lines of the poem that speak of poets moving their hands. It is almost as if the caesura to which Benjamin points is "insistent" or "haunting" (*eindringlich*) because it describes a cut: poetic hands are severed from their bodies. In any case, the entire methodological preface of "Two Poems" is directed toward the impasse in question, for the analysis Benjamin undertakes is searching for nothing so much as a "bump"—a place, in other words, where the unity of myth bumps up against a higher unity, one that is not only nonmythic but cannot be resolved or even only described in mythic terms: "The analysis of great poems will not bump up against myth but will, instead, hit upon a unity generated through the force of mythic elements that strive against one another—this unity as an authentic expression of life" (2: 108). Since life is the sole task of art, and since the fulfillment of this task is truth, an "authentic expression of life" can also be called "the true."

In the concluding paragraph of "Two Poems" Benjamin expands upon his abbreviated account of the "bump" that is bound to occur in the analysis of great works of art: "The consideration of the poetized does not lead, however, to myth but, rather—in the greatest creations—only toward mythic binds and obligations [*mythischen Verbundenheiten*] that are formed in the artwork into a singular, unmythological, and unmythic shape, which cannot be more closely grasped by us [*zu einziger unmythologischer und unmythischer, uns näher nicht begrifflicher Gestalt geformt sind*]" (2: 126).[15] The sudden eruption of an "us" in an investigation that has otherwise avoided the tokens of subjectivity demonstrates the degree to which the author of "Two Poems" finds himself at a loss for words. The absence of a term for the shape of the poetized generated by his commentary indicates the problem to which the introduction of a technical term at the beginning of the investigation ultimately responds: the poetized, as the artistic task, can be separated only abstractly from its fulfillment.

The missing word names something like the shape of truth, as it begins to emerge in Hölderlin's last poems. Nowhere in all of Benjamin's work is a technical term both more necessary and more dangerous: necessary, because there is no such word in the philosophical tradition through which the methodology of "pure aesthetics" is developed; dangerous, because a term of this type would invest the objectivity it names with an unmistakable note of subjectivity.

To a certain extent, a word for the shape of truth has already been found: *the oriental*. There is, as Benjamin admits, little justification for the use of this term in the analysis of "Blödigkeit," but he discerns a certain "eastern" quality in the following image of the carpet, which can be found in a previously discussed line of "Blödigkeit": " Geht auf Wahren dein Fuß nicht, wie auf Teppichen? [Does your foot not stride on the true as on a carpet]." Whereas the carpet ensnares those, like Agamemnon, who overstep their bounds, it unbinds the poet to poetic existence and reveals the distance between the poetized of Hölderlin's late poetry and the mythic category of fate: "so foreign, as from the Eastern world, and yet so much more original than the Greek Parcae" (2: 114). The "oriental" carpet can thus be seen as an image of "the true" taking shape. But *carpet* is still not the term Benjamin seeks, for it is, after all, only an image, not a word that would itself be "the true." In the final paragraph of "Two Poems" Benjamin turns to a poem that is even later than "Blödigkeit," namely "Herbst" (Autumn), which Hölderlin wrote during his so-called madness. This is what Benjamin draws from " Herbst": "the legends" or "the sayings" (*die Sagen*) depart from the earth and return to humanity. "Die Sagen," then, is not so much the sought-after term as its temporary stand-in. At a certain point in the generation of the poetized, the mythic categories that facilitated its description come into such a high degree of tension with one another that a nonmythic—hence "legendary"—unity begins to emerge.

About "die Sagen" only this much can be said with certainty: they are there to be passed on. The categories of the poetized that transcend those of myth must be similarly transitional and could even be called "autumnal" insofar as autumn is a time of transition. But the term to which Benjamin eventually turns has an advantage over *autumn* and even *legend*, in that it suggests a veridical transition, a transition into truth. The term is *Lehre*, which should probably be translated as "doctrine" but which, like *doctrine*, derives from the word for "teaching" or "learning," *lehren*, and also serves at time as a synonym for *theory*, as in, for instance, Goethe's "theory of color" (*Farbenlehre*) or Cantor's "set theory" (*Mengenlehre*). Just as "sayings" must be continually passed on for them to become "legendary," so "doctrine" must be continually taught in order for it to be "doctrinal." The point is not so much the lesson learned as the "fact" of learning. The idea of doctrine can thus be understood as a transformation of what Cohen famously called the "fact of science" (C, 57). For Cohen, modern mathematical science, which derives from the "principle of the infinitesimal method," is the point of departure for epistemo-critique, which then consists in the generative justification of what science has learned but cannot by itself make universally teachable. Doctrine in Benjamin's terms corresponds to the "fact of science" without the doctrinaire position that only the mathematical sciences are so fully unified that they alone can be taught regardless of who is prepared to learn. According to Scholem, Benjamin began to use the term *doctrine* soon after they met in the summer of 1915.[16] By drawing out the poetized of the late Hölderlin to a point where the categories of myth run up against a "higher unity," Benjamin prepares a place for doctrine, which is precisely this: the transition from a lower unity of "mythos" to a higher unity of truth.

As neither the fulfillment of the artistic task nor the attainment of the pure poetized, doctrine is, rather—to use mathematical terms—the ever-decreasing interval between the poetized as limit concept and the limit itself. To use the topographical terms with which Benjamin introduces his program for "pure aesthetics," doctrine is the place where the "sphere" generated by commentary narrows into the "region" of its truth: "a particularly central place [Ort] . . . in which the limit of the poetized has advanced furthermost toward life, at which the energy of the inner form shows itself more powerfully, the more fluid and the more formless the life that is therein meant [das bedeutete Leben]" (2: 122). Here, then, is Dionysian formlessness—not as insight into the underlying fluidity of existence but, rather, as a certain mode of tautological "meaning." When the poetized converges on the pure poetized, a place emerges that indicates what "life"—beyond substance—means. Doctrine is thus a limit concept to the second power, which cannot be called a concept in an unambiguous sense: it is the ever-diminishable interval where the limit becomes recognizable. Conceived in these terms, the poetized of Hölderlin's late poetry acquires a doctrinal character, as it approaches the pure poetized.[17]

§2 Entering the Phenomenological School and Discovering the Color of Shame

A School Divided

At the end of "Two Poems of Friedrich Hölderlin," Benjamin begins to discuss a paradoxical stance that has implicitly guided his analysis: the poet is receptive to the world but remains unaffected by anything in it. Following Hölderlin, Benjamin describes this stance as "sober."[1] In the dissertation he wrote a few years later, *Der Begriff des Kunstkritik in deutschen Romantik* (The concept of art criticism in German romanticism), Benjamin repeats this gesture, as he again refrains from mentioning the term *sober* until the argument has reached its conclusion. Far from being associated with a philistine virtue or a bourgeois value, sobriety designates a condition of nonenthusiasm and nonexcitement: there is no god to whom the existence of the sober one points; and there is no mental state that mediates between inner and outer worlds. There is only the world to which the sober one is receptive without being touched by anything or excited by anyone. Benjamin probably borrowed the term from an image in Hölderlin's two-stanza poem "Hälfte des Lebens" (Half of life; The middle of life), to which his discussion of being "deposited into the middle of life" (2: 125) may also be indebted:

Ihr holden Schwäne.
Und trunken von Küssen
Tunkt ihr das Haupt
Ins heilignüchterne Wasser.

[You swans, graceful
And drunken by kisses

Dunk your heads
Into holy-sober water.]²

In the absence of the image of water the retention of the "holy" in "holy-sober" would be potentially misleading, for it would suggest that there are two opposing kinds of sobriety: a holy one, which would be divinely inspired, and a profane one, which would presumably consist in utilitarian abstinence. For Hölderlin, as for Benjamin, however, there is no such division in sobriety. To the extent that the holy is inviolate, sobriety is always already holy. Just as the swans cannot fully immerse themselves in the naturally consecrated water but can only dip their heads for a luminous moment, so the poet whom Benjamin describes at the end of "Two Poems" exists on the verge of life by momentarily converging with the limit called "life."

Around 1916, while attending the university in Munich, Benjamin's philosophical studies begin to revolve around the problem contained in the word *sober* by translating it into a series of technical terms, beginning with "pure reception" (*reine Aufnahme*) and "pure intuition" (*reine Anschauung*). In a dialogue Benjamin probably wrote in late 1915 or early 1916 under the title "The Rainbow," one of its two characters gives voice to the theorem that underlies these studies: "The absolute exists only in intuition" (7: 21). Nothing comparable to this proposition can be found in "Two Poems." On the contrary, the methodological procedures Benjamin outlines in his program for "pure aesthetics" preclude intuition of any kind: there is no intuitive counterpart to the "limit concept" he constructs under the rubric of "the poetized," and there is no place for "intellectual intuition" even in the lines of Hölderlin's poem that unmistakably allude to this key post-Kantian term. No wonder *The Concept of Art Critique in German Romanticism* repeats the structure of "Two Poems," for what Benjamin says of Friedrich Schlegel in his dissertation corresponds to the manner of thinking that finds expression in his "first major work" (*GB*, 3: 157): "Especially with regard to intellectual intuition . . . [his] manner of thought is distinguished by its indifference to intuitiveness [Anschaulichkeit]" (1: 47).

A major impetus for Benjamin's change of attitude with respect to intuitiveness can be found in the philosophy faculty at the University of Munich, where he encountered a vibrant group of phenomenologists who were taking sides in a conflict over the future of phenomenology that Husserl initiated when, in 1913, he published *Ideas Pertaining to a Pure Phenomenology and to a Phenomenological Philosophy*. Instead of simply going "toward the things themselves" (Hu, 19, 1: 10), as he famously claimed in one of the prefaces to *Logical Investigations*, the philosopher cum phenomenologist is now required to undergo a "radical alteration" in attitude. Husserl invents a number of technical terms to capture the sense of this change, which places the so-called "natural" attitude "out of action," "brackets" it, and thus "turns it off [schalten sie aus]" (Hu, 3: 65). The technical term *Ausschalten* (turn off, shut down), which is drawn from the sphere of modern technology, corresponds to an older one that derives from the practices of Greek skepticism: *epochē*, understood as the "abstention" from judgment as to whether any given statement about the world is either true or false. For Husserl, there is a single yet universal matter of judgment from which the phenomenologist is called upon to abstain: whether there is a world of substantial things lying outside of consciousness. If a motto for *Ideas* could be found that would revise the one with which he launched *Logical Investigations*, the following passage—which Paul Linke prominently cites in an essay published in *Kant-Studien* in 1916 under the title "Das Recht der Phänomenologie" (The law of phenomenology)—is certainly among the best: "Have the courage to receive whatever in the phenomenon is actually appearing" (Hu, 3: 264).[3] The point of the phenomenological reduction lies in courage: instead of assuming that appearances are supported by underlying substances, the phenomenologists are charged with the task of simply seeing what gives itself to be seen.

Far from consolidating the phenomenological movement around the theoretical practice of "turning off" the "natural" attitude, however, *Ideas* created a major division. Among those who were skeptical of Husserl's idea of *epochē* was one of his former students in Göttingen, Moritz Geiger, who once described his first encounter with *Logical Investigations* as a "transformation of my very being."[4] When Benjamin first came into contact with Geiger, he did not make a similar claim; but in a letter to Fritz Radt from December 1915 he came close to doing so—by way of displacement. Benjamin begins the letter with a description of his disappointment with the quality of his classes in Munich, especially those of the art historian Heinrich Wölfflin, in whose "'seminar' (in which he alone speaks)" (*GB*, 1: 297) there is neither teaching nor learning, for the greatness of the objects under discussion—medieval miniatures—

grotesquely contrasts with the small-mindedness of the instructor. But the central paragraphs of Benjamin's letter to Radt are concerned with a momentous encounter with an older student, Felix Noeggerath, who was completing his dissertation in Erlangen on the concept of the philosophical system, in which he seeks to show that Kant's categories of relation can be represented as forms of mathematical series. Noeggerath was apparently drawn to Munich because of Geiger, whose wide range of interests mirrored his own. Benjamin called Noeggerath the "universal genius" and was singularly impressed by his ability to discuss topics as diverse as Kantian critique and pre-Columbian culture: "Next to such conversations, which, for a long time give me something to learn, to a certain extent—whatever else I do, I will not and should not forgot them—nothing, of course, at the university can compare" (*GB*, 1: 301). This, though, implicitly prompts a question, which Benjamin immediately answers: what about the seminar that drew Noeggerath to Munich? "Geiger's seminar has too few hours. He is assaulted by [military] service, and the problems are too difficult to yield much during a seminar" (*GB*, 1: 301). In the case of Wölfflin, the "seminar" in scare quotes exemplifies the contemporary academic situation: great objects, small minds. In the case of Geiger, by contrast, the discrepancy consists in academic form, exacerbated by the militarization of university life. Geiger's seminar cannot handle the breadth of the problems under discussion, and for this reason, above all, he enters into conversations with Noeggerath, who can be seen as Geiger's replacement. Not only are more hours required, so, too, is extra reading, which Benjamin duly assigns himself: "[I] am going through a phenomenological work of his [Geiger's]," Benjamin adds, referring to *Contributions to the Phenomenology of Aesthetic Enjoyment*, "and I am also reading Husserl's difficult, principal groundwork, so as to gain entrance into his school" (*GB*, 1: 301–2).

Nowhere else does Benjamin express a similar desire to enter into a school of thought; on the contrary, he generally keeps his distance from schools of all kind. Among the many reasons Benjamin was attracted to Husserl's school, the following is perhaps the strongest: entering into the school and keeping one's distance from it are not in this case mutually exclusive, for the school is divided against itself. This division is legible in the letter to Radt, as Benjamin refrains from naming which of the two groundworks he is reading: *Logical Investigations* or *Ideas*. In this way, he keeps his distance from any particular branch of the school. In an effort to disambiguate the reference, the editors of Benjamin's collected writings cite the authority of Scholem, who, it seems, should be in a position to know which of the two works was meant: "*Husserl's difficult, principal groundwork*: that can refer only to Husserl's *Ideas Pertaining to Pure Phenomenology* (1913). Since the list that Benjamin kept of the writings he read begins only from the beginning of 1917, nothing more precise can be established" (*GB*, 1: 305). Far from solving the problem, Scholem's remark exacerbates it: if Benjamin could be referring only to *Ideas*, then nothing more precise is require; but of course he could be referring instead to *Logical Investigations*. The first few pages of the "official" list of books and articles to which Scholem refers is missing; but other lists are extant; and in the relevant ones, Husserl's two "difficult, principal" tomes are placed side by side (WBA, MS 506 and 1851v), as if, even in his private notes, Benjamin declined to take sides.[5]

As for what Benjamin learned upon entering into Husserl's school, this much is certain: he learned its doctrine of intuition.[6] In the following definition of intuition, which Scholem preserved in his *Diaries*, Benjamin departs from both the *Critique of Pure Reason* and its post-Kantian successors: "The object of intuition is the necessity of a content that in feeling purely announces itself as perceivable. The perception of this necessity is called 'intuition'" (S, 2: 449). Whereas a self-proclaimed phenomenologist would probably use technical terms such as *a priori* and "stream of consciousness" (*Erlebnisstrom*), Benjamin uses nontechnical ones: *necessity* and *feeling*; but in every other way his definition of intuition—unbeknownst to Scholem, it seems—accords with the phenomenological sense of the term: intuition is not, as Kant proposes, a cognitively indispensable relation to a contingent appearance; nor, as some proponents of "intellectual intuition" maintain, does it consist in the production of the object intuited; rather, intuition is cognition itself—in Benjamin's terms, "the perception of this necessity [of a content]." Around the time that Benjamin defines intuition in this manner, he drafts "On the Program of the Coming Philosophy," which includes inter alia the following remark: "For the development of philosophy, it is to be expected that every annihilation [Annihilation] of these metaphysical elements in the theory of knowledge [which can still be found in the *Critique of Pure Reason*] at the same time points toward a deeper, metaphysically fulfilled experience" (2: 160–61). Among the promising "annihilations" to which Benjamin refers, none perhaps raises greater expectations for the development of philosophy than "the annihilation of the world of things" (Hu, 3: 115), which Husserl first proposes in paragraph 49 of *Ideas*: "absolute consciousness," according to Husserl, is doubtless "modified" by the annihilation of substantial things; but it nevertheless remains "untouched [unberührt] in its existence" (Hu, 3: 115). In Benjamin's terms, drawn from Hölderlin, "absolute consciousness" is altogether "sober."[7] For proponents of the "realist" branch of phenomenology, talk of "world annihilation" goes nowhere: essences can be intuited without such methodological preparations. For Benjamin, the annihilation of the world of substantial things is only a preparation for "metaphysically fulfilled experience." In this way, too, Benjamin can refrain from taking sides in the dispute that traverses the sole philosophical school to which he ever seeks entrance.

Entrance Exam

As part of his "Habilitation" or academic accreditation, Jean Héring, another former student of Husserl's from Göttingen, published a dissertation in 1925 entitled *Phénoménologie et philosophie religieuse* (Phenomenology and religious philosophy), which draws on Husserl's discussion of world annihilation in order to sketch out the lineaments of "religious knowledge" that are latent in transcendental phenomenology.[8] There is no evidence that Benjamin came across Héring's dissertation; but at a crucial passage in the "Epistemo-Critical Preface" to his own *Habilitationsschrift*, he draws on a small treatise that Héring published in the official phenomenological yearbook under the nondescript title of "Bemerkungen über das Wesen, die Wesenheit und die Idea" (Remarks on essence, essentiality, and idea). In the course of developing a theory of ideas, Benjamin adopts the term *essentiality* (*Wesenheit*) from Héring and elucidates truth accordingly (1: 218).[9] Similarly, in convolute N of the *Arcades Project*, as Benjamin incorporates the concept of the dialectical image into his methodological reflections, he draws on Héring's term once again: "What distinguishes images from the 'essentialities' of phenomenology is their historical index" (5: 577; N 3,1). And there is good reason for Benjamin to make use of *essentiality* at crucial moments in the exposition of his own "theory of knowledge," for Héring's treatise deftly responds to the phenomenological problem to which Benjamin paid particular interest while studying in Munich. Simply stated, the problem lies in the fact that phenomenological discourse charges the term *essence* with too many responsibilities. As a result, the object of inquiry succumbs to certain ambiguities, not least of which is the following: the concept of a factual object and its essence are designated by the same term. In 1916, Benjamin attempted to distinguish between essence and concept, and the results of his research represent something like an entrance exam for Husserl's school.

The research paper, "Eidos und Begriff" (Eidos and concept) (6: 29–31), is a unique item in Benjamin's dossier. In no other extant document does he adopt a mode of presentation that is so alien from his characteristic procedures and styles. To a certain extent, it reads as though it could have been written by any student of phenomenology circa 1916.[10] So different is the paper from his other writings of the period that it even violates the stylistic imperative under which he generally operates: "Never use the word 'I' except in a letter" (6: 475). The "I" in this case does not so much consist in the "person" of Walter Benjamin as in the generic phenomenologist who accomplishes the requisite reduction. And there is one further difference between "Eidos and Concept" and Benjamin's other writings of the period: it received a decidedly icy reception from Scholem, who obtained a copy, in 1918, while he was studying mathematics and philosophy at the University of Jena. As it happens, one of Scholem's professors was Paul Linke, who wrote the essay to which "Eidos and Concept" explicitly responds.[11] Upon learning that Scholem was studying under Linke, Benjamin sends him his two-year old paper, and this prompts Scholem to seek entrance into the phenomenological school as well: "I've begun to read the first volume of Husserl's *Logical Investigations*, which, until now, I have been unable to do seriously. Apropos of 'Eidos and Concept' I certainly have much to say. The differences go *much* further than Walter thinks" (S, 2: 140). The differences between eidos and concept may go further than the paper contends—and Benjamin says as much in the accompanying letter (*GB*, 1: 427)—but there is nevertheless something odd in Scholem's reaction: Linke's "Law of Phenomenology" is, broadly speaking, a defense of Husserl's *Ideas* against empiricist misconceptions of its governing motivation. Instead of immersing himself in *Ideas*, however, Scholem turns his attention to Husserl's earlier groundwork, as if there were no difference between the two.

In an effort to confirm his own suspicion that Benjamin's paper does not go very far, Scholem lent his copy to Linke, who, it seems, would be in a good position to grade it. The situation, however, is not quite so straightforward. The opening section of "The Law of Phenomenology" is a defense of the "transcendental turn" that Husserl accomplishes in *Ideas*; by 1918, Linke had turned decisively against the program of transcendental phenomenology in general and *Ideas* in particular. The nucleus of this reappraisal is already apparent in the concluding section of the essay published in *Kant-Studien*, where Linke seeks to demonstrate that many of the objections raised against the idea of the phenomenological reduction are misguided. He concurs, however, in part with one of the objections: "phenomenological insights" cannot always be communicated.[12] As long as one is concerned with the phenomenology of perception, according to "The Law of Phenomenology," the noncommunicability of phenomenological insight poses little danger to its scientific status; the danger appears only when the essence of things like "justice" or "beauty"—and presumably "law," including that of phenomenology—are under discussion.[13] A similar discussion of the noncommunicability of phenomenological insight appears some seventeen years later in the pages of *Kant-Studien*, when Eugen Fink, with Husserl's explicit authorization, once again seeks to defend transcendental phenomenology against widespread misunderstandings.[14] Confronted with the paradox that the phenomenological reduction presupposes itself, which implies that it is impossible to understand the results of the phenomenological reduction unless one has completed it oneself, Finke famously argues in 1933 for a radical transformation of phenomenology, which would finally allow it to pose the basic question with which it implicitly began: the "genetic" question of the origin of the world in transcendental subjectivity. Around 1918, by contrast, Linke becomes convinced that the phenomenological program first advanced in *Ideas* cannot be accomplished and therefore argues for its abandonment. The groundwork he published under the title of *Grundfragen der Wahrnehmungslehre* (Basic questions in the theory of perception) repudiates the "Copernican turn" and proposes, instead, a purely "object-oriented phenomenology." The attempt to disclose the origin of the world leads, in Linke's view, to the philosophical dead end of "intuitional" disarticulation.[15] At the very moment in which Scholem hands Linke a copy of Benjamin's paper, the author of "Law of Phenomenology," in short, no longer acknowledges the rightness of the law he therein defends.

In a diary entry from November 1918, Scholem recounts Linke's wholesale dismissal of Benjamin's efforts: "This afternoon I was at Linke's house in order to discuss 'Eidos and Concept.' His standpoint is the following: it is fundamental to distinguish between idea and essence. Idea = ideal object, therefore simply a group of determined objects, essence = that which is essential for this object. Concept = the nonintuitively represented object, therefore *the object in the intention*. Benjamin's paper concerns only idea and essence, not concept and essence. What Benjamin calls 'eidos' is the idea, what he calls 'concept' is essence" (S, 2: 141–42). Linke's final comments carry a tinge of exasperation, as if Benjamin's exercise was simply beside the point: "The essence of individual objects is *identical* with the object itself, since all characteristics related to an individual object are essential. *Of course*, concept and essence are different" (S, 2: 42). From Linke's perspective, Benjamin fails the entrance exam into the phenomenological school: it is as though he had only made the trivial point that the essence of something should not be confused with its concept, which is known in any case by the fact that *essence* and *concept* are different terms used under different circumstances. As a spectator to Linke's dismissal of his friend's paper, Scholem, it seems, developed a permanent conviction that Benjamin was not particularly talented in the sphere of phenomenological research. As he writes some fifty years later, "[Benjamin] gained an indistinct notion of *Logical Investigations* during his time in Munich."[16]

What is missing from Linke's assessment, however, is the problem with which Benjamin begins: of course, concept and essence are different; but how does the concept emerge in contrast to the eidos, which is immediately given, if—and this is crucial—it is not from a process of abstraction? The starting point of the analysis is doubtless *Logical Investigations*, specifically the second investigation, where Husserl laboriously demonstrates the inadequacy of abstraction theories of concept formation; but the investigation does not then produce a positive result, which would disclose the distinction between the essence of something and its corresponding concept. At a certain point in the "little treatise" (*GB*, 1: 343) that Benjamin wrote in 1916 under the title "Über Sprache überhaupt und über die Sprache des Menschen" (On language as such and on human language), he briefly resumes the investigation and tentatively suggests that abstraction is rooted in "the judging word, in judgment" (2: 154). The principal point of reference in this cryptic comment is probably Cassirer's *Concept of Substance and Concept of Function*, which presents abstraction theories of concept formation as correlates of metaphysical interpretation of concepts that are themselves rooted in the subject-predicate sentence.

The paper Benjamin wrote in response to Linke around the same time concludes with a recommendation that is as indebted to both Husserl and Cassirer: develop a theory of concept formation that does not fall prey to the theories of abstraction that they successfully refute. In this context, Linke's equation of the concept with "the nonintuitively represented object" simply bypasses the problem by way of a purely negative term: *nonintuitively*. In the letter to Scholem accompanying a copy of his paper, Benjamin indicates that he takes "little joy" in the results of his inquiry into the difference between concept and eidos but that he is nevertheless confident that he is moving in the right direction: "Linke's view that there is no need for a theory of the concept because concepts are eidetically given is—in this form at least—untenable" (*GB*, 1: 427–28). This does not mean that "Eidos and Concept" proposes its own theory of concept formation, which would not rely on the idea of abstraction and would not therefore succumb to the critiques of Husserl and Cassirer; rather, it seeks to show, even if only in highly preliminary form, that such a theory must supplement eidetic intuition in order for the specificity of both eidos and concept to be either seen in intuition or grasped by a concept. The task at hand can be thus described in relation to *Ideas*: the point of the phenomenological reduction lies in securing the eidetic character of eidetic sciences, beginning with the science of phenomenology itself; but insofar as it lacks a satisfactory theory of concept formation, the project is in danger of creating ambiguities. The requisite theory would then occupy an unsettling position within the phenomenological project, for, as a theory, it is not given and cannot be presented as a dimension of the eidetic science; but insofar as it concerns the formation of precisely those concepts that correspond to the essences under investigation, it cannot be detached from it either. The problem, in short, is to develop a theory of concept formation that belongs to phenomenology, even if it cannot itself be a product of phenomenological insight.

The starting point of the investigation Benjamin undertakes consists in reducing the factual world to the sphere of consciousness, thus allowing the eidos of an object to appear, so that it can be distinguished from its corresponding concept. The example he adduces is drawn from Linke's essay: "I imaginatively present [vergegenwärtige] the eidos of this red inkblot before me. In doing so, I abstract from the fact that it has a position in this *real* temporal continuum and at this *real* location.... The fact that the location in space, the time point at which the inkblot necessarily must 'really' be found is inessential; this fact is abstracted, it succumbs to the reduction" (6: 29–30). The concept of the inkblot, by contrast, requires that it be *of* the very inkblot to which it refers. Of course, higher-level concepts can be developed as well; but at issue in Benjamin's paper is, as it were, this "of course," which presupposes that concept formation be based on the object in view:

> I form the concept of this inkblot: therein it is essential for this concept, that is, it belongs with its content that this inkblot exists at this location of real space in one viz. several units of real time. In themselves, of course, a concept like an essence (in accordance with their essence) is timeless; but it belongs to the concept of this inkblot that it exists at this point of *real* time, of *real* space. In my view, the singular-factual is essential for the concept [of the inkblot]; but for the essence it is precisely inessential. . . . The concept is also grounded on its object; it is simply a concept "of" this object—and it can even be a concept of this singular-factual thing if this, *its* object, is a singular-factual. But an eidos of a singular-factual object is never an eidos of what makes it singular-factual. (6: 30–31)

Benjamin concludes his paper by noting that, whatever theory of concept may result, only a "minor role" will be accorded to the idea of "generality" (6: 31). More pressing than the idea of generality is that of identity, which generates the problem under consideration, since the word for the eidos word and the word for the corresponding concept "can refer to something identical [können sich auf ein Identisches beziehen]" (6: 31). When Linke responds to Benjamin's paper by claiming that the eidos of an individual object is "identical" to the object, far from refuting it, he is only directing attention to the problem with which it concludes. A few years later Héring proposes to call an essence that is identical to the thing of which it is the essence an "essentiality" (*Wesenheit*), which is not given in intuition but is, instead—and this is the passage Benjamin quotes in his *Habilitationsschrift*—the product of a long and potentially fruitless search for its elements.[17]

Around the time that Benjamin lends Scholem a copy of his response to Linke's essay, he also sends him, with similar misgivings, a set of eleven "Thesen über das Problem der Identität" (Theses on the problem of identity), in which he is no longer concerned with the difference between concept and eidos but is, instead, attentive to the exacting point where they meet, despite their "entirely different structures" (6: 31): the word for the concept and the word for the essence refer to "something identical" about which, it seems, nothing further can be said—except perhaps that it must be eminently wordlike, that is, linguistic.¹⁸ A basic schema of the various questions posed by "Eidos and Concept" and "Theses on the Problem of Identity" moves toward a common problem:

$$\text{concept} \qquad \text{essence} \qquad (6\!:\!21)$$

A schema of this kind appears among the papers Benjamin drew up around 1920, as he sought a suitable topic for his *Habilitationsschrift*. In the same folder Benjamin sketches a number of other schemata, all of which are considerably more complex than the triangle above; but from a certain perspective, the triangle is more revealing than the succeeding schemata, for it indicates precisely what is missing, namely the "thing" to which the word would presumably refer. Benjamin thus replaces the traditional semantic triangle, in which the "thing" stands at the top, while "concept" and "word" stand below, with one in which the thesis of thinghood is "turned off," so that the difference between concept and essence, on the one hand, and the identity of the word, on the other, can come to light.

Scholastics

It comes as no surprise, then, that Benjamin proposes to develop the schema sketched above into a *Habilitationsschrift* on the scholastic theory of language, for the scholastics—especially the school of Duns Scotus—immersed themselves in the triangle through which the basic structure of meaning is represented: the nature of the thing is analyzed under the rubric of *modus essendi* (mode of being) and is treated in the science of metaphysics; the nature of the concept is analyzed under that of *modus intelligendi* (mode of thinking) and treated in the science of psychology; and the nature of the word comes under the term *modus significandi* (mode of meaning), which Thomas of Erfurt analyzed under the rubric of *grammaticae speculativae* (speculative grammar) in a treatise that was generally printed in Duns Scotus's *Opera omni* and was thus misattributed to the "subtle doctor" until the early 1920s, when the renowned scholar of medieval thought Martin Grabmann finally corrected the historical record.¹⁹ Even outside of those interested in "medieval spiritual life," to quote the title of Grabmann's masterpiece, scholastic thought enjoyed a certain renaissance in the early part of the century, prompted in part by the emergence of phenomenology, as Benjamin himself notes (2: 163), for in the school of Duns Scotus, as in that of Husserl, the structure of *intentio* could be examined in its own terms, without concern for naturalizing theories of mental activity. In a brief sketch for a *Habilitationsschrift* on Scotian theory of language, Benjamin works his way into a position where he can replace the traditional semantic triangle thing-concept-word with the one that "brackets" the *modus essendi*:

> If, according to the theory of Duns Scotus, references [Hindeutungen] are founded on certain *modi essendi* in accordance with what they mean [bedeuten], there naturally arises the question how something more universal and more formal than its *modus essendi* and thus the signifier [Bedeutendes] can separate from the signified [Bedeutete] in order to function as a fundament of the signifier. And if one is able to abstract from the complete correlation between signifier and signified with respect to the question of foundation, so that the circle is avoided, then the signifier aims for the signified and at the same time rests on it.—This task is to be solved by considering the domain of language [Sprachbereich]. Insofar as a linguistic dimension [Sprachliches] can be removed from the signified and thus gained [as a dimension in its own right], it is to be designated as its *modus essendi* and therefore as the fundament of the signifier. The domain of language extends itself as a critical medium between the domain of the signifier and that of the signified. So that the following can now be said: The signifier aims for the signified and is at the same time grounded on it with regard to its material determination—not, however, in an unlimited way; rather, only with regard to the *modus essendi*, which language determines. (6: 22)

If, as Benjamin concludes, language determines the *modus essendi*, then the thing no longer occupies the top of the semantic triangle and cannot form the basis of either the *modus intelligendi* or the corresponding *modus significandi*; rather, the latter takes precedence over the former, and metaphysics becomes as much a science of language as of being. The traditional doctrine of categories must therefore be fully revised, so that it becomes a theory of meaning as well. Along with accounting for the basic concepts of being, it must bring out the categories of meaning (*significare, Bedeutung*), which, in turn, determine—tautological formulations are almost unavoidable here—the different modes in which meaning means a meant-thing. In preparation for such an ambitious project, Benjamin draws up a three-part list of more than one hundred texts under a title that is reminiscent of his earlier paper on eidetic intuition, "Word and Concept" (WBA, 1851–53). Among the items on the list, he includes primary texts such as Duns Scotus's *Opera omnia* and a recent edition of *Grammaticae speculativae*, which was still attributed to Scotus; scholarly inquires into medieval modes of thought such as Grabmann's *Gegenwartswert der geschichtlichen Erforschung der mittelalterlichen Philosophie* (Contemporary value of historical research into medieval philosophy) and Karl Werner's *Die Sprachlogik des Johannes Duns Scotus* (Duns Scotus's logic of language); Husserl's *Logical Investigations* along with his *Ideas*; well-known inquiries into the genesis and structure of the categories, such as Adolf Trendelenburg's *Geschichte der Kategorienlehre* (History of the doctrine of categories) and Eduard von Hartmann's *Kategorienlehre* (Doctrine of categories); works on the foundations of logic, including Hermann Lotze's *Logik* and Emil Lask's *Lehre vom Urteil* (Theory of judgment); and near the top of the list, a work that stands out by virtue of the fact that its nondescript title points in the direction of Benjamin's own project, namely Martin Heidegger's *Die Kategorien- und Bedeutungslehre des Duns Scotus* (Duns Scotus's doctrine of categories and meaning), with which he was "habilitated" in Freiberg under Rickert's direction.

Upon first reading Heidegger's *Habilitationsschrift*, Benjamin was unimpressed. As he tells Scholem in a letter from January 1920, it makes no significant contribution to the problem that his own work would address:

> It is incredible that someone can be habilitated with such a work, which requires for its completion *nothing* more than a lot of study and the command of scholastic Latin and which, despite all of its philosophical make-up, is at bottom nothing more than a piece of good translation work. The undignified sycophancy of the author toward Rickert and Husserl does not make the reading any more pleasant. Philosophically the linguistic philosophy of Duns Scotus remains undeveloped and thereby leaves behind no small task. (*GB*, 2: 108)

Around the time in which Benjamin criticizes Heidegger in these terms, he writes "The Task of the Translator." In this context, the assertion that Heidegger's work is little more than a "piece of good translation" cannot function as a wholesale condemnation, especially since "The Task of the Translator" takes over, perhaps with help from Heidegger, the Scotian term *modus significandi* and translates it as "mode of meaning [Art des

Meinens]" (4: 14). Benjamin was impressed enough with *Dun Scotus's Doctrine of Categories and Meaning* to add another of Heidegger's texts to his list of works to be consulted—in this case, an obscure article on "new research in logic" that Heidegger published in a journal for "Catholic Germany" before the two of them perhaps encountered each other in Rickert's seminar on Bergson and the metaphysics of time.[20] And a year after Benjamin dismisses Heidegger's *Habilitationsschrift*, he rescinds his original assessment and, as a result, abandons his own plans for a *Habilitationsschrift* on Scotian speculative grammar: "The text of Heidegger perhaps does reproduce that which is most essential in scholastic thought for my problem—in an entirely nontransparent way, to be sure—and even the genuine problem can already somehow be detected in connected with it" (*GB*, 2: 127).

By "my problem," Benjamin probably means the task of evading the "circle" he identifies in his sketch of the project on Duns Scotus: "The signifier aims for the signified and at the same time rests on it." Heidegger's argument proceeds along similar lines, beginning with his translation of *modus essendi* as "givenness," which results in a corresponding reordering of the traditional semantic triangle: "On the leading thread of a givenness (*modus essendi*), which, for its part, is only a givenness as something known (in the *modus intelligendi*), the forms of meaning (*modi significandi*) are read off" (He, 320–21). What Benjamin describes as Heidegger's "sycophantic" relation to Husserl is comprehensible from this perspective: following the discussion of "pure grammar" in the fourth of the *Logical Investigations*, he presents *modi significandi* as phenomena in their own right.[21] In other words, they are given in their own manner of givenness and cannot be seen as such when they are interpreted as derivatives of self-subsisting things. For a "modist" such as Thomas of Erfurt, *modi significandi* are something like the underlying principles of well-constructed discourse.[22] For Heidegger, by contrast, they constitute the prejudgmental stratum of meaning, which is difficult to capture precisely because "capturing" a phenomenon generally involves judgment. This is probably what Benjamin means when he speaks of the "genuine problem," for in the following passage from Heidegger's work—which is marked by a certain tension, if not by an outright contradiction—the problem comes to light: "This a priori organized connection of meanings into semantic complexions does not suffice to constitute what we call valid meaning. In the meaning-complexions as such, as they are organized through the *modi significandi*, the truth-value, which is tied to judgment, is not yet realized. But insofar as the valid sense of judgment, which is expressible in sentences, is determined through the structure of such modes of meaning, something is realized already within the mere realm of meaning" (He, 336–37).[23]

It is not as though the *modi significandi*, according to Heidegger, simply make up a stratum of meaning that has not yet been realized in judgment, which, for its part, constitutes the telos of meaning-related acts. For *something* is already realized in the *modi significandi*, even in the absence of judgment: a "mere" realm of meaning lying below or alongside the realm proper. Following Rickert, Heidegger calls the prejudgmental stratum of meaning "sense."[24] Yet the character of this enigmatic stratum of sense is by no means clear—which justifies Benjamin's remark about the "nontransparency" of the analysis. In the conclusion to his *Habilitationsschrift*, amid much talk of "living spirit," Heidegger presents the enigma in the following terms: "how [does] the 'unreal,' 'transcendent' sense [Sinn] guarantee us true reality and objectivity" (He, 406). During the time in which Benjamin decided against pursuing a *Habilitationsschrift* on scholastic theories of language, he also abandoned another long-standing project, which concerned the intuition or perception of color. From around 1915 until the early 1920s, he produced a large number sketches, remarks, aphorisms, fragments, and at least one dialogue on this theme. At first glance, the two projects seem to have nothing to do with each other: the first is ontological-linguistic and largely directed toward medieval modes of thinking; the other is aesthetic with a certain inclination in the direction of modern art. If, however, as Benjamin asserts, something of Heidegger's treatise captures what is "most essential" and this "something" can be identified with the enigma of what Heidegger, following Rickert, calls prejudgmental "sense," then the two projects converge at a very precise point: certain colors, especially those of fantasy, are *modi significandi* that cannot be derived from any *modi essendi* and are wholly independent from their corresponding *modi intelligendi*, which would be the relevant color concepts. As Benjamin decided against working out the first project, the other second one lost its retrospective foundation.[25]

In his studies of color and fantasy Benjamin repeatedly emphasizes the stance of receptivity: "color must be seen" (6: 109, 119); "color can only be received" (6: 118).[26] This is true not only of the colors that present themselves as properties of certain substances but of the colors of fantasy as well. Because the colors of fantasy cannot be constructed according to a concept and because they are in a certain sense nonsensible—otherwise they would not be colors "of" fantasy—they are comparable to space and time, as Kant discusses them in the "Transcendental Aesthetics" of the *Critique of Pure Reason*. Just as space and time are both pure and nonconceptual, so, too, are the colors of fantasy. And just as a merely discursive intelligence, according to Kant, would not be able to understand the difference between outer and inner if space were not already given, so a mind that is not already receptive to the colors of fantasy can form no conception of what they look like. From some perspective, then, a revision of the "Transcendental Aesthetics" would include an exposition of color. But whose perspective? A simple answer runs: from the perspective of those to whom such colors appear, namely colors that correspond to nothing in reality and whose concepts cannot therefore be constructed. Whenever "nonthings" of this kind appear, they exhaust themselves in their appearance, and for the same reason, they can be called "pure phenomena," which, as such, are the objects to which a "pure phenomenology" must turn. While "Two Poems of Friedrich Hölderlin" draws from Kantian thought and phenomenology, with emphasis on the former, Benjamin's studies of color do the same thing, with emphasis on the latter.

Goethe and Children

From late in the second decade of the twentieth century to the early 1920s Benjamin created a bibliography of some twenty-five texts that address the question of color (WBA, MS 530). In contrast to the lists he produced under the general title of "Word and Concept," the one he wrapped around the theme of color is remarkable for its heterogeneity. The list includes, among other items, Anton Marty's *Frage nach der geschichtlichen Entwickelung des Farbensinnes* (Inquiry into the historical development of the sense of color), which expands upon Ludwig Geiger's provocative papers on language and color; studies of color and fantasy conducted either from a Darwinian perspective, such as Grant Allen's *Color-Sense*, or from the perspective of empirical psychology, including Emil Lucka's *Phantasie: Eine psychologische Untersuchung* (Fantasy: A psychological investigation); a study of blushing from a psychoanalytic perspective; art- and cultural-historical reflections such as Viktor Goldschmidt's *Farben in der Kunst* (Color in art) and Luisa von Kobell's *Farben und Feste* (Colors and festivals); a study of ecclesiastical colors; a translation of Leonardo da Vinci's *Treatise on Painting*; and a lexicon of symbolism and mythology. The list is not altogether shapeless, however, for it is framed by the following texts: Alfred Peltzer's *Ästhetische Bedeutung von Goethes Farbenlehre* (Aesthetic significance of Goethe's theory of color) and Hans-Gerhard Gräf's nine-volume edition of *Goethe über seine Dichtungen* (Goethe discussing his poetry). The study of color and fantasy is thus enclosed within the framework of Goethe's famous *Farbenlehre* (Theory of color).

The place of Goethe on Benjamin's list comes as no surprise, for no study of color in a German-language context can ignore for long the looming enigma of the *Theory of Color*, dedicated as it is to the destruction of Newton's reputation in the field of optics. And with regard to the theme of color *and* fantasy, the name of Goethe rises to even greater prominence, for the *Theory of Color* begins with a consideration of "fantastical" colors: "It is fair to place these colors first," Goethe writes in the first paragraph of its first section, "because they belong entirely, or in a great degree, to the subject—to the eye itself."[27] Goethe calls the colors in question "physiological" because they correspond to both the nature of investigated object and to that of the investigating subject. In other words, physiological colors are an immanent phenomenon, without any defining relation to an exterior object—whether it be a material medium or a colored surface. For this reason, however, physiological colors are fleeting; they appear only in transition, as the eye begins to adjust itself to a new visual environment. As Goethe notes, again in the first paragraph of his study, "because they [physiological colors] are too evanescent to be arrested, they were banished into the region of phantoms, and under this idea have been variously described"—as, for instance, *colores imaginarii, colores phantastici*, or *Scheinfarben* (apparent colors).[28] When Benjamin begins an inquiry into the colors of fantasy, he is thus revisiting the point of entrance for one of the major monuments of modern German culture.

And yet, as manuscript 530 also suggests, Benjamin retains a certain distance from Goethe: the text itself is not on his list, only studies of it are there. The *Habilitationsschrift* Benjamin would eventually publish begins with one of Goethe's descriptions of the "primordial phenomenon" (*Urphänomen*) in the *Theory of Color* (1: 207); but for the most part, in the studies of color themselves, Goethe's work plays only a muted role. A review that Benjamin published in the same year as his *Habilitationsschrift* gives some indication as to why he would keep his distance from a theory of color that might otherwise accord with his own. However much the proponents of the *Theory of Color* may want to reclaim it for "higher culture"—or, in the case of the theosophist and pedagogue Rudolf Steiner, whose study of Goethe's theory of color is also on the list, reclaim it for "higher" spheres of existence—it cannot be abstracted from its polemical setting: the refutation of Newton's *Optics*. The book under review is a new edition of the *Theory of Color* with an introduction by one of Steiner's disciples, who tries to distance Goethe from his anti-Newtonianism:

> In the hundred-year debate about Goethean optics there is a definite, decisive question that should no longer be suppressed: Is Goethe's physical theory of color fundamentally different from Newton's; that is, under certain circumstances can it stand independently from Newtonian theory? Or is the contrary true; that is, must one of them be true, while the other is false, and vice versa? And if Newton's theory is really not an argument against Goethe's, and it is accurate to say that physics "cannot make a judgment about Goethe's *Theory of Color*" [as the author of the introduction does]—if, in short, it is "not competent in this question," then fairness dictates that one emphasize in response that Goethe himself was entirely unclear about his relation to Newton, whom he discussed in drastic terms, calling him the "chief of the Cossacks." (3: 149)[29]

The formulations that Benjamin cites in this review do apply, however, to his own studies of color: physical science is "not competent" to judge its claims. In this way, his studies maintain as much distance from Goethe's theories as they do from Newton's. This can be put in another way: Benjamin's reflections on the phenomenon of color not only fail to get beyond the opening paragraph of the *Theory of Color*; they never even quite reach that point. Goethe contrasts the "physiological" colors with "physical" ones, which result from interaction with material media and "chemical" colors that are fixed onto objects. The primary point of contrast for "physiological" colors is, however, their "pathological" counterparts, which come under discussion in the appendix to the first section. A "healthy" eye, according to Goethe, experiences physiological colors in transitional moments; a "defective" eye, by contrast, undergoes certain color pathologies, such as seeing colors where there are none and being attracted to especially strong or vivid colors. The final paragraph of the appendix makes it clear, though, that the difference between defective and healthy is less a matter of nature than of *Bildung* (culture, education, self-formation). Just as Goethe's *Wilhelm Meisters Lehrjahre* (Wilhelm Meister's years of apprenticeship) is the prototypical bildungsroman, so is his theory of culture a paradigmatic *Bildungslehre*: "Finally, it is also worth noting that savage nations, uneducated people [ungebildete Menschen], and children have a great predilection for lively colors; that certain colors make animals fall into a rage; that cultured people [gebildete Menschen] avoid lively colors in their dress and their surroundings, and appear to distance themselves thoroughly from them."[30] The visual norm lies in the eyes of adult, "gebildete" Europeans, whose education instills a noticeable aversion to striking varieties of color.[31]

To the extent that one attains culture, according to Goethe, one shies away from any colors that would overly emphasize its color-character. Both children and "human beings in their natural state" (*Naturmenschen*) find themselves attracted to energetic and vivid colors, and they likewise display an unmistakable "inclination toward the motley [das Bunte]. But"—Goethe immediately adds, as though he had caught himself in a conundrum—"the motley really emerges when colors in their greatest energy are placed together without harmonic balance."[32] Benjamin may not have had this particular passage of the *Theory of Color* in mind when, around 1915 or 1916, he began a study under the title "Die Farbe, vom Kinde aus betrachtet" (Color, considered from the perspective of the child); but the study can nevertheless be considered a riposte to Goethe's remarks on "pathological" colors. Goethe and Benjamin agree that there is a motley character to the manner in which children see things; but motliness—which Benjamin would later investigate under the rubric of "montage"—cannot be ascribed to harmonic imbalance precisely because, as he claim in a contemporaneous study of the theme, color gives "no support for a doctrine of harmony. . . . Among colors, number is only the expression of infinite possibilities, which are to be comprised only systematically" (6: 109). For similar reasons, Benjamin declines Goethe's invitation to tabulate the correct number of primary colors. Instead of reflecting on primary colors, Benjamin emphasizes the primacy of the colors that the child sees, which are unavailable to those who, as it were, survive their youth. And children are not, as Goethe claims, so much "attracted" to motliness as indifferent to the substantial character of things, which means that their visual world cannot fail to be motley, since it is not defined by the solid forms through which it deduces underlying substances: "The colorfulness [Farbigkeit] of a child's picture proceeds from motliness [Buntheit]. . . . Motliness does not affect in an animal manner [affiziert nicht animalisch] because it constantly originates from the unbroken fantasy-activity of the child's soul" (6: 111).

The opposition between Goethe and Benjamin in the matter of motliness could scarcely be more sharply drawn: according to the former, a motley appearance results from a "harmonic imbalance" among "chemical" colors; according to the latter, motliness is as close to a purely "physiological" color as possible. And just as the "physiological" colors, according to Goethe, are fundamentally fleeting, so is motliness in Benjamin's account, for childhood is itself evanescent, even as the origin of the phenomenon in question is emphatically "constant." What Goethe says of "primordial phenomena" at the end of the section on "diotripical colors of the first class" is valid verbatim for motliness as seen by the child: "nothing in appearance lies beyond them [nichts in der Erscheinung über ihnen liegt]."[33] There is neither an independent substance nor a deductive law that explains what is thus sensed. As described in Benjamin's study, however, motliness is not simply one primordial phenomenon among others; it is a primordial phenomenon to the second power, an *Ur-urphänomen* that is so primordial that the subject who sees it must be equally primordial—closer to an uneducated and uncultured "natural human being" (*Naturmensch*) than to a potential coworker in the field of visual science. The old adage from "the Ionian school" that Goethe adapts for his own purposes, "like is known only by like [nur von Gleichem werde Gleiches erkannt]," can likewise be applied to Benjamin's studies of color.[34] The name of the motley-like subject is not so much "the child" as its "soul"—"its" because the word *child* is neuter in German and because, in addition, Benjamin, who is elsewhere concerned with gender difference, makes no such distinction in this case.[35] Even more exactly, however, the name of the still neutral subject is "unbroken fantasy-activity," which removes the soul of the child from its "animal" affectivity, while at the same time making it receptive to the pure play of appearances. Benjamin thus continues in "Color," as he gropes toward a conclusion: "Because, however, it [unbroken fantasy-activity] sees purely, without letting itself be psychically perplexed, it [the child] is something spiritual [etwas Geistiges]" (6: 111).

With this remark, Benjamin returns to the claim with which his study begins: "Color is something spiritual [etwas Geistiges], something whose clarity is spiritual or whose blending is nuance, not fuzziness" (6: 110).

To say of color that it is "something spiritual" means, above all, that its "sense" is not a matter of sensation. The same goes for the child, whose "unbroken fantasy-activity"—as opposed to a "faculty" or "power" of imagination, which must be activated—"constantly" grants it access to the pure phenomenon of motliness. To say of color that it is nonsensory does not mean, however, that it should be seen as something intellectual. In "Two Poems of Friedrich Hölderlin," Benjamin uses the terms *geistig* (spiritual) and *intellektuel* (intellectual) almost interchangeably. In "Color," by contrast, he makes an exacting distinction: "intellectual" derives from the unity of self-consciousness, whereas everything "spiritual" is always already its own unity—hence, the emphasis on "something" (*etwas*). The objectivity of this "something" is correspondingly different from the objectivity of things that owe their origin to sensible-intellectual synthesis: "The objectivity [Gegenständlichkeit] of color does not rest on form; without empirically affecting intuition, it goes directly to the spiritual object through the isolation of seeing. It supersedes the intellectual combination of the soul and creates pure mood [reine Stimmung] without thereby giving up the world" (6: 111). A "pure mood," unlike its affective counterparts, does not result from excitations that are themselves measured in accordance with a certain self-estimation. One grows angry, for instance, because one feels belittled by someone. Since "something spiritual" does not derive from the unity of consciousness and is accordingly unrelated to any units of self-measurement, the mood it creates is as far removed from anger as it is from pride and cannot be grasped as a synthesis of the two.

All of this can be summarized in the following formula: "something spiritual" can be received only by someone who is equally spiritual, and the "pure receptivity [reinen Empfänglichkeit]" (6: 111) that constitutes the "soul" of this neutral one consists in unqualified openness to a spiritual world that is itself "moody." This formula could be made even more precise with the replacement of the term *something spiritual* by Rickert's term *sense*: every color then appears as a certain "unreal" unity that gives itself as such only so long as the recipient is characterized by a correspondingly pre-predicative condition. This, however, is essential: such a condition cannot be adopted at will. It is for this reason, above all, that it belongs to the sphere of childhood, which cannot be "recovered." For the same reason, the sphere is closed to intuition; there is no evidence for it, least of all the evidence of remembrance, insofar as the latter can be activated at will.

A paradox thus arises: "childhood" designates the neutral sphere in which pure phenomena exhaust themselves in giving themselves; but everything that falls under this designation must be correspondingly constructed—constructed, that is, in and by fantasy, now understood as an intermittent "force" or "power" that can be realized at will.

The paradox generates two major consequences. On the one hand, any construction of a sphere that owes its origin to fantasy must also be a deconstruction of the very same sphere. The term for "deconstruction" that Benjamin develops in this context is *Entstaltung* (deformation), which describes the work of fantasy in the absence of its ownmost sphere: "A moment of constructivism is proper to all fantastical formations [Gebilden]—or (spoken from the perspective of the subject) a moment of spontaneity. Genuine fantasy, by contrast, is unconstructive, purely deforming [entstaltend]—or (seen from the perspective of the subject) purely negative" (6: 115). On the other hand—and more importantly for the project initiated in "Color, Considered from the Perspective of the Child"—every claim about the sphere of childhood is subject to a certain degree of conjecture, and none are more uncertain, none less murky, than those that direct attention toward its end. Thus, Benjamin ends the fragment with the following remark: "Children do not feel ashamed, for they have no reflection, only vision" (6: 112). In a sketch written a few years later, under the perfunctory title "Erröten in Zorn und Scham" (Reddening in rage and shame), he makes the exact opposite claim, which perhaps indicates why he would abandon "Color" in midsentence: "Highly developed shame feeling among children. That they so frequently feel shame is connected with the fact that they have so much fantasy, particularly in their earliest maturity" (6: 120).

The Color of Shame Versus the Call of Conscience

To use a term that Goethe employs in his attack on Newton's optics, the phenomenon of shame is the *experimentum crucis* of Benjamin's studies of color.[36] The contradiction between the last sentence of "Color, Considered from the Perspective of the Child" and the final remark of "Reddening in Rage and Shame" indicates at the very least that Benjamin reconsidered the question as to whether children are exposed to the experience of shame. In the earlier study, shame is presented as a function of reflection: the absence of self-reflection makes it impossible for children to feel shame. In the later fragment, by contrast, the feeling of shame is a function of fantasy: the more energetic the latter, the more frequent the former. And in a dialogue entitled "The Rainbow," which Benjamin probably wrote between the two sketches—and which is discussed at length in the following chapter—a change in weather marks a change of attitude: "Children dwell entirely in innocence, and in blushing they themselves return to the existence of color. In them fantasy is so pure that they are able to do so.—But look, it stopped raining. A rainbow" (7: 24). In no other passage is Benjamin more explicit about the "reductive" character of shame, which leads children back to a sphere where they are what they see: the neutral "subject-object" of motley vision.

It is, of course, impossible to determine what exactly led Benjamin to change his mind about the phenomenon of shame. Perhaps it had something to do with the his reflections on the collapse of the youth movement at the beginning of the First World War, which, for Benjamin, was intimately associated with the suicide of Fritz Heinle. Brief citations of Heinle's poetry routinely appear in Benjamin's color studies, especially the following line, which can be found immediately before the passage from the "Rainbow" dialogue quoted above: "If I were made from material, I would color myself [Wäre ich aus Stoff, ich würde mich färben]" (7: 24; cf. 6: 121). The child who changes its color represents the "material" of Heinle's poetic world—in contrast to which there is the sight of youth shamelessly marching in step with the Wilhelmine regime. It was also in this period that Benjamin sought entrance into Husserl's school and thereby obtained a certain phenomenological *Bildung*, which allowed him to replace terms such as "something spiritual" with meaning-related equivalents. Thus, at the end of a brief yet decisive study entitled "Über Scham" (On shame), which begins by explicitly taking issue with a passage from Goethe's *Theory of Color*, the phenomenon under consideration acquires a name that both derives from and transcends the terminology Husserl first developed in *Logical Investigations*: not the feeling of shame itself but, rather, its corresponding coloration is called "expressionlessly signifying appearance [ausdrucklos bedeutende Erscheinung]" (6: 71).

The passage from the *Theory of Color* to which Benjamin draws attention at the beginning of "On Shame" concerns the perfection of human physiognomy as compared with that of the ape. Benjamin may have been drawn to this passage because it is once again concerned with motliness—now as a term that refers to the phenomenon of the seeing subject rather than to that of object seen: "The following remark from Goethe leads to the innermost meaning of the redness that, with shame, creeps over the human being: 'If in the ape certain naked parts appear motley, with primary colors, this simply shows how distant such a creature is from perfection, for one can say: the more noble the creature, the more everything material [alles Stoffartige] is assimilated; the more essentially its surface coheres with its interior, the less primary colors can appear on it. For, precisely where everything is supposed to make up a perfect whole, something specific cannot be separated off here and there'" (6: 69).[37] The same passage from the *Theory of Color* may have also—and at the very same time—attracted the attention of another avid reader of Goethe, namely Franz Kafka, who published a story in Martin Buber's journal *Der Jude* that is narrated by an ape who is literally "reddened" when hunters shoot him in the hip and then again in the face. The name he receives as a result, Red Peter (*Rotpeter*), is a shameful reminder of the primary color that, by appearing on the primate's skin, makes him imperfect in his natural imperfection.[38] Benjamin, for his part, does not further consider the figure of the ape. Nor does he appear to give any thought to the variety of colors appearing on the skin of human beings. But the absence of any reflection on skin colors cannot be taken as a sign of a thoughtless racialism that simply assumes that the "normal" color is the—impossible—absence thereof.[39] For Benjamin goes after the core of Goethe's remark, which is, for its part, the fundamental premise of racialist discourse: namely, the outer appearance of human beings, especially the colorations of their skin, expresses their inner disposition.

This is, of course, precisely what "expression" (*Ausdruck*) designates: the interior presses outward. According to Benjamin, the coloration of shame not only does not express the inner disposition of the one who undergoes shame; rather, it does away with the very interiority that makes expression possible: "The redness of shame does not tarnish the skin; no inner division, inner disintegration appears therein appears on the surface. It announces absolutely nothing inward. If it were to do so, it would again truly be occasion enough for new shame, the shame of the human being thus discovered in his fragile soul; instead—as it is in truth—it extinguishes with its redness all reason for shame, everything interior" (6: 69). The phenomenon of shame is not therefore one phenomenon among others; it is the phenomenon that discloses a sphere of outwardness that runs counter to the immanence of absolute consciousness. Those who show their shame withdraw from appearances, without thereby finding refuge in a metaphysical region above or beyond all phenomena: "The redness of shame does not well up from the interior . . . but, rather, gushes from outside onto the one who is ashamed and extinguishes the disgrace [Schande] and simultaneously withdraws him from disgracing [Schändern]. For the dark red with which the shame douses him draws him under a veil and withdraws him from the gaze of human beings. The one who feels shame [Wer sich schämt] sees nothing but is also not seen" (6: 69–70).

The technical language of phenomenology falters in view of this phenomenon, which can only be described as the phenomenon of radical de-phenomenalization. The first of Husserl's *Logical Investigations* begins by distinguishing between expression and indication, under the presumption that a sphere of pure expression, without any residual indication, can be attained in the form of interior monologue. For Husserl, an indication communicates something to someone under certain condition; by contrast, an expression means something—so much so that meaning and expression are correlated terms: "It belongs to the concept of an expression to have a meaning. . . . A meaningless expression is therefore, properly speaking, no expression at all" (Hu, 19, 1: 59). The coloration of shame would appear at first glance a kind of indication, even if only in an auxiliary sense of the term, where it is applied to "natural" events that indicate the existence of something other than themselves. Just as smoke indicates the presence of fire, so blushing indicates that the blusher is ashamed; and just as the fire does not itself mean anything by producing smoke, so those who feel shame do not mean to show their feelings—on the contrary. And this, for Benjamin, is the essential point: those who undergo shame mean to conceal themselves. If the coloration of shame can be called "something spiritual," then it undermines the terminology with which Husserl launches the phenomenological program, for it consists in a meaning that is without expression and that can be confused, as a result, with a "natural" indication. As the final remarks of "On Shame" emphasizes, moreover, the phenomenon under investigation splits the viewing subject, so that its faculty of understanding is at loss to understand what the "soul" sees:

> The color of shame is pure: its red is neither colorful [Farbiges] nor color [Farbe] but coloration [Färbendes]. It is the red of transition from the palette

of fantasy. For that authentic, purest [eigentliche reinste] coloring light is none other than the colorful, many-colorful color of fantasy. To it there is appropriated [eignen] the colors in which a being appears without being an expression of an interior. And only this colorful appearance is pure, and for this reason it is incomparably powerful: not in terms of its effect on the understanding, to which it betrays nothing but, rather, its effect on the soul, to which it says everything. Expressionlessly signifying appearance [ausdrucklos bedeutende Erscheinung] is the color of fantasy. Expressionlessly signifying appearance of transition [Vergehens] is the redness of shame. (6: 71)[40]

The dismantling of the terminology with which Husserl begins *Logical Investigations* goes hand in hand with the identification of the motivation for the "turning off" of the "natural" attitude. To be affected by something, to be excited by someone—these not only elicit shame-feelings; they are inherently shameful.[41] Those who show that they are overcome with shame, however, recede from phenomenality and return to—without actively remembering—the time of childhood, when there may be shame but there is nothing to be ashamed of. The moment of shame can be understood as the index that there is still "unbroken fantasy-activity." And the evanescent character of the appearance in question makes it—in Goethe's terms—the "physiological" color par excellence. The coloration of shame is the "natural" reduction of the "natural" attitude, which can be accomplished only by those who have no intention of doing so: it gives evidence of pure outwardness, which can be called "death" and which solicits terms like "mortification."

The coloration of shame is thus comparable to the call of conscience, as it appears in the book Heidegger would publish under the programmatic title *Being and Time* a decade after his *Habilitationsschrift* on Scotian speculative grammar. And there is good reason for the comparability of the two phenomena: neither Benjamin nor Heidegger accepts the premise on which the phenomenological reduction is founded, namely, that the "natural" attitude is genuinely natural; and yet neither of them adheres to the "realist" branch of the phenomenological school, which seeks to supplant *Ideas* by reinitiating a purely "object-oriented phenomenology." On the contrary, for both Heidegger and Benjamin, the reduction of the "natural" attitude has already taken place in everyday activity. In an introductory paragraph to "Schicksal und Charakter" (Destiny and character), Benjamin goes so far as to invoke the idea of reduction by reducing the term to its original sense as "lead backing": "Between the human being at work and the outer world, everything is reciprocal action [Wechselwirkung]; their circles of action interpenetrate each other; their representations may be ever so different, but their concepts are inseparable. . . . The outside in which the active human being finds himself can in principle be reduced [zurückführt], however much one wishes, to his inner; his interior, however much one wishes, to the outside. Indeed, his interior can be seen in principle as his outside" (2: 173). For the "human being at work" (*wirkenden Mensch*), the "natural" attitude is always already "placed out of action." Far from positing a world of things that affect consciousness and to which it reacts in return, there is only the "working" situation, and the distinction between interior and exterior is purely functional, not substantial. Benjamin's brief exposition of the "human being at work" so perfectly captures Heidegger's meticulously crafted analysis of the phenomenon of the workplace in which *Dasein* by and large finds itself that it would not be an exaggeration to suggest that he may have been prompted to analyze the phenomenon of the workplace as a consequence of coming across Benjamin's essay in the pages of the avant-garde journal *Argaunauten* in 1921.[42]

In any case, the lines of the comparison are clear: just as, for Benjamin, the exposition of the "human being at work" is a preliminary part of the larger project of determining the relation between destiny and character, so, for Heidegger, the analysis of the phenomenon of the workplace prepares the way for a full determination of the relation between being and time. Caught up in the workplace, *Dasein* does not conceive of itself as subject affected by objects but tends, instead, to understand itself in term of what it actually does. Since, however, the existential character of *Dasein* consists in its sheer possibility, without any underlying or overarching "actuality," it thereby covers up its own being and finds itself in the movement of "falling." Heidegger goes in search of a movement that would run counter to the fallenness that consists, above all, in trying to remain secure, holding on to whatever presents itself as handy. In this way, *Being and Time* repeats the basic structure of *Ideas*: both begin with a description of an "everyday" attitude and proceed to describe its reduction. Instead of returning the "natural" attitude to its origin in absolute consciousness, however, *Being and Time* leads *Dasein* back to its "ownmost" possibility, which is called "death." By exposing itself to the possibility of its impossibility, *Dasein* is drawn out of its self-actualizing tendency and led back to its existential character. But—and this is decisive—one cannot intuit one's own death; it does not appear as such. Heidegger therefore looks for a phenomenon that vouches for a nonphenomenon: "*Dasein* needs evidence of a being-able-to-be-oneself that it already is according to its possibility. What in the following interpretation is claimed to be such evidence is known in the everyday understanding of *Dasein* as 'the call of conscience.'"[43] The call of conscience thus has the same status in *Being and Time* as the sphere of inner monologue in *Logical Investigation*. In both cases, there is no residue of unfulfilled intention, nothing merely "indicated," nothing beyond pure "expression."[44] Conscience is therefore certain: *das Gewissen ist gewiß*. The certainty of such "evidence" is irreducibly different than the self-certainty of the ego. Instead of using *conscience* in this context, Benjamin prefers less ego-associated terms, such as "surety" (*Bürgschaft*) and "guarantee" (*Gewähr*). For Heidegger, the phenomenon of non-egological certainty recalls *Dasein* to its existential character—sheer possibility, without any substance or function. Since, however, *Dasein* has always already actualized itself as someone in particular, it always finds itself acting contrary to its own existential character: "And this means: *Dasein* as such is guilty [schuldig]."[45] To the extent that *Dasein* is receptive to the nothingness to which it is called, it resolves to be how it always already finds itself: guilty for being someone. As Heidegger then adds, in resolving to be guilty, it sees through itself. Just as the call says nothing, it sees that there is nothing "there" for it to see: "Resoluteness gives *Dasein* its authentic transparency."[46]

To Heidegger's assertion of transparency—perhaps the most ambiguous of all the technical terms introduced into *Being and Time*—Benjamin's studies of color proleptically respond: it is not in self-transparency that the self "sees" its ownmost possibility of not-being-able but in the phenomenon of self-coloring. To the "silence" of the call, Benjamin similarly responds with the "expressionless" character of this coloration. And to the "nothing" that *Dasein* says to itself in the "call of conscience," Benjamin responds with the "everything" that the "soul" receives in the absence of understanding. Whereas the "call of conscience," according to Heidegger, is a matter of the "moment" (*Augenblick*), the coloration of shame is altogether transient: its time is always "now." The identification of the "call of conscience" as the witness for the paradoxical phenomenon of total de-phenomenalization makes it possible for Heidegger to make a transition from the analysis of authentic temporality to an exposition of its historical counterpart, in which *Dasein*—shamelessly perhaps—finds its "destiny" and therein its "people" (*Volk*).⁴⁷ For Benjamin, by contrast, the coloration of shame shows how the world would appear if there were nothing in the world to be ashamed of.

Toward a Messianic Reduction

The color under consideration in "On Shame" wipes away the disgrace of whomsoever is thus ashamed. This is the reason for its transience: it does not simply come and go, as do "physiological" colors; it goes away by erasing its very motivation. Nothing is therefore required for the restitution of the status quo ante, least of all a judgment about the one who feels ashamed. More to the point: a judgment such as "so-and-so has done something wrong, as evidenced by this visual marker" is not only *not* equivalent to the sight of shame; it makes it impossible to see the "given" as such. The coloration of shame and the faculty of judgment are thus mutually exclusive: the first eludes the second, and the second can at most interpret the first. The "natural"—which is to say, in Benjamin's terms, mythological—interpretation reads like this: the one who experiences shame is guilty of doing something shameful. If Heidegger and Benjamin do indeed proceed along a similar line of phenomenological inquiry, then the one who officially entered into Husserl's school by becoming his assistant takes a monumentally wrong turn when he interprets the phenomenon of de-phenomenalization in terms of guilt. The coloration of shame is "pure," as Benjamin repeatedly emphasizes, because it frees those who experience it—and also those who see it being experienced as such—from this very experience. Guilt, by contrast, is irreducibly murky; as a result, it excites the faculty of judgment, which makes distinctions in both senses of the phrase. Whenever the faculty of judgment is excited, there can be no phenomenon of shame, only a deepening of guilt.

As a consequence of encountering Kierkegaard's *Concept of Anxiety*, Benjamin drew up a series of notes on the problem of "original" or "inherited" sin (*Erbsünde*) that, according to Scholem, was to serve as the basis for a longer project.⁴⁸ As with his intention to develop his color studies into a major work, little came of this plan. Its traces, though, are legible in the densely argued paragraphs of "Destiny and Character." Whatever distinguishes the concept of destiny from that of character, they share one essential trait: each of them refers to a "connection" (*Zusammenhang*) that is not "immediately visible" (2: 174).⁴⁹

Because neither of these "connections" appears in its own right, both concepts belong to the order of the "sign" (*Zeichnen*) and its "designation" (*Bezeichnete*), in contrast to that of the "signifier" (*Bedeutende*) and the "signified" (*Bedeutete*). In other words, neither destiny nor character is pure phenomenon. Any application of the concepts of destiny and character is therefore a function of the "natural" attitude, which not only presupposes that consciousness and the world of things affect each other but takes for granted that nothing can be immediately given as a result. Whoever adopts a "natural" attitude overlooks phenomena and posits systems of decipherable signs instead. Benjamin does not directly discuss destiny and character in terms of the "natural" attitude; but as he turns from his discussion of destiny to a corresponding consideration of character, he comes very close: "It is no accident that both orders are connected with interpretative practices and that in chiromancy character and destiny coincide altogether. Both concern the natural human being, more exactly, nature in the human being, and precisely this announces itself in the signs of nature, either those that are there by themselves or those that are given through experimentation" (2: 176).

Far from being a given, the nature in the human being is a product of interpretation, and the first thesis of interpretation consists in taking the relation between self and world to be in need of interpretative practices. Of the two modifications of this "natural" attitude under discussion in Benjamin's essay, destiny is farthest from the sphere of phenomenality. Not only is the concept of destiny "connected" with certain practices of interpretation; it refers to a "connection" that, as a "net," ensnares anything which would otherwise have the power to act on its own—that is, every living being. Benjamin accordingly defines destiny as the "guilt connection of the living [Schuldzusammenhang des Lebendigen]" (2: 175). The murky character of this connection excites judgment, which is itself always under the jurisdiction of a rule or law: "It was not in law [das Recht]," Benjamin adds, "but in tragedy that the head of genius arose for the first time out of the fog of guilt [Nebel der Schuld]" (2: 174). The head of the tragic hero does not dispel the fog but only makes its fogginess apparent "for the first time." Guilt calls for an epochal construction of the historical process for this reason: unlike the coloration of shame, guilt says nothing "to the soul" (6: 71); it can be captured only in the form of a construction that distinguishes one epoch or worldview from another in terms of its relation to a pervasive murkiness. As for the specifically "moral infantilism" (2: 175) of the tragic hero, who knows he is better than the gods but cannot find the words to express his knowledge, it is fundamentally different from the infantilism of the child, who cannot say what he or she sees. In "Destiny and Character" Benjamin does not discuss colors, considered from the perspective of the moral infant; but in his concluding discussion of character he briefly alludes to a theory he otherwise keeps out of the public eye: "The vision of character, however, is liberating under all forms: it is connected to freedom, as cannot be shown here, by way of its affinity to logic.—The character trait is therefore not the knot in the net. It is the sun of the individual in the colorless [farbenlosen] (anonymous) heaven of humanity, which casts the shadow of comic action" (2: 178).

According to a set of notes that Benjamin presumably wrote in conjunction with his reading of *Logical Investigations*, "logic analyzes with a view toward judgments of meaning" (6: 10). To the extent that the "character trait" in comedy is similar to the constitutive marks or "characteristics" of a logical term, which exhaustively determine its meaning, the art of comedy is akin to the science of logic. Unlike destiny, which never departs from the order of the sign, character thus begins to merge with meaning—which is to say, in Husserl's terms, with expression; in Benjamin's terms, with the name. For the same reason, moreover, the "character trait" acquires a relation to the phenomenon of color that may be entirely negative but is a relation nevertheless. The absence of color and the absence of the name are, each in its own way, negative modes of meaning, which constitute the threshold to its sphere. The conclusion of "Destiny and Character" moves in a similar direction: toward a twofold theory of meaning, one part of which would be concerned with the idea of the name, the other with the paradoxical color of colorlessness, comparable to the redness of shame, where the total transparency of existence would be achieved.

Nowhere in Benjamin's extant writings can such a formulation be found. Only in rare instances, almost all of which are scholastic, does Benjamin draw on specifically phenomenological terms such as *reduction* and *eidos*. Thus—to give only two examples—in "Eidos and Concept," from 1916, he speaks of a reduction in so many words; and in a curriculum vitae from around 1928, he describes the mode of procedure that guides his *Habilitationsschrift* as an "eidetic way of taking appearances into consideration" (6: 219). Outside of such academic contexts, however, he strenuously avoids technical terms in accordance with the stylistic imperative that he succinctly articulates in the "Epistemo-Critical Preface" to his *Habilitationsschrift*: "The introduction of new terminologies . . . is worrisome within the philosophical domain. Such terminologies—an unfortunate naming, in which intention takes a greater share than language—betrays the objectivity that history has given the principal coinages of philosophical reflection" (1: 217). Given that the term *reduction* was introduced into philosophy around 1913, there is little chance that it can be placed among the "coinages" that history has sanctioned. The same is obviously not true of *eidos*, which stems from Plato. And it is equally untrue of the word that corresponds to *reduction* in Plato's lexicon: *anamnesis* or "remembrance." Just as the phenomenological reduction makes it possible for phenomenologists to "intuit" essences as they give themselves within the limits of their givenness, so *anamnesis* brings philosophers back to the place-time in which they could "see" the ideas that constitute the "really real." Only insofar as one carries out the phenomenological reduction does one begin to be a phenomenologist; similarly, only by remembering what presents itself on the "plain of truth" (248b) does one become a lover of wisdom.

Benjamin and Scholem read Plato's *Symposium* together in the original when they shared a villa for a few days in August 1916 (S, 1: 389). And in a set of notes that Benjamin probably wrote around the same time as "On Shame," under the title "Zu einer Arbeit über die Schönheit farbiger Bilder in Kinderbücher" (For a work on the beauty of colorful pictures in children's books), he recasts the conclusion to "Color, Considered from the Perspective of the Child" in terms of the Platonic name for the phenomenological reduction: "If there is anything like Platonic anamnesis, it takes place among children, whose illustrated picture book [Anschauungsbilderbuch] is paradise. They learn how to remember [Sie lernen am Erinnerung]. . . . They learn in the remembrance of their first intuition [Anschauung]. And they learn about the motley, for the home of remembrance without longing [Sehnsucht] is in the fantastical play of colors, which can remain without longing because it is unclouded [ungetrübt]" (6: 124).[50] All of this follows, in a certain sense, from Benjamin's other colors studies; but he immediately retracts his initial proposal concerning *anamnesis* and adds another term, which points in a different direction altogether: "To this extent, then, Platonic anamnesis is not really [eigentlich] the entirely peculiar form of remembrance among children. It [anamnesis] is not without longing and regret, and this tension toward the messianic [Spannung zum Messianischen] is the proper effect of authentic art [eigentliche Kunst], whose perceiver not only learns from remembrance but also from longing, which it [art] satisfies too early and therefore too slowly" (6: 124).

According to "On the Beauty of Colorful Pictures," the child does not need to be brought back to paradise—defined, according to the earlier version, as that situation in which there is no "amalgamation in the object of experience as a result of incitement [Anregung]" (6: 112). For insofar as children are receptive to the motley dispersion of things without being excited by any one of them, they already exist within a paradisal landscape. In other words, children are sober. In still other words, there is no murkiness in their perceptual world, only motliness. To the extent that an illustrated picture book is paradisal, then, it reduces children to their own condition—"reminding" them, so to speak, who they are, namely, children, who see from the perspective of the child. Remembrance of this kind, which is unlike any other kind of memory, takes place in the absence of longing precisely because there is nothing to long for. By contrast, anamnesis in the proper sense of term incites both regret and longing. Benjamin does not then suggest that authentic works of art remind their viewers of the paradisal vision that characterized their childhood. In other words, he precisely does *not* claim that a work of art "turns off" the so-called "natural" and allows appearances to give themselves as they are—as unclouded, unmurky, and nonadumbrated as a rainbow. If the coloration of shame can be said to "reduce" its viewers in this way—a question that must be left undecided, since the sketch makes no mention of shame—it is only because this phenomenon is precisely *not* a work of art. Which means, in short, that the requisite reduction, here presented in Platonic terms, cannot be accomplished by those who intend to accomplish anything. Nevertheless, and for this very reason, the failure of anamnesis creates or makes palpable—there can be no principal distinction here—a "tension" that would otherwise go unnoticed: "tension toward the messianic."

Benjamin says nothing further about the messianic in this context. And in a certain sense the sudden appearance of the term can be interpreted as a sign of an insuperable impasse: in colloquial terms, the sphere of children and that of adulthood never converge. Inscribed into anamnesis in the proper sense of the term is a certain amnesia: there is no remembrance, that is, of the original intuition that, according to Benjamin, children remember when they look into an illustrated picture book. The term *messianic* emerges out of this impasse. Works of arts produce a paradoxical effect upon their viewers: their authenticity lies in "satisfying" the longing for the paradisal condition, which is constitutively without tension; and in doing so, a supplementary tension— "toward the messianic"—is either created or disclosed. The satisfaction afforded by "authentic art" is by no means false; it really does satisfy the longing for paradise; but it does so at the wrong time and therefore with the wrong timing: "too early and therefore too slowly." The supplementary tension is therefore always too weak, without sufficient force to propel the viewer across the divide that separates the divergent spheres. The satisfaction of art is thus the source of intense frustration: the region of paradise has not only not been remembered, it is even more impenetrable than ever. The question, then, is not whether a greater tension toward the messianic can be generated but how the weakness of this tension is related to the stretch of time, even decreasing, in which it becomes recognizable.

§3 "Existence Toward Space"

Two "Rainbows" from Around 1916

The Canon of Art

Benjamin begins his notes "For a Work on the Beauty of Colorful Pictures in Children's Books" by referring to the only color study of his student years that he apparently completed: "See the dialogue on the rainbow" (6: 123). Probably written in late 1915 or early 1916, while Benjamin was studying in Munich, "Das Regenbogen: Gespräch über die Phantasie" (The Rainbow: Dialogue on fantasy) can be considered the predicate of his further studies of color and fantasy, including a supplement he wrote under a similar title, "Der Regenbogen oder die Kunst der Paradieses" (The Rainbow; or, The art of paradise).[1] The question around which the dialogue revolves develops directly out of "Color, Considered from the Perspective of the Child": does the same "unbroken fantasy-activity" give rise to the colors of fantasy and works of art? Reflection on this question takes the form of a dialogue to the extent that its author is of two minds about the answer: he wants to affirm that fantasy is the source of both the visual field of the child and the authentic work of art; but he also wants to draw a sharp distinction between the colors of fantasy and those of any actual picture. In the "Rainbow" dialogue a character named Margarethe gives voice to the first view, which rests on an intimation that there is only a single fantasy-activity, while her friend Georg, a painter, draws a distinction between the pure colors of fantasy and those of artistic creation. Inasmuch as a rainbow appears to be a natural work of art, its unexpected arrival near the end of the dialogue represents a silent reconciliation of the two voices. That the reconciliation fails—perhaps because of its silence—becomes apparent in the supplementary "Rainbow," which implicitly repudiates the promise that the rainbow in the dialogue appears to announce: the promise, namely, that natural beauty can recall the paradisal condition in which the viewer is receptive to the world without being affected by anything in it.

The dialogue begins by implicitly resolving a problem into which "Color, Considered from the Perspective of the Child" issues: what happens to fantasy when one departs from the sphere of childhood? If, as the fragment claims, "unbroken fantasy-activity" is indeed continuous, it goes underground, as it were, and generates another dimension of vision, which in waking life is called "dreaming." As long as one is dreaming, the "natural" attitude is naturally "turned off." And in the case of Margarethe's dream, which occasions the dialogue, there is so little left of the "natural" attitude that everything she sees presents itself exactly as a pure phenomenon, including herself:

> MARGARETHE: It is early in the morning, I was afraid to disturb you. And yet I could not wait. I want to tell you a dream before it fades away.
>
> GEORG: How delighted I am when you come to me in the morning—because I'm then entirely alone with my images and do not expect you at all. You've gone through the rain, which has refreshed you. Now tell.
>
> MARGARETHE: Georg—I see that I cannot. A dream doesn't allow itself to be said.
>
> GEORG: But what have you dreamed? Was it beautiful or terrible? Was it an experience, one with me?
>
> MARGARETHE: No, nothing like this. It was entirely simple. It was a landscape. But it glowed in colors; I have never seen such colors. Even painters know nothing of them.
>
> GEORG: They were the colors of fantasy, Margarethe.
>
> MARGARETHE: The colors of fantasy, so it was. The landscape shimmered in them. Every mountain, every tree, leaves: they had infinitely many colors in them. Indeed, infinitely many landscapes. As if nature vivified itself in a thousand-fold innate-ness [Eingeboren-Sein].
>
> GEORG: I know these images of fantasy. I believe that they stand within me when I paint. I mix the colors, and I then see nothing but color. I'd almost say: I am color.
>
> MARGARETHE: So it was in my dream. I was nothing but seeing. All other senses were forgotten, vanished. Even I myself did not exist, nor my understanding, which discloses things from the images of the senses. I was not a viewer, I was only viewing. And what I saw were not things, Georg, only colors. And I myself was colored in this landscape. (7: 19–20)

Benjamin is not simply ignoring Freud's *Traumdeutung* (Interpretation of dreams) when he produces a dialogue that is so evidently uninterested in dream interpretation; rather, he is replacing a theory of unconscious desires with a theory of fantasy-activity. Both the unconscious and fantasy-activity are continuous; neither of them knows anything of maturation or acculturation; and each announces itself primarily in dreams, since sleeping diminishes the self-censoring mechanism that Margarethe identifies with her "understanding."[2] Yet there is one major difference between Freud's elaborate theory of unconscious desires and Benjamin's preliminary account of fantasy-activity: for Freud, dream imagery is always distorted; for Benjamin, by contrast, as expressed in Margarethe's description of her fast-fading dream, it is sheer clairvoyance. In this regard, Benjamin is closer to Nietzsche, who in the opening section of *The Birth of Tragedy* casts the dream as the "Apollonian" phenomenon par excellence.[3] To the Nietzschean idea of the dream as the sphere of sheer plasticity, Benjamin counters with the dream of pure color. Thus, Margarethe is "there" in her dream only as one of the infinitely many colors she sees. Georg, for his part, understands what she says she sees because similar colors can be found in his palette, even if not in the images he paints. When he does begin to paint, however, he experiences a similar ecstatic standstill of his understanding, which does away with all substantial elements and leaves only "pure properties" that are "permeated" (*durchdrungen*) with his residual self: "What you describe is like ecstasy [Rausch]. Remember what I told you about that strange and exquisite feeling of drunkenness which I know from earlier times. I felt myself altogether weightless in these hours. Above all, I perceived only what let me be in things—their properties, through which I permeated them" (7: 20). Up until this point Margarethe and Georg are of one mind; but as soon as she poses a question, their conversation turns into a dialogue:

> MARGARETHE: Why have I never found in those images made by painters such glowing, pure colors, the colors of the dream? For the place from which they come, their source—fantasy, and which you compare to ecstasy, the pure reception in self-forgetting—that is the soul of the artist. And fantasy is the most inward essence of art. Never have I seen this more clearly.
>
> GEORG: Even if it were the soul of the artist, it is not yet for this reason the essence of art. Art creates. And it creates objectively, that is, in relation to the pure forms of nature. Think about this—and often you have done this with me--: to the forms. Nature creates according to an infinite canon,

which grounds infinite forms of beauty. They are forms, they all rest in the form, in the relation to nature. (7: 20)

The dialogue is generated on the basis of a nuanced distinction between a voice that allows for no ultimate distinction between artist and artwork and a voice that develops a theory of art on the basis of this distinction. For Georg, no matter how much artists forget themselves in the creation of their art, they remain distinct from their work, which has its source in an altogether objective "canon." The conversation between Margarethe and Georg, then, consists in the latter's attempt to make a case for the existence of the canon without succumbing to certain aesthetic theorems both he and Margarethe reject insofar as they leave no room for the experience that initiates their conversation—the experience, once again, of finding oneself permeating the very space of perception. The first of the rejected theorems is among the oldest: "Art imitates nature." In response to Georg's equation of what he calls "the forms" with nature, Margerethe asks: "Do you mean art copies nature?" To which he immediately responds: "You know that I do not think so. It is true that at bottom the artist always only wants to grasp nature; he wants to receive it purely, to know it formally. But the inner, productive forms of reception rest in the canon" (7: 20). The basic elements of a theory of art are thus in place: "fantasy-activity" is itself purely receptive, without any form-imposing character; as such, it is the essence of the artist; but in order for the essence to express itself in any particular artwork, the receptivity of the artist is compromised by form, and the formation of an artistic mode of receptivity comes to depend on the so-called canon.

The burden of Georg's theory of art thus falls on the term *canon*; but it is a burden that he lightly shrugs off. Earlier conversations with him have made Margerethe familiar with his use of the term, so he does not have to stipulate what it means. What "canon" means to the reader of the dialogue emerges in its course and thus remains relatively opaque. Nevertheless, it is not wholly esoteric, and Benjamin is not the only thinker of the period to make the idea of the canon into a foundational element of aesthetic theory. Benjamin participated in Geiger's seminar on Kant's *Critique of Judgment*, and one of the topics that presumably came under discussion was Hermann Cohen's recently published *Aesthetics of Pure Feeling*. Georg's initial thesis about the nature of art is an almost verbatim quotation from Cohen's epistemo-critical revision of the *Critique of*

Judgment. First, Cohen: "the principal problem forms around art-creation [Kunstschaffen] itself."[4] Now, Georg: "Art creates [Die Kunst schafft]. And it creates objectively, that is, in relation to the pure forms of nature" (7: 20). According to Cohen, the following proposition, drawn from the *Critique of Judgment*, forms the "cornerstone of Kant's grounding of aesthetics" and represents the point of departure for his own theory of art: "No objective principle of taste is possible."[5] Without an objective principle, however, there can be no object of art, and without such an object, according to Cohen, "there is no purity."[6] For the sake of his own system, then, he goes in search of an aesthetic object that would correspond to the moral and epistemological ones that the previous volumes of his "system of philosophy" identify: "Can a pure object be isolated, as demanded by systematic aesthetics? Is there a proper object apart [eigenen, aparten Gegenstand], which can be generated as a pure object from a corresponding groundwork, as required by the problem of aesthetics?"[7]

Cohen's answer to this question passes through the idea of the canon. In a section of the *Critique of Judgment* on the "ideal of art," Kant briefly refers to Polycleitus's famous statue of a spear bearer, which acquired the title of *The Canon* as it became the "standard" against which other statues were assessed and from which sculptors began to learn their art.[8] The canon, in this sense, is the artistic counterpart to "doctrine." Kant has nothing good to say about Polycleitus's work, which he derisively describes as "the rule" and places in the company of a similarly "rule-bound" representation of a cow. Restoring the Greek term—which Kant suppresses, perhaps because it performs a systematic function in the *Critique of Pure Reason* (K, A 795–831; B 823–59)—Cohen uncovers a "contradiction" in Kant's discussion: "Human form, according to Kant, is alone supposed to be capable of being the ideal. In the canon of Polycleitus, however, the human figure is presented. And yet the canon is not supposed to be the ideal."[9] The contradiction that Cohen identifies—if it is one—derives, in his view, from the uncertain foundation on which the *Critique of Pure Reason* is based: it takes its point of departure from receptivity, which naturally presupposes a thing-in-itself that transcends the sphere of pure thinking. In the third *Critique* this uncertain foundation forces Kant to propose that "free beauties" (K, 5: 229) are pure, whereas works of art, especially images of human beings, solicit teleological considerations—hence the devaluing of an aesthetic "canon." He does not consider art from the perspective of creators, who base their creative work on a second-order receptivity to the canon of their form of art. For the artist, according to Cohen, the canon is equivalent to the given—but it is precisely not a "raw given," comparable to mere sensation. On the contrary, it is itself an object of art, which, for this reason, comes close to being the "proper object" on which a systematic aesthetics can rest.

Cohen, however, ultimately declines to pursue this path: a canon is a rule; but it is only a relative rule, not the absolute regularity or "lawfulness" (*Gesetzlichkeit*) that generates the epistemological object in the form of pure knowledge and the ethical object in the form of the pure will. If the canon of art were lawfulness pure and simple, then there would be no place in art for genius. Like the *Critique of Judgment*, Cohen's major aesthetic treatise presents creation primarily in relation to the genial artist, who repudiates mere rules in favor of the singular lawfulness that discloses itself solely in the newly created work: "Here the sharp distinction between law and lawfulness must be maintained. The genius [Genie] is a revelation of lawfulness; therefore he reveals new laws. The artist who does not reach the height of genius does not so much injure the laws as lawfulness."[10] Thus, despite Cohen's attempt to separate himself from Kant on the basis of their respective evaluations of the canon, he ultimately agrees that it can assume only an ancillary function in systematic aesthetics. To this, Georg—and Benjamin as well—respond by way of a certain silence: the canon so completely saturates the space previously assigned to the figure of the genuis that the latter term is altogether elided, even if there is still room for the Muses, in whom the canon becomes communicable. Only insofar as artists are receptive to the canons of their art can they create an object of art that is genuinely objective—and not an expression of either "lived experience" or in Cohen's terms, "pure feeling," which consists at bottom in "love for humanity." The dialogue on the rainbow is dedicated to Grete Radt. In a letter to her brother Felix, Benjamin recalls Cohen's reflections on genius by uncompromisingly rejecting the term: "The genial [Genie] is not important but, rather, the genius [der Genius]. In the incomparable word *genius*, whose meaning has been forgotten since Hölderlin, Wilhelm von Humboldt, and Goethe, there lies in the clearest possible manner that purity of productivity which flows only from a clear consciousness of its objective sources [sachlichen Quellen]. The genial [Genie] remains problematic in a work and in creation: 'holy-sober' is the genius [Genius]" (*GB*, 1: 298–99).

What Benjamin calls an "objective source" in the letter to Felix Radt appears under the rubric of "the canon" in the dialogue dedicated to his sister. A "clear consciousness" of the objective source or canon consists in pure receptivity or "holy sobriety." And the latter is so far from being categorically distinguished from "ecstasis" (*Rausch*) that they converge: whenever one is receptive to the canon, one no longer maintains oneself as an individual unity of consciousness that is affected by particular things in the world; rather, the self permeates and is permeated by what it receives. And this total interpenetration is, in turn, the general character of the canon. To a certain extent, then, the term *canon* not only displaces the concept of the "genial" artist but also replaces that of "the organic," which presumably came under discussion in Geiger's seminar on the *Critique of Judgment*. Understood as the unity of part and whole, such that each part reciprocally affects every other part, thereby constituting a unified whole whose unity transcends the transcendental unity of apperception, the organic doubtless bears a certain resemblance to the canon, and insofar as Kant's famous thesis about genius is valid—"genius is the talent (natural endowment) that gives the rule to art" (K, 5: 307)—it can be reformulated in organic terms: genial works are thoroughly organic. But there is nevertheless a major distinction between the organic and the canon: the former is a matter of reciprocal causation among constituent parts, whereas the elements of the latter so thoroughly permeate each other that there is no place for causal connectivity of any sort.

Only in one place can the categorial relation of ground and consequence be applied to the canon: it is the basis of any work that presents itself as a work of art, where *of* means *from*. The task of the artist, then, does not consist in discerning a "standard" of art from any given canon; rather, the artist must enter into the canon and remain there—*in* the canon, once again understood as a particular mode of total interpenetration, comparable to the dream of a colorful landscape in which the dreamer disappears into the colors she sees. In "Aphorismen zum Thema Phantasie und Farbe" (Aphorisms on the theme of fantasy and color) Benjamin briefly establishes the distinction between creating according to a canon and being in it: "The intuition of fantasy is a seeing within the canon, not according to it; therefore purely receptive, uncreative" (6: 109). Georg applies the same formula to his painterly existence: "that ecstasis which rushes through our nerves during moments of the greatest spiritual clarity, the consuming ecstasy of creating—this is the consciousness of creating in the canon, in accordance with the truth we fulfill" (7: 21–22). Creating according to a canon is uncreative. It is for this reason that Cohen, following Kant, ultimately contrasts such "mechanical" creativity, which merely follows rules, with "spontaneous" genius. By contrast, for Georg, creating according to a canon is insufficiently *un*creative. Being-within-the-canon is so thoroughly uncreative that it requires complete passivity on the part of the artist—with the paradoxical result that the genuine work of art cannot appear as such.

Just as Cohen distinguishes relative laws from lawfulness in general, so Georg draws a distinction between form and law: the first is relative to nature, whereas the latter derives solely from the absolute creator. Receiving the laws themselves, without relation to any form in which a law becomes accessible to a certain kind of eidetic intuition, is the ideal mode of artistry. It is as far from spontaneous creativity as possible. The aesthetic theory Georg proposes in "The Rainbow" culminates in the Neoplatonic idea of divine emanation, in which the creator is likened to a self-sustaining fountain, whose negative mode of creativity consists in allowing whatever it receives to flow onward, so that it may be received into a lower basin, which likewise overflows into a still lower one ad infinitum. Each basin of this continuous fountain of fantasy represents a particular form of art. And the entire structure is its canon, into whose waters artists can plunge under the condition they receive the flux without interrupting or diverting its course: "This reception out of fantasy is no reception of the model [Vorbild] but, rather, of the laws themselves. It would unite the poet with his figures [Gestalten] in the medium of color. To create entirely out of fantasy would mean to be divine. It would mean to create entirely from the laws, immediately and free from all the relation to them by means of forms [Formen]. God creates out of an emanation of his essence, as the Neoplatonics say, for this essence would be nothing other than fantasy, from whose essence the canon arises" (7: 24).[11]

Georg places everything he says about a purely receptive mode of creation in the subjunctive mood, for there is no evidence that creation of this kind could ever take place. And this is no accident: nothing, strictly speaking, can be the "product" of fantasy, least of all one of his pictures, for whenever fantasy ceases to flow, it ceases to be what it is: continuous flux. To the extent that every form of art is comparable to a basin in the ever-flowing fountain, each one can be taken as a representative of art as a whole; each canon of art, in turn, can stand for the canon of art in general. And Georg can present all art in terms of his own. Georg therefore assumes a kind of "canonical" synesthesia, in which each color is comparable to some form of art; being within the canon of painting is likewise akin to seeing oneself as a color of the rainbow; and all of these self-seeing colors share the defining feature of its existence: unattainability, which ultimately consists in being untouchable. Like the rainbow, pure art, which would consist in creation from pure receptivity, is no illusion; it does really appear; and although, like the end of the rainbow, it can never be reached, it is not thereby an infinitely receding goal, much less an illusion that could somehow be dispelled.

The Problem with Nature

By making pure color into the canon of art, the aesthetics that Benjamin works out in "The Rainbow" takes leave of sensation altogether—and is therefore no longer an "aesthetics" in the original sense of the term. If he were to emphasize this term, he would do so in the tradition of Hegel, who complains about its inapplicability to the field of art and yet, for reasons of terminological continuity, retains it nonetheless.[12] Benjamin, however, is as concerned with the nature of sensation as was Alexander Baumgarten, who first introduced the term into philosophical discourse.[13] For Baumgarten, sensation is a "lower" mode of cognition than intellect; but it could achieve a perfection of its own, which each work of art seeks to present; for Benjamin, the more a sensation can be disentangled from both the understanding and the imagination, the more receptive it is, the less creative it becomes, and the closer it approaches the fountain of pure fantasy, where there is no longer any distinction between sensation and the thing sensed. According to both Georg and Margarethe, the outstanding sensations in this respect are smells, tastes, and color perception. It is possible to "imagine" or "make up" new smells, new tastes, and new colors only by creating new things, which are then smelled, tasted, or perceived anew. And as Georg proceeds to explain, uncreative sensations are localized on the face, suggesting that they should be considered more prominent, if not "higher," than other senses for this reason:

> Of course, I see that there arises in the face a particular region of human sense to which no creative capacity corresponds: color perception, smell, and taste. See how clearly and sharply language designates this. It says the same thing about these objects as about the activities of the senses themselves: they smell

and taste. But from their colors: they look like [sehen aus]. For one never says anything of the kind about objects when one wants to designate their pure form. Do you get an intimation of the secret, deep region of spirit that begins here? (7: 22)

In response, Margarethe goes one step further—or better yet, takes a step backward into the "region" where the phenomenon of pure color originates:

> Hasn't this intimation come earlier to me than to you, Georg? Still, I want to lift colors purely from the secret realm of the senses. For, the more deeply we climb up that second realm of receiving senses, to which no creative capacity corresponds, the more vexed its objects become in terms of their substantiality, and the less the senses are permitted to sense pure properties. One cannot receive them for themselves alone, with a pure, detached sense but, rather, only as a property of a substance. But color originates, for this reason, in the most inward region of fantasy, because it is only a property; in nothing is it substance or does it even refer to substance. Therefore only this can be said of it—that it is a property, not that it has a property. For this reason colors have become the symbols for those who are bereft of fantasy. In color the eye is purely turned toward the spiritual [rein dem Geistigen zugewandt]; color spares the path of whomever creates in nature by means of form. In pure reception it allows sense immediately to encounter the spiritual, upon harmony. A viewer is entirely in the color; to see it means to immerse the gaze in a foreign eye, where the viewer is swallowed—in the eye of fantasy. Colors see themselves; pure viewing is in them, and they are its object and organ at the same time. Our eye is colorful. Color is generated from viewing and colors pure viewing. (7: 22–23)

If the defining feature of metaphysical speculation consists in proposing an identical subject-object, then Margarethe almost formulates a speculative thesis. But the "almost" is important: only the senses are under discussion, not the intellect, much less the synthesis of intellect and sensibility in the form of an intuitive intellect or an intellectual intuition. It is even erroneous to say that "the" senses come under discussion, for Margarethe distinguishes purely receptive senses from their receptive-spontaneous counterparts. She thus distinguishes the senses from themselves and identifies receptive ones with two mutually exclusive "domains": in the first domain, sensations are represented as belonging to things that affect the subject; in the second, they appear as pure properties, which belong to no underlying substance and can therefore be attributed to fantasy alone. A kind of double vision thus traverses perception: colors appear as properties of things or as pure properties. In the latter case, colors appear only to one who is neither excited nor aroused by them; rather, it is "purely turned toward the spiritual." Georg expands on this "turning" by declaring that the sight of color per se is nothing less than the "overcoming of the viewer [Überwindung des Sehenden]" (7: 23): an overcoming that not only does *not* draw the viewer into the sphere of the supersensible but, on the contrary, makes the viewer into a viewed phenomenon. Drawing on the double sense of the term *Aussehen* (look like, look out)—which Benjamin does repeatedly in his color studies—the two characters emphasize the same point: colors look at those who look at them.

From the beginning of the dialogue onward—with the notable exception of his excurses on the canon—Georg takes Margarethe's words as the point of departure for his own reflections. His last long speech is exemplary in this regard. Beginning as an expansion of Margarethe's description of color perception, it turns into a reflection on the nature of natural things: "And all these things about which you speak are only various sides of one and the same color of fantasy. It is without transitions and yet plays in countless nuances; it is moist, blurring things in the coloring of their contours. It is a medium, pure property of no substance, motley and yet monochromatic, a colorful fulfillment of *one* infinite by fantasy. It is the color of nature, of mountains, trees, rivers, and valleys—but above all of flowers and butterflies, of the sea and the clouds. The clouds of fantasy are so near because of color" (7: 25). All the elements of a "transcendental aesthetics" in both senses of the term come into play—along with an impediment to the integration of these elements into a *single* doctrine that would reflect the "*one* infinity" about which Georg speaks. The impediment lies in the word *nature*. It is scarcely a surprise that the discussion of receptivity should issue into an enumeration of natural things, each of which can serve as an exemplum of what cannot be made. In seeing them, creativity is reduced to mere receptivity. And for this reason, as Georg says apropos of clouds, they are "so close." But close to whom? Georg does not say. In particular, he does not say "us," nor even "us artists." In response to Margarethe's reprise of "Color, Considered from the Perspective of the Child," Georg is speaking of children, who are close to clouds precisely their color perception is as yet "unclouded" (*ungetrübt*) and their vision is, as a result, "dispersed" (*zerstreut*). Nevertheless, he remains silent about the "subject" of the vision. And his earlier thesis is of no help in this context. To say that the viewer is "overcome" means very little, for the colors of the natural things under discussion are not "pure properties without substance." On the contrary, they are the colors of "mountains, trees, rivers, and valleys." Georg is drawn into an impasse marked by the trope of enumeration precisely because of what he himself established at the beginning of the dialogue: nature cannot be received *purely*. It can be received only in its infinite variety of forms. To the extent that clouds are themselves lacking in form, they may approximate pure data of nature, which can be received without the mediation of forms; but Georg gets no closer to identifying the subject of the vision under discussion—beyond, perhaps, the suggestion that this subject is as cloudy as the object it sees.[14]

In the midst of this impasse Georg interprets the appearance of the rainbow: "And the rainbow is to me the purest appearance of this color that thoroughly spiritualizes and animates nature, leads its origin back [zurückführt] to fantasy, and makes it into a silent, intuited archetype [Urbild] of art" (7: 25). *Urbild* can be translated in a more colloquial manner as "primordial image," which appears only to primordial eyes; and *zurückführt* (leads back) can be translated in a more technical vocabulary as "reduces."[15] Georg's claim can thus be reformulated in the following way: a certain color reduces nature to the sphere of pure fantasy, in which there is no distinction between viewer and nonviewer, on the one hand, and viewer and viewed, on the other. But—and this reproduces the question above—the color in question does not itself appear; rather, it is the substrate of the colors that appear as the rainbow. Because of its invisibility, the color is accessible only by way of interpretation, the sign of which appears in the emphatic "to me." To Georg, the appearance of the rainbow means the reduction of nature to "the spiritual," that is, to meaning pure and simple. If the viewer were reduced along with nature, there would no longer be any place for the dative "me" who sees something as the appearance of something else. In sum, Georg subjectively interprets the rainbow as the absence of all subjective interpretation.

Art, for Georg, does not imitate nature; but the reduction of nature to the continuous fountain of fantasy that can be discerned whenever certain natural phenomena appear—and above all, when the skies brighten and a rainbow appears—takes precedence over the reduction of the so-called "natural" attitude that occurs in the sphere of art. As a result of Georg's expansive conclusion, in which an invisible color functions as the substrate of appearances, art is in danger of being rendered functionless. Even worse, it may represent an obstacle to the sight of the phenomena in which nature is "led back" to its origin. And this leads to a schematic version of the impasse that led Georg to interpret the rainbow in the first place: if something is created, it cannot be a matter of pure receptivity, and because the essence of beauty lies in pure receptivity, creation and beauty are mutually exclusive. Georg tautologically brings the dialogue to a close by demanding that discussion come to an end—presumably so that the sight of the rainbow, in contrast to its imputed significance, will break the impasse with which the dialogue concludes: "The ground of all the beauty that alone appears to us in pure reception lies in fantasy. It is beautiful, indeed it is the essence of beauty that we can do nothing else but receive the beautiful, while the artist can live only in fantasy and immerse himself in the archetype. The deeper the beauty enters into a work, the deeper it is received. All creation is imperfect; all creation is unbeautiful. Let us be silent" (7: 26).

Generating Space

The opening sequence of other "Rainbow" is worlds apart from the conclusion of the dialogue. So different are the two "Rainbows"—despite a certain similarity in theme—that Benjamin may have modeled his own work on the two poems of Hölderlin he that analyzed a few years earlier: "Dichtermuth" (Poetic courage) is to "Blödigkeit" (Infirmity) as "The Rainbow: Dialogue on Fantasy" is to "The Rainbow; or, The Art of Paradise." Whereas the first in each pair looks beyond art, the second moves in the direction of a self-contained solution to the task of art. The second "Rainbow" is so little concerned with nature that it does not even have room for the appearance of a rainbow. And in place of a dedication to a friend, Grete Radt, the second "Rainbow" announces its fragmentary and pseudoepigraphical character: "from an old manuscript" (7: 562). Finally, the first sentence of the second "Rainbow" affirms the exact opposite of what Georg declares at the end of the first one. Because beauty is solely a function of receptivity, according to Georg, works of art are always in a way unbeautiful. According to the opening sentence of the other "Rainbow," by contrast, there is something dubious about any application of the term *beauty* to nature: "It is a difficult question: whence comes the beauty of nature? For it must [be] entirely different than the beauty of painting and the plastic arts, because these are not the imitation of natural beauty" (7: 562).

By asserting a categorical distinction between the beauty of art and that of nature, "The Rainbow; or, The Art of Paradise" denies the underlying thesis of the like-named dialogue: all beauty owes its origin to the sphere of fantasy. The univocity of the term *beauty* is thereby rendered problematic. In the same stroke, however, art acquires a new function: that of generating space. Benjamin is here working within the parameters established by Kant and Cohen. Works of art are subservient to neither cognitive nor moral ends, and yet this independence does not mean that works of art exist solely for their own sake, still less that they exist only for the particular delight of the viewer. Rather, works of art serve an aesthetic function. This proposition is nontautological because of the ambiguity of the term *aesthetics*. That works of art serve an aesthetic function means that they make the formal conditions of appearance apparent. The second "Rainbow" retains the term "spiritual" (*geistig*) from the first and presents the concept of beauty accordingly: beauty consists in "spiritual appearance" (*geistigen Erscheinung*). In negative terms, *geistig* qualifies any phenomenon that can be received without thereby inciting motion or exciting emotion. The condition of nonaffective receptivity is possible only if the phenomenon exhausts itself in so appearing. In other words, there cannot be anything beyond, below, or behind appearances. In still other words, the phenomenon must be its own meaning. Whenever meaning is given independently of any meaning-giving act, there is beauty. And it remains problematic whether such an event can happen in the absence of art:

> Space is actually the medium of generation in art, and it is creative only to the degree that it productively presents what is spiritual about space. Space has no spiritual appearance other than in art. In the plastic arts space is generated [erzeugt] in a certain manner, and one will not doubt that this happens. Space is made an object in relation to its dimensionality. Natural space, in accordance with vision, is undimensional, dull, a nothing, if it is not enlivened in an empirical and lower manner." (7: 562)

The argument of the fragment appears to be circular: plastic arts are supposed to generate the space that is itself the medium of generation in art. A similar appearance of circularity prompts Kant to begin the "Transcendental Aesthetic" by declaring that space cannot be generated in thought: if it were not always already given, it could never be thought, and because it *can* be thought—otherwise, it would make no sense to undertake a discussion of space—it must be immediately given, hence a "pure intuition" (K, A 25; B 39). The second "Rainbow" elaborates on this argument by means of Cohen's concept of "generation" (*Erzeugung*), in which production and giving "evidence" (*Zeugnis*) are indistinguishable: space cannot be generated in thought; but a work of plastic art can generate it by making it evident. Only insofar as space is generated can *it*—rather than something *in* space—appear as such, that is, as a "form of appearance" (7: 563). Things in multidimensional space are given only in adumbrations, to use the term Husserl introduces into *Ideas*, as he seeks for a criterion to distinguish the givenness of consciousness from that of reality. (H, 3: 95–98).[16] The term *spiritual appearance* applies to works of plastic art because they make the "transcendental" character of space apparent: they are wholly given because they generate the very space "in" which they are seen. In the "Rainbow" and related studies color is called "spaceless [raumlos]" (6: 25; 7: 25) for this reason: to the extent that colors are "something spiritual," they do not generate the space in which they appear but are, instead, identical to the moment of their appearance.

Once it is recognized that a work of art generates the space in which it is located, it goes without saying that there is no limit to the number of dimensions that can be generated: "Given this restriction, three-dimensional space is inconceivable" (7: 562). The function of plastic works is defined in this manner: they disclose the space in which they appear—so much so that it would be a mistake to say that they occupy any given space, which would be accessible to intuition in the Kantian sense. The term *disclosure* is similarly misleading, for it suggests that the dimensions of the space that make themselves knowable in a work of plastic art were somehow "there" in advance of the work. According to the "Rainbow" fragment, this is not the case: the dimensions of the space that is disclosed through the work of art do not exist prior to the work; they are "there" only insofar as they "spiritually" appear, and they can do this only insofar as the dimensions unfold in the work. With respect to the plastic arts, therefore, the fragment concludes: "Plastic art [Plastik] has to do with a space [einem Raum] that emerges via generation and is therefore as conceivable as it is unlimited" (7: 562). "A space" (*ein Raum*) turns into space in general (*der Raum*) whenever it is taken for a condition of appearances that itself recedes from appearances.

Concentration

In comparison to the first "Rainbow," the second is a model of stylistic ascesis. It contains no extravagant digressions, nor does it include any esoteric doctrines. Still less does it make mention of any lived experience. Indeed only in one respect are the two "Rainbows" bound together: the second one develops the theory of the painterly "plane" (*Fläche*) that Georg proposes in the first. As a pseudo-epigraphical fragment, the second "Rainbow" thus appears as though it were the ancient source of Georg's theory of painting. Both "Rainbows" in any case reject the supposition that the function of painting lies in reproducing what is seen. In keeping with its ascetic character, the fragment says nothing about the canon of art or the ecstasis of the artist; rather, it soberly concerns itself with the question: what is the plane of a painting? Its answer, in brief, runs as follows: it is what Husserl's *Ideas* professes to be, namely an "introduction to pure phenomenology." The painterly plane is the place in which there is no distinction between appearances and the thing-in-itself or between signifying intention and fulfilling intuition. Viewers see exactly and only what gives itself to be seen within its self-defined limits of givenness. In this way, viewers are led directly into what they see. Instead of describing the disappearance of the viewer into the thing viewed, however, the fragment is concerned solely with the condition for the possibility of this disappearance, namely that the space "construed" by the painterly plane is infinite, which thus absorbs all finite viewings. In colloquial terms, it is inexhaustible, even though there is never anything new to see: "The spatial nature of things develops itself in the plane, and it does so nonempirically, concentrated. Not the dimension but, rather, the infinitude of space is construed in painting. This happens via the plane, in such a way that things do not absolutely develop their dimensionality, nor their extension *in* space but, rather, their existence *toward* space [nicht ihre Ausdehnung *im* Raume, sondern ihr Dasein *zum* Raume]" (7: 563).

The painterly plane is the "space" in which space itself develops. As such, the plane can be called the spatialization of space, which is "prior" to space, understood as the a priori form of outer intuition. The infinitude of space appears "on" the plane precisely because the plane and the things "on" the plane inseparably appear and appear inseparable. The plane of the painting thus distinguishes itself from every other surface: it is not simply a material condition but also a transcendental condition of whatever appears "on" it. And the fragment has a name for the criterion of this distinction: "concentration," which makes a plane into the "space" of art. Thus, "beauty rests on concentration" (7: 562), and the term *concentration* takes over the function Georg assigned to "the canon": it is the objective source of "spiritual appearance." In the absence of concentration, appearances are interpreted as the attributes of an underlying substance. Concentration is so entirely a matter of the object that it does not preclude—and perhaps even invites—the opposite on the part of the viewer, namely the "distraction" or "dispersion" that Margarethe discerns among children: "The color perception of children is itself dispersed into color" (7: 25). Because things are "concentrated" on the plane, they appear as themselves, distinct and individuated, by virtue of a certain unfolding or development. Like the space construed by the plane, development is potentially endless. The basic structure of the plane is both paradisal and paradoxical: it immediately appears along with all of the things "on" it, and yet there is always more "there" to see, despite the fact that nothing is hidden from view. As the things that already appear begin to appear, they do so, as both the first and the second "Rainbows" emphasize—not "in" but "toward" space, that is, toward *the* space, *der Raum*, which is nevertheless not the homogeneous space in which exterior objects affect an inwardly constituted subject. Things on the painterly plane unfold in the direction of paradisal space where, as Georg says, the viewing subject is "overcome" (7: 23).

From the perspective of "The Rainbow; or, The Art of Paradise," in sum, painting is the art of paradise, distinguished from the rainbow—as well as all other natural phenomena—by virtue of its depth, in which the paradoxical character of the painterly plane is once and for all resolved. In the central sections of his *Contributions to the Phenomenology of Aesthetic Enjoyment* Geiger seeks to uncover the different strata of depth that disclose themselves in conjunction with aesthetic phenomena.[17] To Geiger's subject-oriented phenomenology Benjamin poses an objective counterpart. Depth is not a dimension of space but, rather, the nature of its infinitude. The plane of a painting gives rise to depth because it shows everything, including itself, and yet what it shows can be infinitely developed:

> Depth yields [ergibt] infinite space. The form of concentration is thereby given; but this now requires its fulfillment, the satisfaction of its tension, a presentation of an infinite that is in itself no longer dimensional and extended. Objects demand a form of appearance that is grounded purely on

their relation to space, a form of appearance that does not express their dimensionality but rather their contural tension (not their structive [struktive] but, rather, their painterly form), their existence in depth. For without this, the plane does not come to concentration; it remains two-dimensional and gains only a graphic, perspectival, illusionistic depth—not, however, a depth as nondimensional relational form of spatial infinitude and object. The requisite form of appearance, which constitutes this, is color in its artistic significance. [Diese verlangte Erscheinungsform, welche dies ausmacht, ist die Farbe in ihrer künstlerischen Bedeutung]. (7: 563)[18]

Whereas the "Rainbow" dialogue begins with a discussion of color, the "Rainbow" fragment—which could henceforth be called "The Art of Paradise in Place of the Rainbow"—begins to break apart where color comes under discussion. And whereas the first "Rainbow" presents color as the problem that sets the dialogue into motion, the second makes color into the solution demanded by the objects of art themselves. As for the nature of the problem that color solves, it can be stated as follows: concentration is the primary condition for beauty understood as "spiritual appearance"; not only the plane but also the things "on" the plane achieve a certain degree of concentration, for otherwise, though they will doubtless appear, there will be no more to see. In colloquial terms, they will appear flat; in Benjamin's own terms, they may appear, but their appearance would not be "spiritual." Concentration is therefore the "objective" side of the phenomenological reduction: to the extent that the subject is reduced, the things it experiences are reduced, that is, concentrated; by contrast, when the subject adopts the "natural" attitude, the appearance of depth results from the lived experience of expansive, perspectival viewing. The very structure of the objects on the plane—to the extent that there is a plane in the first place—"demands" that they be deep. No matter how often a viewer looks at them, there must be more to see, and this "more" expresses itself in a straining or striving.[19] The object thus acquires a peculiar tension: its finitude runs counter to the depth into which it wants to be released. It is as though the plane, far from being flat, bends in response to the tension thus generated. Here, then, is the function of color in painting: it satisfies this tension of the object, insofar as it is not exactly *another* space but, rather, like space, a "form of appearance." The object thus gains its depth without losing its defining features. It—and therefore the painterly plane as a whole—finds itself in paradisal satisfaction. Anyone who looks into the object is identical to that object, which is itself structured in accordance with truth: the signifying intention of one form of intuition, namely space, is intuitively fulfilled by another, which is called "color."

But—and this is where the fragment starts to breaks off—the fulfillment of the tension take places only as long as color can indeed be secured as a form of appearance in its own right. Only in this way does color gain its artistic meaning and the object its satisfying depth. As the manuscript proceeds to explain, however, it is by no means clear that the color can be seen in its artistic meaning without some degree of artifice—without, for instance, a stipulation in whatever form that "here stands art." If such a stipulation is required, however, color gains its artistic meaning only when someone says that it is meaningful, which indicates at bottom that it is not. And who can doubt that the painterly plane must be so designated? It is for this reason that the painting requires those *parerga*, such as frames, clothing, and decoration, to which Jacques Derrida draws attention in his analysis of the *Critique of Judgment*.[20] That Benjamin is thinking along the same lines can be discerned from the seemingly out-of-place sentence with which the fragment abruptly ends—the sole sentence, moreover, that mentions the earlier "Rainbow": "Clothing and decoration not mentioned in the dialogue" (7: 563). If this were not enough, the fragility of the color solution becomes even more apparent in a comment appended to the manuscript: "Painterly color cannot be seen for itself" (7: 564).

Because painterly color does not appear as such, it cannot appear "on" the plane without the plane somehow saying—not showing—that it is painterly. The tension of the object, which "longs" for its depth, can thus be satisfied only if the solution is designated as art. But the designation *art* runs counter to the art that is thereby designated. And so the tension not only remains; it can even be said to intensify: there is no paradisal satisfaction, which is to say, no satisfaction in the form of paradisal art. If there is to be a release of the tension that Benjamin identifies in the second "Rainbow," it can be neither in nature nor in art. The tension can be perhaps called "messianic" in contrast to the paradisal character of its satisfaction. This term doubtless says very little in this context; but it at least suggests two further forms of appearance: time and language.

Missing the Mark

Time and language meet space and color in what Benjamin calls the "the mark" (*das Mal*). The term does not appear in either of the "Rainbows" despite what Benjamin writes at the beginning of his notes "For a Work on the Beauty of Colorful Pictures in Children's Books": "See the dialogue on the rainbow. There, color applied onto watercolors is distinguished from the color of painting that makes the mark [das Mal]" (6: 123). Unless there is a yet another dialogue entitled "The Rainbow"—or the copy that Giorgio Agamben received from Herbert Blumenthal-Belmore is somehow edited or otherwise incomplete—Benjamin here misremembers his own text.[21] Or perhaps he only indicates to himself what is missing from the "Rainbows" as well as from the rainbow itself. In any case, as he proceeds to explain, the basic traits of the mark in contrast to watercolor lies in language and time: "Whatever appears in the mark speaks connaturally via perception to the entire metaphysical essence of the human being, and fantasy speaks in the mark without being detached from the necessarily morally codetermined and valid longing [Sehnsucht] of the human being" (6: 123). It is in the mark, then, that the continuity of "continuous fantasy-activity" continues beyond childhood and into waking life: as a voice, so to speak, from a sphere that cannot be remembered.

The mark is missing from "Rainbows"—or at least missing from the "Rainbows" that Agamben discovered and Suhrkamp Verlag published. The term prominently appears, however, in an addendum or corrective that Benjamin wrote in response to a letter he received from Gershom Scholem in August 1917. Having visited a major exhibition of contemporary painting organized by Herwarth Walden under the title of "Sturm" (Storm), Scholem tells Benjamin about his encounter with the works of Braque, Chagall, Kandinsky, Kokoschka, and Picasso, among others.[22] The letter has been lost; but its content can be reconstructed from Scholem's *Diaries*. In a small treatise, "Über und gegen den Kubismus" (On and against cubism), Scholem launches a polemic against Picasso, who, in his view, fails to see that color destroys the redemptive promise of cubism, which consists in resolutely "dismantling" (*zerlegen*) space. Scholem thus expresses a desire for a purer version of cubism, which would take an ethical stance and thus renounce any lingering infatuation with color: "Color should not be an element in this world in which space is dismantled" (S, 2: 32). Of particular importance in this regard is Picasso's *Woman with a Violin*, which allows its viewers to see through its own kitsch quality, derive from its use of color, and thereupon recognize the essence of space laid bare: "In the vertical middle of the world stands the woman, the fixed and formed essence of space; the center of the violin goes through her spiritual center; the music is brought into deep connection with the space that is dismantled" (S, 2: 31–32).

In response to Scholem's polemical account of Picasso's recent paintings, Benjamin writes a polemical letter not so much against Scholem as against the pseudophenomenological language in which he expresses his views. It is not as though Benjamin—who, as he explains to Scholem, regrets that he will not have a chance to see the exhibition in person—is a proponent of Picasso's version of cubism; indeed, as he notes, he concurs with Scholem's general assessment of Picasso. But upon encountering his own theory of color applied to specific paintings, Benjamin demurs. Not only does Scholem fail to recognize the difference between a painting and a geometrical demonstration; he also imputes to painting the *Wesensschau* (intuition of essence) that belongs to philosophy:

> Purely polemically, I only want to say to you that, without even sketching out an ordering of cubism, I take your characterization of it to be false. You consider the quintessence of cubism to be "communicating the essence of the space that is the world through dismantling." It seems to me that this definition contains an error concerning the relationships between painting and its sensible object. In analytic geometry I can, to be sure, produce an equation for a two- or three-dimensional figure in space without thereby overstepping the analysis of space; but in painting *Woman with a Fan*, for example, I cannot paint in order thereby to communicate space through dismantling. Rather, the communication must under all conditions concern only the "woman with a fan." On the other hand, it is probable that painting really has nothing to do with the "essence" of anything, for then it would collide with philosophy. I am not now in a position to say anything about the relation between painting and its object; I believe, though, that it is a matter of neither imitation [Nachbildung] nor essence-cognition [Wesenserkenntnis]. (*GB*, 1: 395)

Scholem's vision falls short in two ways. On the one hand, perhaps as a result of his mathematical training—in the same letter Benjamin asks him to evaluate a paper by Ernst Barthel about the applicability of non-Euclidean geometry to physical space (*GB*, 1: 391)—he misconstrues the relationship between the painterly plane and the dimensional space of analytic geometry; and on the other, perhaps because he fails to see that "artistic content and spiritual communication are entirely the same" (*GB*, 1: 395), he overlooks a painterly problem that cubism accentuates: how does a painting acquire a name, which would be precisely *its* name, not a label that someone imposes upon it, even if this "someone" happens to be the artist? The final sections of a systematically inclined fragment Benjamin wrote in response to Scholem under the title "Über die Malerei oder Zeichen und Mal" (On painting or sign and mark) are directly concerned with this question. At the opening of the subsection on the sign, Benjamin alludes to the first objection he raises against Scholem's discussion of cubism by explicitly postponing any discussion of the relationship between the geometrical and the artistic line. Similarly, near the beginning of the subsection on the mark, Benjamin alludes to the second of his objections—that the relation between a painting and its object should not be confused with the *Wesensschau*. In this case, however, he does not postpone a discussion of the relation between painting and eidetic intuition; rather, he borrows a decisive term from Husserl's *Ideas* in order to capture the "temporal meaning" of what he calls the "absolute mark" (*das absolute Mal*). Just as the intuition of essence requires that the so-called "natural" attitude be "turned off" (*ausgeschaltet*), so in the presence of the absolute mark "the resistance of the present to the future and past is switched off [ausgeschaltet]" (2: 605).

It is no surprise that Benjamin should emphasize the temporal character of the mark, since the same word, *mal*, means "time" in phrases such as "for the first time" and in equations such as "$2 \times 2 = 4$." Equally unsurprising is the fact that he places the mark on the side of language, since the term from which it is categorically distinguished, "the sign" (*das Zeichen*), is always associated, for Benjamin, with "improper meaning [uneigentliche Bedeutung]" (6: 10) and therefore with a negative mode of language. According to the sketch "On Painting," the opposition between the "absolute sign" and the "absolute mark" is "metaphysically of *immense* importance." Each of them is associated with a Kantian form of appearance and placed in a certain relation to the "person" understood as mask bearer that assumes a certain position in a legal or social order: "The sign appears to have a more pronounced spatial relation and more relation to the person, whereas the mark . . . has a more temporal and personal-expelling meaning" (2: 604). As Benjamin notes, however, the "fundamental distinction" between sign and mark lies in the manner of their appearance: "the sign is impressed [wird aufgedrückt], whereas the mark emerges [hervortritt]"— to which Benjamin immediately adds, "this indicates that the sphere of the mark is that of a medium" (2: 605). The medial character of the mark means that its emergence is immediately recognizable as such. There is no need for a supplementary sign that would say something to the effect that "this is a mark as opposed to a sign." The sketch "On Painting" thus dissolves the impasse into which issues the exposition of the painterly plane in the second "Rainbow": a plane does not require a supplementary sign or a parergon such as a frame in order to designate its painterly character; rather, painting is composed of two opposing elements, the combination of which makes it possible for a painting to be *one* painting, regardless of where it is located in the space that is available to the "natural" attitude. The two elements are the mark and its "composition," which likewise "transcends" it. By virtue of this structure of immanent transcendence, the image enters into the sphere of "nameability [Bennenbarkeit]" (2: 607). To the extent that a painting is nameable, regardless of whether it acquires a name or not, it can do without either a sign or a frame that would designate *its* plane as painterly.

The dissolution of the impasse that Benjamin encounters in "The Rainbow; or, The Art of Paradise" thus depends on the cogency of his inquiry into the absolute mark. Although there is no reason to doubt that Benjamin considered the distinction between the absolute sign and the absolute mark to be of singular consequence, especially for a nascent theory of myth, there is some doubt as to whether the term *absolute* can be attributed of the mark, especially since, as Benjamin explains, the phrase *absolute mark* is a pleonasm: "There is no opposition between mark and absolute mark, for the mark is always absolute and is in its appearance like nothing else" (2: 605). The emergence of a mark is therefore a phenomenon in the genuine sense: the mark gives itself within its own limits of givenness. It is, in colloquial terms, the postchildhood version of color, considered from the perspective of the child—except that its "essence" (*Wesen*), as opposed to that of color, lies in myth. And if, as Benjamin suggests in similar contexts, "ambiguity [Zweideutigkeit]" (1: 183) characterizes myth in general, then the mark, as a mythic counterpart to color, cannot emerge absolutely—which means, however, that it is doubtful whether any mark can emerge as such. Doubts about the mark find a certain corroboration in each of the examples that Benjamin adduces for its putatively incomparable appearance: "Altogether striking is how the emergence of the mark on the living is so often bound up with guilt (blushing [Erröten]) viz. innocence (cicatrices of Christ); indeed, even when the mark appears on something lifeless (sun spots in [August] Strindberg's *Advent*), it is often an admonishing sign [Zeichen] of guilt" (2: 605). To the extent that a mark functions as a sign, it ceases to be a mark, and this event of the mark becoming a sign makes blushing into an index of guilt rather than into a coloration of shame.

The exposition of blushing in terms of guilt indicates the degree to which something is missing from Benjamin's account of the mark. It is perhaps for this reason that the mark is missing from both of the "Rainbows," which look toward the innocent spheres of fantasy, on the one hand, and the art of paradise, on the other. There is no sign of a term that would transcend the mark—only one into which the mark falls once it no longer emerges. In the absence of a third term, it cannot escape becoming contaminated by the sign against which it is defined. Of course, this is precisely what Benjamin locates in painting: the mark-transcending element of "composition" that makes a painting nameable. The argument requires, however, that there be a correlate to a composition for each mark. Every mark must solicit a "compositional" element that makes it amenable to a "higher power" (2: 607). A "composed" mark would no longer simply be a mark, and yet it would also be the only absolute one. What is missing from Benjamin's account of the mark is therefore akin to the missing term to which he refers at the end of "Two Poems of Friedrich Hölderlin": a term for the transcendence of myth that emerges from the countermythic tendencies of the "mythos" that takes shape as the poetized. Nothing of this sort appears in "On Painting." With the theory of the mark, Benjamin attempts to arrive at a generally accessible counterpart to the kind of colors that are visible only to the child. When he turns to the theme of the mark in his notes "For a Work on the Beauty of Colorful Pictures in Children's Books," and provides the missing term, the sphere of childhood is as irrecoverable as ever: the "composed" mark is the place where the "tension toward the messianic" (6: 124) would crystallize. Such a thesis says almost nothing, however, for the "composed" mark can emerge only when there is such a thing as cicatrices.

§4 "The Problem of Historical Time"

Conversing with Scholem, Criticizing Heidegger in 1916

Scholem's Decisions

The cubist paintings that Scholem encountered at the "Sturm" exhibition in the summer of 1917 gave him a chance to propose a negative correlate to Benjamin's theory of color: when cubists renounce color, they show the essence of space by means of its "dismantlement." The value of these paintings, in Scholem's view, extends beyond their confirmation of the theory of color he adopts from Benjamin, for, as he notes, they unexpectedly corroborate a theory he had begun to develop in the previous year: "Cubism is the artistic expression of the mathematical theory of truth" (S, 2: 33). Among Scholem's various interests, as they are reflected in his diaries of the period, mathematics and its metaphysical foundations occupy almost as much attention as Zionism and its Jewish sources. He understood himself accordingly as a divided soul, who must make a decision along the lines Kierkegaard described in his pseudonymous authorship: either remain within the sphere of the ethical in the form of rigorous mathematical studies or leap into the religious sphere that corresponds with the vision of Zion. Along with Nietzsche, whose *Also sprach Zarathustra* (Thus spoke Zarathustra) he finds singularly impressive (S, 1: 51), Kierkegaard provides a model of philosophical authorship, upon which a new era of Judaism, cognizant of the past yet oriented toward the future, can be constructed: "Kierkegaard is the palace from which the rock of the Talmud is erected.... He is the mediator between God and Israel" (S, 1: 224). Something in Kierkegaard's *Stages on Life's Way* so moved him that he experienced "physical disgust" with the individual named "Gerhard Scholem" (S, 1: 238–39) and began to call himself "Gershom" as a result. Gershom Scholem, unlike Gerhard, has made an existential decision in favor of Zionism (Gershom) without having to renounce the appeal of pure mathematics (Scholem).

Around 1916, however, Scholem finds himself in a position where he simply must make a decision—not, however, a decision for or against mathematics. Rather, he must make a decision for or against Martin Buber, whose version of "cultural" Zionism had hitherto influenced his own. In early 1916 Scholem invited Buber to address a Jewish youth-group and introduced him in glowing terms: "Martin Buber, who will come under discussion this evening, is, along with Ahad Ha'am, the most significant personality of contemporary Zionism and one of its strongest spiritual potencies" (S, 1: 111). And there is no wonder that Scholem originally found himself drawn to the figure of Martin Buber. According to a widely circulated series of lecture that Buber gave shortly before the war, "the Jew" is the divided soul par excellence, particularly the modern, "Western" Jew, who is forever being drawn away from the wellspring of genuine religiosity and thus absorbed into the "hustle and bustle" of modern life.[1] Only by way of nonrational "lived experience" (*Erlebnis*), according to Buber, can the division be overcome. By the summer of 1916, Scholem assigns himself the task of overcoming Buber's version of overcoming the division that characterizes his own life, for otherwise—this is the reason the decision can no longer be avoided or delayed—he will not be able to enter into more intimate conversation with Benjamin, whom he had met in the previous summer. In a diary entry from August 23, 1916, Scholem presents his situation in characteristically Kierkegaardian fashion: "One says, either 'I have *experienced* my Jewishness' . . . or: 'I have *seen* Zion. For this is something *entirely* different: vision or lived experience [Vision oder Erlebnis]" (S, 1: 386). Confronting this either/or—as opposed to the decision between mathematics and Zionism—Scholem shows no signs of hesitation: *Erlebnis* is now a seen as a version of aesthetic dissoluteness, while "vision" represents something like ethical rigor. Just as, for Benjamin, "pure seeing" means that viewers see themselves colored into a landscape of pure color, so, for Scholem, the "vision" of Zion means that viewers see themselves as already on Mount Zion. In the same diary entry in which he speaks of vision, Scholem acknowledges a debt to Benjamin; but the debt goes no further than the language in which his thoughts are clothed: "Not that I have learned from Benjamin; on the contrary, I have thought precisely the same for months, and only on a single point has it now also become clear in linguistic form: in the denial of the value of 'lived experience'" (S, 1: 386).

Unlike Scholem, Benjamin was never attracted to Buber's work, much less to its cardinal concept of *Erlebnis*. Among the reasons Benjamin may have been drawn to the seminars of Moritz Geiger was the latter's similar distaste for "Erlebnis" discourse, even when it was used by Husserl.[2] Sometimes Benjamin expresses his attitude toward Buberian *Erlebnis* subtly, as when, for instance, while presiding over a major youth organization, he invites Buber in 1914 to a discussion of his recently published *Daniel: Gespräche von der Verwirklichung* (Daniel: dialogues on actualization), which, as it turns out, he did not bother to read (*GB*, 1: 218–19). At other times, his distance takes the form of almost palpable repugnance, as, for instance, in the following passage in a letter to Ludwig Strauss from 1912: "Perhaps it is necessary at the beginning that I repeat how my position with respect to Judaism is formed. Not through a Jewish experience [Erlebnis]—through no lived experience whatsoever. Rather, [it is formed] simply through an important experience [Erfahrung]: whenever I turned my ideas toward the outside, in the spiritual and the practical, in almost all cases Jews came to meet me. . . . My experience brought me to this insight: Jews represent an elite in the host of the spiritual" (*GB*, 1: 75).[3] When Buber solicits a contribution from Benjamin to the journal he launches in 1914 under the provocative title *Der Jude*—which says, in effect, that the Jews of Germany no longer need to hide themselves under assimilating labels like "Mosaic Confession"—Benjamin indicates that he would welcome a forum to work out the thoughts that he outlines in his letter to Strauss: "The problem of Jewish spirit is one of the largest and most persistent objects of my thoughts" (*GB*, 1: 283). But after Benjamin has a chance to read the first issue of *Der Jude*, he revokes his offer and writes, instead, a remarkable letter to Buber in which he categorically declines to participate in the journal as it is currently constituted.[4]

As Benjamin indicates at the beginning of the letter he sends Buber in July 1916, his initial response to the first number of the journal was largely determined by his antipathy to the position taken by all of its contributors toward "the European war" (*GB*, 1: 325); but, as he proceeds to explain, this antipathy is only a dimension of a deeper aversion to the journal's misconception of the relation between language and action. The contributors to *Der Jude* assume that "political" language is primarily a means, the purpose of which consists in providing sufficient motivations for a proposed action. For Benjamin, by contrast, language immediately achieves a certain effectiveness when the schemata of cause-and-effect and means-and-end are eliminated: "When I disregard other forms of effectiveness—other than poetry and prophecy—it still always seems to me that the crystal-pure elimination of the unsayable [Unsagbaren] is the closest form in which to be effective in language and therefore effective through language" (*GB*, 1: 326).[5] In the opening passage of the letter, Benjamin indicates that he has spoken with "Mr. Gerhard Scholem" about his decision with regard to *Der Jude*. And when Benjamin invites Scholem to visit him and Dora Pollak at her current husband's estate outside of Munich in August 1916—Dora and Walter were married the following spring, soon after she divorced Max Pollak—the first order of business, so to speak, is a public reading of the letter that crystallizes their contra-Buberian friendship. Quoting passages of the letter from memory, Scholem summarizes its imperative in the diary entry he wrote soon after his exciting visit: "Benjamin demands of [Buber] and his journal that their words should be directed at the 'core of innermost muteness'" (*S*, 1: 384; *GB*, 1: 326).

The Afternoon Conversation

In August 1916 Scholem, having distanced himself from Buber's language of lived experience, arrives at the large home of Max Pollak, who is away on business, or so he is told, and there he stays for three rain-soaked days. Benjamin and Scholem are both under the threat of being inducted into the war they equally detest, and in accordance with Benjamin's implicit guidelines for discussion, there will be no talk of the war or other "current events." Dora, for her part, participates in the conversation; but there is little record of what she says, except when she defends Hegel against the accusation that he tries to deduce the world from the principles of his logic and when she makes fun of Scholem for expressing an interest in the work of Ernst Mach (*S*, 1: 387–88).[6] Many topics come under discussion; but one topic is of particular importance, for, resuming the conversation that first brought the two of them together, it consumes an entire afternoon.[7] Our knowledge of this conversation stems entirely from Scholem, who twice recorded some of Benjamin's remarks—first in a series that the editors of Benjamin's collected writings call "Aphorismen" (Aphorisms), and again in a diary entry entitled "From a Notebook Walter Benjamin

Lent Me, 'Notes Toward a Work on the Category of Justice'" (S, 1: 402).[8] Most of the notes taken from the borrowed notebook are concerned with a theory of justice—which comes under discussion in the final chapter of this volume—but at the end of the notes, as an appendix of sorts, Scholem transcribes the remark with which their afternoon-long conversation began. We are certain of the accuracy of this report because Scholem gives it twice—first a few days afterward and then some sixty years later.

The reconstructed version of the conversation that Scholem presents in *Walter Benjamin: The Story of a Friendship* runs as follows: "We discussed it [the philosophy of history] for an entire afternoon in connection with a difficult remark of his [Benjamin's]: the series of years are doubtless countable but not numerable. This led to [the meaning] of *course* [Ablauf], *number, series, direction*. Whether time, which is certainly a course, also has a direction. I said whence we can know that time would not behave like certain curves, which have a continuous course at every point but at no particular point would have a tangent, that is, a determinable direction. We discussed whether years, like numbers, are exchangeable, just as they are numerable."[9] This account of the conversation is perplexing—and not only because some of Scholem's sentences are uncharacteristically ungrammatical. First of all, the passage is traversed by an outright contradiction. Benjamin's opening remark is difficult indeed: "the series of years are doubtless countable [zählbar], but not numerable [numerierbar]."[10] As the conversation continues, however, it turns out that years are numerable and perhaps also exchangeable. Among the many questions raised by the passage, one is particularly pressing: why does Benjamin's "difficult remark" give way to a discussion of the word *Ablauf*, here translated as "course"? This question is prompted by the striking difference between *Ablauf* and the other terms under discussion, each of which is a major element of mathematical reasoning. *Ablauf*, by contrast, cannot be considered a mathematical term at all.[11] Benjamin may have brought the term into the discussion, for at a certain point in "Two Poems of Friedrich Hölderlin" he draws attention to its ambiguity: in the first poem under consideration, the "course" (*Ablauf*) of the analogy that links the movement of poet with that of the sun is "interrupted" without returning with "total intensity" to the poet; by contrast, in the later poem, "nothing but the existence of the poet is given in the run-off [Ablauf]" (1: 122).[12] In this context *Ablauf* means at once the entire "course" and its outcome.

Yet, nothing about this use of the term suggests any connection with "number, series, direction"—or mathematics in general. For this reason, it can be supposed that Scholem brought the term under discussion, perhaps in response to certain "epistemo-critical" theorems of Felix Hausdorff, one of the most important and innovative mathematicians of the era. In 1914 Hausdorff published a groundbreaking study of set theory in which he developed the idea of topological dimension with which his name has henceforth been associated. Years earlier, however, he had published a number of literary and philosophical writings under the name of Paul Mongré. There is no question that Scholem knew that Mongré was Hausdorff's pseudonym, for he notes as much in a diary entry from 1914 (S, 1: 179), where he briefly describes his attraction to Hausdorff's first philosophical treatise, *Sant' Ilario: Gedanken aus der Landschaft Zarathustras* (Sant' Ilario: Thoughts from Zarathustra's landscape). In the figure of Hausdorff-Mongré Scholem perhaps sees an image of his own divided soul: Hausdorff is the rigorous mathematician who never strays from the subject matter at hand, whereas Mongré is the wide-ranging philosopher who, like Scholem circa 1914, locates the source of philosophical reflection in the mountainous regions of "Zarathustra's landscape." Scholem may have communicated his interest in the mathematician-philosopher to Benjamin, who includes *Sant' Ilario* among a list of works to be read (WBA, MS 1850r). The term *Ablauf*, as it happens, plays a prominent role in Hausdorff's second philosophical work, *Chaos aus kosmischer Auslege: Ein erkenntniskritischer Versuch* (Chaos from cosmic selection: an epistemo-critical essay), beginning with the following terminological stipulation:

> The investigation of time to which we will turn can be divided into two major parts. In our idea of time two heterogeneous ideas are connected with each other in a manner that can be easily separated. We will call them "time content" and "time course" [Zeitinhalt und Zeitablauf]: the first consists in a continual series of world-states, the material substrate of time; the second consists in an enigmatic formal process, through which every world-state experiences the transformational series of future, present, past. By "world-state" I mean a fulfilled stretch of time [erfüllte Zeitstrecke] of length null; by moment [Augenblick] I mean an empty stretch of time of length null.[13]

The primary point of Hausdorff's "epistemo-critical essay" consists in working out a hyperbolic version of transcendental idealism in which the "cosmo-centrism" of Kantian critique reveals itself to be as short-sighted as the geo-centrism of pre-Copernican metaphysics. Showing the ideality of space, for Hausdorff, is considerably less difficult than demonstrating the ideality of time, whose unidimentionality and unidirectionality appear to be necessary in contrast to the three-dimensions of generally reversible space. Above all, Hausdorff seeks to show that the "stretch of time" cannot be represented in terms of a straight line and proposes, instead, the image of a "temporal plain" (*Zeitebene*) within which the limited "time line" of the individual expands into an ellipse with neither a beginning point nor an endpoint that would correspond to the moments of birth and death.[14] When Hausdorff writes under the name of Mongré, he largely avoids mathematical terminology, but the final sections of *Chaos from Cosmic Selection* adopt enough mathematics that the work begins to appear as a "mathematical theory of truth"—or, better yet, as a mathematical theory of nontruth, since chaos is the final word. In 1916, Scholem, by contrast—and perhaps in response to Hausdorff—begins to experiment with the opposite thought, which rise above "Zarathustra's landscape," namely the thought that a "great differential equation that expresses the world" (S, 1: 390) can be discovered. In his diary from August 1916 Scholem prefaces his account of the afternoon-long conversation with Benjamin by presenting this grand equation as the solution to, and therefore the redemption from, the problem of the world, which is otherwise dominated by myth. This, for Scholem, is the messianic mission of mathematics: to express the world in a form that is wholly immune from myth. Above all, the equation captures the course of time. And thus emerges a question: what, after all, does "course" mean in the context of associated mathematical terminology?

The notes that Scholem records in his diary a few days after the conversation with Benjamin give certain indications of an answer, even as they diverge at crucial points from the version he would publish many years later: "We spent an entire afternoon discussing a very difficult remark: 'the series of years is indeed countable but not numerable.' Which led us to [the meaning of] course [*Ablauf*], series of years, and as the final starting point, direction. Is there a direction without a course? 'Direction is the differential measure of two straight lines,'" Scholem adds, probably in response to Ferdinand Frobenius's lectures on analytic geometry that he attended in the summer semester of 1915.[15] He continues: "This is a thought complex that I very much want to think about again.

Doubtless, time is a course [Die Zeit is wohl ein Ablauf]; but does time have a direction?" (S, 1: 390). The question provides a perspective from which the sense of the term *Ablauf* can be determined: it designates as a continuous "processes" or "run" (*Lauf*) that can "run out" and thus "expire." The drafting of a straight line is a geometric representation of such a "run," but as Hausdorff emphasizes, contradicting Kant, rectilinear motion is by no means the only representation of the "cursive" character of time: "For it is a thoroughly metaphysical assertion that time is like a straight line; perhaps it is a cycloid or something else, which has no direction at many points. (Where there are *no* tangents.)" (S, 1: 390). The parenthetical comment indicates the line of argument that brought the enigmatic quotation under discussion. The direction of a curve can be determined in two ways: by means of parameterization—this is probably what Scholem means by "differential measure"—and in terms of its tangent lines.

The account of the "cursive" character of time that appears in Scholem's diary runs in the opposite direction to the one he provides in *Story of a Friendship*; but the question around which both accounts revolve remains the same. And the source of this question can be identified with precision: it does not derive from Hausdorff, whose *Grundzüge der Mengenlehre* (Basic features of set theory) Scholem had not yet had a chance to study (S, 2: 263) but, rather, from certain mathematical inquiries conducted by Konrad Knopp, a less innovative but still prominent mathematician of the period, under whom Scholem studied ordinary differential equations in the spring semester of 1916 and whose lectures on the theory of functions he probably attended the following semester.[16] During the years in which Scholem was studying mathematics, Knopp was engaged in a program of research into a loosely connected set of mathematical problems that had as yet no formal title. In 1916 he published a paper on so-called Cantor sets, which are linear perfect yet nowhere dense; that is, they are in some sense equivalent to the complete set of points on the linear continuum, even though there are an infinite number of gaps on the corresponding line.[17] In the following year Knopp published a more substantial paper on certain curves that are everywhere continuous yet nowhere differentiable—precisely the type of curve to which Scholem directs Benjamin's attention in their August conversion: "(Where there are *no* tangents)."[18] And in the following year, as a summation of his contribution to this unnamed field of mathematical research, Knopp published a paper on continuous yet nowhere differentiable functions in which he proposed a "simple procedure" for their construction.[19]

The importance of continuous yet nondifferentiable functions for the development of nineteenth-century analysis can scarcely be overestimated. Karl Weierstraß—who is often described as "the father of analysis" and who was, incidentally, Husserl's *Doktorvater*—surprised his students and colleagues in the 1860s when he constructed a trigonometric function that was so "pathological" that it had no tangent at any point.[20] A curve corresponding to this function is unimaginable in the exact sense of the term: every point consists in a sharp turn. After Weierstraß's initial demonstration, a number of similar functions were constructed, including, for instance, a curve that includes an isosceles triangle between any given points and a curve that fills the space of a square. Hausdorff developed the idea of topological dimension for precisely such cases of apparently paradoxical dimensionality. After proposing his "simple procedure" for constructing continuous yet nondifferential functions, Knopp made no further contributions to the unnamed field; when he left Berlin for Königsberg in 1919, he returned to his earlier fields of research for which he had already gained a considerable reputation. Around the time when Scholem published *Walter Benjamin: The Story of a Friendship*, in the 1970s, the field of mathematics to which Knopp contributed during his last years in Berlin years received a formal name. With assistance from Hausdorff's *Basic Features of Set Theory*, Benoit Mandelbrot, a Jewish-Polish-French-American mathematician and polymath, called the objects in question "fractals," and the field of research accordingly acquired the name "fractal geometry."[21] As reported by Scholem, then, the question toward which he directed the conversation with Benjamin could be formulated as follows: does the course of time have a fractal shape? In his diary entry from August 1916 Scholem says "perhaps," whereas in his book from 1975 he inclines toward "no."

Despite incomplete and at times contradictory documentation, the course of the afternoon-long conversation between Scholem and Benjamin can thus be reconstructed in broad outline: Benjamin's "difficult remark" on the discrete years of history gives way to a reflection on the continuous course of time. The field of mathematics under discussion accordingly changes from number theory to analytic geometry, as Scholem proposes a series of curves that could capture the "cursive" character of time, the least satisfying of which would be the drafting of a straight line. In response, then, to the claim that historical years are countable but nonnumerable, there emerges a corresponding claim that the course of time is continuous but nondifferentiable. The conversation then returns to the field of number theory: "Also among the series of numbers there is the same problem [as the one described in the original remark]," Scholem notes: "But in this case, numbers [Zahlen] somehow bear in themselves numerals [Nummern]; properly speaking, however, they are not numerable, since numerability presupposes exchangeability, and that is true neither of numbers nor of years: they are in no way exchangeable" (S, 1: 390). Although this highly abbreviated theory of number coheres with Benjamin's nascent theory of language—where ordinal numbers correspond to proper names—for the elderly scholar of the Kabbalah, the theory is so strange that he changes the story for his *Story of a Friendship*: "We discussed whether years, like numbers, are exchangeable, just as they are numerable." Stranger still are the final remarks in the diary account, which are completely elided from the subsequent book: "Beyond this [the nonexchangeability of numbers and years], it became clear that direction is a determination in relation to the following dimension. . . . Very difficult things also emerge with respect to the direction of crooked lines through the insertion [Einsetzung] of the tangent direction as curve direction, which is doubtless arbitrary yet satisfying to the human spirit and is something that appears to be self-evident" (S, 1: 390). Whatever else may be said of these difficult remarks, which probably stem from Scholem's study of parametrization, it is clear that they do not advance the "difficult remark" with which Benjamin began their afternoon conversation in August 1916.

Mathematical Studies

Yet Benjamin may have invited Scholem to the Pollak residence for precisely this reason—so that they could resume their original conversation about the nature of the "historical process" in conjunction with a new "mathematical" thesis about the structure of chronology.[22] Scholem should have been a productive participant in a discussion of this topic because he was, after all, taking courses with the likes of Frobenius and Knopp. Benjamin, for his part, was not so well prepared; but he was by no means uninterested, and Scholem was not his only mathematically inclined acquaintance. The "universal genius" (*GB*, 1: 291) Felix

Noeggerath—whom Benjamin met while taking courses with Moritz Geiger, himself a skilled philosopher of mathematics, who would soon publish an acclaimed axiomatization of Euclidean geometry—wrote a dissertation on the concepts of system and synthesis, which recasts the relational categories of Kantian critique in terms of the mathematical concept of seriality and which, in addition, includes a rather pedestrian appendix on the applicability of non-Euclidean geometry.[23] Benjamin also came into contact with at least one truly genial mathematician, Robert Jentzsch, who was a poet and a friend of Friedrich Heinle and who wrote a dissertation on power series that has been recognized as a minor masterpiece of its genre.[24] As for Benjamin's own preparation for higher mathematics, he considered himself competent enough to accompany Scholem to Henoch Berliner's seminar on number theory during the brief time in which the two lived together in Bern. In conjunction with his interest in this area of mathematics, he posed questions to Scholem about convergent series (*GB*, 2: 44 and 2: 58) and, for a time, considered asking Käte Holländer to tutor him in mathematics (*GB*, 2: 111).

Above all, though, Benjamin was the nephew of a major mathematician of the period, Arthur Schoenflies, whom he "often met" and "got along with very well" (*GB*, 3: 382), and whose mathematical papers he would often consult (WBA, MSS 1850 and 1857).[25] After studying with Weierstraß in Berlin, Schoenflies become a prominent proponent of set theory, whose reports on its development were widely used by German mathematicians, until they were permanently superseded by Hausdorff's *Basic Features of Set Theory*. One of Schoenflies's reports on set theory briefly describes its genesis: "Set theory as a science arose at the precise moment when Cantor introduced denumerability [Abzählbarkeit] as a well-defined mathematical concept, undertook the division of infinite sets according to their power [Mächtigkeit] and showed, in particular, that algebraic numbers form a countable set, whereas the continuum is not countable."[26] The distinction upon which Benjamin wants to develop a theory of historical time is akin to the one from which set theory is generated. It is perhaps for this reason that, as Scholem notes in his diary, Benjamin experienced one of Berliner's lectures on number theory as "uncanny" (S, 2: 225).[27]

Scholem, for his part, makes nothing of this affinity. Concepts drawn from set theory play no role in the reflection on mathematics contained in the published version of his diaries—which does not mean, of course, that he was unfamiliar with them.[28] Scholem is adverse enough to the mathematics of the infinite that when he first attends Knopp's lectures on differential equations, he is thrilled to hear the following words: "exact version (better: exclusion) of the infinite" (S, 1: 267). Even if the infinite is not altogether excluded from Scholem's reflections on mathematics—especially after his conversations with Benjamin in August 1916 and in light of his growing interest in the work of Bernard Balzano—nowhere does he reflect on the difference between the "power" of denumerable sets in contrast to that of nondenumerable ones. Perhaps Cantor's use of Hebrew letters for transfinite cardinals was a source of concern, especially since, as he doubtless knew, Cantor was not Jewish. In any case, his aversion to new theories of the infinite is of a piece with a more general tendency, which can also be discerned perhaps in his relation to Judaism. Despite a certain rhetoric of radicalism, which requires that the roots of the current crisis be exposed, his primary interest lies in discovering overlooked connections to a vibrant past in the midst of a confounding present. The scientific version of this tendency is nowhere more evident than in a little treatise he wrote in July 1916, under the title "Allerlei vom mechanistischen Weltbild" (Potpourri regarding a mechanistic world-image).

At this pivotal moment in the development of the mathematical sciences—when Einstein had just formulated the differential equations that would henceforth be known as "the general theory of relativity" and thereby displaced Newton's *Principia* by freeing physics from the strictures of Euclid's *Elements*—Scholem's midsummer "Potpourri" presents classical mechanics as the model of scientific inquiry; he goes so far as to propose that the combination of the *Elements* and the *Principia* forms the paremeters within which a purely alogical mysticism can finally meet the pure rationalism of mathematics.[29] The resulting combination, devoid of all traces of a falsely mystical organicism, prepares the opening through which the messiah comes: "Out of the work of Euclid and Newton the Bible of this new world would be formed, complete with an introduction by the cognizing mystic or the new Novalis." And Scholem concludes his "Potpourri" on the following note, in which his divided soul finds, as it were, the highest possible synthesis: "Mysticism is eternally in danger of forgetting mathematics. This is the original sin of mysticism; the new mechanistic world-image would be the decisive step in advance of the final synthesis. . . . The aforementioned mathematical mystic or mystical mathematician—he will certainly be a Jew. He will be the messiah" (S, 1: 353).

As the summer of 1916 gives way to the fall, and Scholem begins to view Benjamin as a "new Novalis," he does not then see his own role as that of reminding the mystic of a certain mathematical rigor. Instead, he pursues a theory of his own, which he begins to call "the mathematical theory of truth." Three months after the seminal conversation in August, as Knopp's lectures on the theory of functions get underway, Scholem finds himself in a position to draw up a preliminary version of the theory. Its point of departure is the traditional definition of truth as *adequatio* or *Übereinstimmung* (correspondence), which Scholem mathematicizes by interpreting *adequatio* in functional terms: "If truth is dependent on correspondence with being [das Sein], it follows that truth is a *function* of the entity [Seienden]" (S, 1: 416). Thus emerges the equation: "To being *a* there belongs the truth of this entity = $f(a)$" (S, 1: 417). If it is assumed that there is only one being, "only *one* thing," then there is only one truth, "since *x* is not a variable." This means, however, that "*being has no derivatives*, therefore no tangents," which leads, in turn, to the unhappy conclusion that "truth is not really a function because it has no interval" (S, 1: 417). In order for there to be a "mathematical theory of truth," then, the opposite assumption must be made: "there are many existent things, entirely in Plato's sense" (S, 1: 417). On the basis of this assumption, the following ontological theorem can be ascertained: "$x + \Delta x$ means the heightened being of a thing in relation to x; Δx is the increase in being" (S, 1: 417).

Thus Scholem finds himself in a position to define a hitherto missing term, namely concept, which corresponds to *intellectus* in the traditional definition of truth. To say of truth that it is the correspondence with the thing means that it is not the correspondence of the thing with its concept. The "mathematical theory of truth" takes shape as a Cartesian coordinate system in which being and truth are the *x* axis and the *y* axis, respectively. Concept or thought is then the derivative of function—which is to say: "The concept of a thing can be designated as the direction of its truth toward God" (S, 1: 417). As for the concept of truth, it is identical with itself, since "the *concept is the tangent of truth*, and the straight line its own tangent at *every* point" (S, 1: 417). As if he were replaying in his own mind the conversation in August, Scholem immediately adds: "Probably, though, the relation is still more difficult and truth is *not* a straight line but, rather, a parabola or something similar, with which the intensification of being approaches truth more *quickly*" (S, 1: 418). Scholem's sketch of the "mathematical theory of truth" abruptly concludes with questions similar to the ones that appeared to govern the conversation in August, now framed in terms borrowed from Benjamin's "Two Poems of Friedrich Hölderlin," which Scholem had a chance to read in the interim: "Is truth continuous? Is it always differentiable, that is, does every thing have concept and every truth an inner form? What is an entire function of being? *Does the* historical *allow itself* to be *investigated* as a limit concept? This would be the truest doctrinal book of logic and the theory of science, which only the messiah will write, the most objective and metaphysical book of world literature" (S, 1: 419). None of the questions Scholem thus poses, however, specifically responds to the "difficult remark" that began the conversation out of which the "mathematical theory of truth" emerged. Among the questions Scholem declines to pose in this context, the following are perhaps the most pressing: Does the distinction between countability and numerability correspond to the difference between cardinals and numerals? In what sense is numerability related to denumerability? And to what extent can the nonnumerability of historical years be understood as an expression of their "power"?

Other things are missing as well. In his *Story of a Friendship* Scholem claims that he still possesses a document in which he records in greater detail the part of the conversation that concerns the relation of numbers to years, but it is nowhere to be found.[30] In all of the written records, moreover, Benjamin's "difficult remark" is accompanied by another comment, which cites the title of a little treatise that Georg Simmel published in 1916: "The problem of historical time," Benjamin writes, making use of Simmel's title, "is already posed by the peculiar form of historical chronology [eigentümlichen Form der historischen Zeitrechnung]" (2: 601; cf. S, 1: 402). And in one of the folders that contain these two closely related remarks about the problem of historical time, Scholem preserves a note about the structure of the ensuing theory: "The theory obviously does not relate to reality; rather, it must cohere with language [mit der Sprache zusammenhängen]. Here lies an objection to mathematics [gegen Mathematik]" (2: 601–2). Tiedemann and Schweppenhäuser are slightly suspicious of the folder in which this remark appears, and accordingly decide to change its title from "Remarks" to "Aphorisms" (2: 1411). The emendation is probably unwarranted, since both Benjamin and Scholem regularly spoke of their remarks as "remarks"; and, in any case, there is nothing wrong with the word *remark*. Nevertheless, there is something suspicious about the previously quoted "aphorism," for, after all, who ever said that mathematics is related to reality? Certain forms of Platonism consist, of course, in the thesis that mathematical objects are the ultimate elements of reality; but the offending party in this case is Platonism—not mathematics.

If the editors of Benjamin's papers were to look for someone who claims that mathematics is itself related to reality, they need look no further than the scholar who preserves the remarks in question. Scholem's "mathematical theory of truth" is predicated on the traditional understanding of truth as *adequatio*. Benjamin apparently responds to Scholem's theory by asserting that the theory of historical time must cohere not so much with language as with the *theory* of language. For this reason, the final words of the remark should probably read: "gegen die mathematische." And the sentence as a whole would thus run: "Here lies an objection to the mathematical [theory of truth]." Construed in this manner, Benjamin's judgment could scarcely be more damning. Scholem may have failed to record it accurately for this reason. Or Tiedemann and Schweppenhäuser may have misread the remark—and displaced their suspicions onto the title of the folder as a whole. In either case, the result would be the same: Benjamin would be saying, in effect, that the "mathematical theory of truth" is seriously misguided, perhaps even as misguided as Simmel's *Problem der historischen Zeit* (Problem of historical time), about which he says the following in a letter to Scholem from 1917: "a completely wretched fabrication that, after the faculty of thinking goes through many contortions, incomprehensibly utters the silliest things imaginable" (*GB*, 1: 409).

As for Benjamin's attempt to reflect on the nature of mathematics on his own—without help from any of his more mathematically knowledgeable friends—there is no question that he found himself at a loss, for he admits as much in a letter to Scholem from October 1916. The very same letter announces that his failure has not prevented him from completing a major work, indeed a "little treatise," which develops the theory of language that he first sketched in his letter to Buber from the previous July. Benjamin began to write the treatise as a response to a situation in which he found himself "unprepared for the infinitely difficult theme of mathematics and language, that is, mathematics and thinking, mathematics and Zion" (*GB*, 1: 343). The title of the resulting treatise sounds systematic enough: "On Language as Such and on Human Language." Yet the title also indicates that the "little treatise" is missing something essential: mathematics as such, and perhaps even human mathematics, *die Mathematik des Menschen*, if there is such a thing.[29] The absence of mathematics in any case means that "On Language as Such and on Human Language" cannot pursue either of the two related problems to which Benjamin refers in his letter to Scholem: those of thinking and Zion. Nothing further needs to be said at this point about the problem of thinking. It should be clear, though, that in the context of the conversations between Benjamin and Scholem the word *Zion* is shorthand for "the problem of historical time."

Criticizing Heidegger

In the same letter to Scholem Benjamin thus returns to the place where "the problem of historical time" first materialized. He adds no further thesis; but he does make a few comments about an essay by his fellow classmate in Rickert's seminar on Bergson's theory of time, namely Martin Heidegger, who published "The Concept of Time in Historical Scholarship" in conjunction with his appointment to the University of Freiburg.[32] In Benjamin's opinion—which will later be echoed in his initial response to Heidegger's *Habilitationsschrift* on Scotian speculative grammar[33]—the success of such routine academic work is something of a scandal: "This essay documents in an exact manner how *not* to go about this matter. A terrible work, one that you [Scholem] perhaps will look into, if only to confirm my sense that not only what the author says about historical time (which I can judge) is non-sense but that his discussion of mechanical time is also wrongheaded, as I surmise" (*GB*, 1: 344).[34] Despite a certain similarity between Benjamin's assessment of "The Concept of Time in Historical Scholarship" and his evisceration of Simmel's *Problem of Historical Time*, there is nevertheless a major difference: Heidegger's essay is not a sheer "fabrication." By showing in an *exact* manner how *not* to go about posing the problem, it gives direction by way of negation to the problem at hand. And there is reason to suppose that Heidegger shared Benjamin's assessment, for at a crucial juncture in his essay—as he passes from his analysis of time in the physical sciences to an exposition of the concept of time in historical scholarship—he describes a direct path to the problem, which, however, he is unwilling to take until the end:

> In historical scholarship [as opposed to physical science], the path that leads from its goal to the function of its concept of time and thereupon to the structure of this concept appears as a detour. Historical scholarship can achieve its

goal much more easily and quickly if we only consider the fact that there is a particular auxiliary discipline in the methodology of historical scholarship, an auxiliary discipline that actually concerns itself with the determination of time in historical scholarship: historical chronology. Here the peculiarity [das Eigentümliche] of the historical concept of time immediately comes to light. Why this path is not taken can only be explained in the conclusion. (He, 426)

The lines of inquiry pursued by Heidegger and Benjamin so closely converge that they express themselves in the very same terms. For Benjamin, historical chronology is the "peculiar form" (*eigentümlichen Form*) in which the problem of historical time poses itself; for Heidegger, historical chronology leads immediately to the "peculiar character" (*das Eigentümliche*) of the concept of time in historical scholarship. Heidegger ultimately decides against following this path and accepts the methodological directives of Rickert, who recommended his appointment.[35] When Benjamin condemns Heidegger for proceeding in *exactly* the wrong direction, he is presumably taking issue with this decision. As in his *Habilitationsschrift*, Heidegger is not so much wrong as obsequious. And when at the end of his essay he returns to the "auxiliary science" of historical chronology, the direct path to the problem can be ascertained simply by appending a "not"—or "not necessarily"—to each of his assertions:

> Year numbers [Jahreszahlen] are convenient numerical marks [Zählmarken]; still, considered in themselves, they are senseless [ohne Sinne], since for any number, another number could be equally substituted, if one only shifts the inception of the counting. But precisely the beginning of chronology shows that this time-reckoning always starts with an historically significant event (the founding of the city of Rome, the birth of Christ, Hegira). The auxiliary discipline of historical scholarship, historical chronology, is therefore only meaningful [bedeutsam] for the theory of the historical time concept from the perspective of the *beginning* of time-reckoning. (He, 432)

Much could be said about the ecumenical character of these concluding remarks, especially their noticeable silence about Jewish chronology, which begins with the beginning. It is as though the creation of the world must somehow be excluded from Heidegger's reflections on the significance of significant historical events. In any case, from the perspective of Benjamin's assessment of the essay—specifically, that its discussion of historical time is "non-sense" (*Unsinn*)—only one point must be made: year numbers are *not*, as Heidegger claims, "senseless." More exactly, they are not necessarily "convenient numerical marks," which can be replaced by other numbers if only a different starting year is stipulated. Conversely, as soon as *any* year is designated as the start, the subsequent numbers must be "convenient markers," the meaning of which depends on the particular temporal power—whether it be Roman, Christian, or Islamic—that converts a stipulation into a convention. For this reason, however, the year numbers are not so much *historical* as they are exponents of the relevant regime. If, by contrast, the beginning of the count is indeed the beginning, then historical years *can* be numbered. In this case, year numbers enjoy the same status as proper names, according to the formulations of the "little treatise" that Benjamin wrote in lieu of completing his inquiry into the "infinitely difficult theme of mathematics and language." Negatively spoken, year numbers have nothing to do with stipulation or convention. The following remark thus appears among Benjamin's papers of the period: "Historical years are names" (6: 90).

In conjunction with his contemporaneous attempt to solve Russell's paradox—which is discussed at the beginning of the next chapter—Benjamin draws an insuperable distinction between "judgments of designation" and "judgments of predication." Only in the case of the latter can one speak of "meaning" (*Bedeutung*) in the "proper" (*eigentlich*) sense of the word (6: 10). Terms that owe their origin to "judgments of designation," by contrast, are only improperly meaningful, for they mean only what they are said to mean. It is no accident that Benjamin's first and primary example of a "judgment of designation" is drawn from the field of mathematics: "*a* designates the *BC* side of a triangle" (6: 9). For mathematical proofs begin with the stipulation of what their signs are henceforth supposed to mean. Benjamin does not take sides on the scholastic debate between nominalists and realists as to whether the resulting theorems are themselves affected by the semantic conventions in which they are expressed. But it is nevertheless certain that, in his view, one form of applied mathematics, namely historical chronology, is incapable of freeing itself from the "judgment of designation" with which it begins. The stipulation of the first-year number is the foundation of historical chronology. Years can be counted as soon as someone says which one is first; but as soon as someone *says* that it is first, regardless of whether it is the beginning, year numbers are no longer meaningful in the proper sense of the word.

There is no evidence that Scholem responded to Benjamin's request that he read Heidegger's essay and determine whether its account of mechanical time is as wrongheaded as its discussion of historical time.[36] At the end of the diary entry in which he proposes the "mathematical theory of truth," however, Scholem does make the following note to himself: "The essay on historical time is very ridiculous and unphilosophical. Benjamin is right in his judgment" (S, 1: 418). With this remark, the conversation begun in August comes to an end—without a conclusion. Nevertheless, the following line of argument can be extrapolated from Benjamin's remarks on "the problem of historical time" in conjunction with his critique of Heidegger's "Concept of Time in Historical Scholarship." Anyone with the requisite arithmetic competence can count the years, beginning with whatever year he or she wishes. Prisoners are thus said to mark the days of their imprisonment. And just as the marks made by prisoners are meaningful for them, and for them alone, so are the numbers with which one counts the years—unless, of course, a regime makes some chronological system into a convention, which rules the lives of those under its sway. Heidegger inadvertently hits upon the "problem of historical time" when he writes in parentheses: "the founding of the city of Rome, the birth of Christ, Hegira." The diversity of historical chronologies is comparable to the dispersion of languages after the collapse of the Tower of Babel. It is not as though the years are nonnumerable because humanity as a whole has failed to agree on a single chronological system; rather, there is no such thing as "humanity as a whole" until the years are properly numbered—at which point there are no longer any years to count. When, in short, years are countable, they cannot be numbered; when they are numbered, there are none to count.

The theory that develops out of Benjamin's remarks on the problem of historical time thus coheres with the equally difficult theory of language that he develops in his "little treatise." "Historical chronology" is the name for one of the primary points at which meaning (language) and designation (mathematic) converge. An "historically significant event"—to quote Heidegger against himself—must be designated as such in order for a chronological system to begin; but the "judgment of designation" that distinguishes the meaningful event from all others deprives it of its meaningfulness. There is only one way that the event with which time-reckoning begins can *remain* meaningful: it must be taking place, again, this year. Every year would therefore be the first year as long as the year is historical—and not simply stipulated as such by whomever assumes power. Alongside the statement that "historical years are names," Benjamin adds a fragmentary remark: "The series of historical years" (6: 90). The remark could be expanded along the lines suggested above: the series of historical years is not only convergent; it is hyperconvergent, because it never diverges from the numerical value with which it begins. To find a formula for such a series is no small feat.

Another Detour

So it goes with the problem of "mathematics and Zion." With regard to the coordinated problem of "mathematics and thinking," Benjamin left some scattered notes, which were probably intended as the basis for the letter to Scholem that he could not complete. At the center of these notes is a "magic circle of language" with the following four moments: "God creates" at the top; "The thing is called" in the middle, left side; "The human being knows" in the middle, right side; and at the bottom, "Mathematics thinks" (7: 786). The word *thinking* doubtless sounds good; but as its position at the nadir of the circle indicates, it is not necessarily so. Indeed, thinking may be positively evil. Benjamin does not say so in his preliminary notes; but it becomes an essential element of the work that most fully develops the lines of inquiry he first sketched in 1916, namely his *Habilitationsschrift* on the *Origin of the German Mourning Play*, the opening sentences of which contain in miniature the starting point for his reflections on the relation of mathematics to languages: mathematics belongs to a lower order than language, for, whereas it acquires only a "signum of genuine knowledge," languages "intend truth" (1: 207).[37] It would perhaps be better to say, however, that the theme of his *Habilitationsschrift* develops out of the impasse he encountered in pursuing—without reliable assistance—the "difficult remark" with which the conversation in August 1916 began. For immediately after Benjamin says of historical numbers that they are proper names, he adds a remark that discloses an alternative path to the "problem of historical time." The advantage of this path is immediately evident: it goes around the "infinitely difficult theme" of mathematics and language. And the path can be justified on the basis of what Benjamin enigmatically calls the "Kantian typic" (2: 160). According to the "metaphysical exposition" of space and time with which the *Critique of Pure Reason* begins, the structures of these two "forms of appearances" are equivalent; similarly, for Benjamin, the structure of historical time corresponds to that of historical space: "The problem of historical time must be grasped in correlation with that of historical space (history on *stage* [*Geschichte auf dem* Schauplatz])" (6: 90).

Just as Heidegger introduces a methodological digression into his analysis of the concept of time in historical scholarship, so Benjamin takes a detour in his attempt to discover the "peculiar form" in which the problem of historical time poses itself. In his case, however, it is not a detour dictated by a desire to occupy an academic position; it is, rather, a detour through the idea of dramatic form, with particular attention to the form in which history acquires a thoroughly spatial character. The sketch from 1916 entitled "Tragedy and *Trauerspiel*" begins with a few remarks about the distinction between historical and mechanical time that are not as far removed from Simmel's *Problem of Historical Time* as Benjamin's remark to Scholem would lead one to suppose. And in the first section of the work that developed out of his sketches from 1916, Benjamin demonstrates the degree to which the "time-space [Zeit-Raum]" (1: 313–14) of the baroque *Trauerspiel* rigorously excludes time in the "proper" sense of the word; instead, it admits a place only for its "parasitical" counterpart (1: 313), which consists in the endless repeatability that he had earlier discussed under the rubric of countability. The theme of "mathematics and Zion" is thus effectively bypassed. The corresponding problem of "mathematics and thinking" reappears in the final section of the *Trauerspiel* book as an inquiry into the characteristics of the purely subjective thinker, who, by virtue of his subjectivity, falls into the abyss of evil. The language of the melancholic "ponderer" (*Grübler*) shares an essential trait with that of the mathematician: it always begins with "judgments of designation." Just as the mathematician arbitrarily assigns meanings to certain marks, so does the ponderer willfully endow things with "meaning," which only indicates that they lack any. Unlike the mathematician, however, the ponderer never proceeds beyond the beginning. The null-language of purely subjective thinking is not therefore the mathematical sign but, rather, in Benjamin's words, "allegorical expression," which could henceforth be called the missing element of the intended system: human mathematics or *die Mathematik des Menschen*. And the purely subjective thinker, in turn, can be likened to the mathematician who is always only beginning, saying in ever different ways, for example, "a designates the BC side of a triangle."

For Scholem, the messiah must be a mathematician. The notes and remarks that Benjamin drew up in the latter half of 1916 point in a different direction: messianic completion is nonnumerable because its unity is of a higher "power" than that of any countable unit. But this difference says nothing against mathematics—only against its beginning, which historical chronology forever reiterates.

§5 Meaning in the Proper Sense of the Word

"On Language as Such and on Human Language" and Related Logico-Linguistic Studies

Russell's Paradox

As he sketches a plan for an article rebutting Arthur Schoenflies's paper "Über die logischen Paradoxieen der Mengenlehre" (On the logical paradoxes of set theory), Gottlob Frege ends on a note of despair, which is enlivened by the expectation that his own *Begriffsschrift* (concept-notation) will remain untouched: "Concept and object, *nomen appellativum, nomen proprium*. . . . Russell's paradox cannot be solved in Schoenflies' way. . . . Set theory convulsed. My concept-notation in large part not dependent on it. (In contrast to other similar attempts.)"[1] Around 1900, Bertrand Russell discovered a contradiction or paradox that he widely publicized in the following form: "The set of all sets that are not members of themselves is both a member and not a member of itself."[2] In an appendix to the *Principles of Mathematics* Russell tentatively proposes the "doctrine of types" as a remedy: individual objects constitute type 1; sets of individual objects, type 2, sets of those sets, type 3, and so forth. As long as this hierarchy is rigidly maintained, it is impossible for a set to be a member of itself: "It is the distinction of logical types that is the key to the whole mystery."[3] In a relatively nontechnical article for the 1906 yearbook of the German mathematical association, Benjamin's maternal great-uncle Arthur Schoenflies provides a significantly simpler solution to the paradox, which consists at bottom in demonstrating that the set in question cannot be brought into the science of mathematics: "We ask . . . whether there is a set that contain itself as an element. *This is impossible.* Every attempt to form or imagine a set of this kind must fail."[4] By clarifying the meaning of the term *Menge* (set), Schoenflies attributes the origin of the paradox to a logico-linguistic confusion:

> The Russellian set M of all sets μ, which do not contain themselves as an element, is therefore nothing other than the "set of all sets." This is really mere words; but we immediately conclude that these words cannot really represent a concept free of contradictions. . . . The "set of *all* sets" would in any case represent a set that contained itself as an element, which is indeed impossible. In this way, the source of Russell's paradox is disclosed.[5]

In response to Schoenflies's solution, which relies on a distinction between phrases that are scientifically meaningful and those that are made up of "mere words," Frege reiterates the fundamental distinction on which much of his work rests: "The article by S[choenflies] occasions me to make the following remarks, in which I repeat much that I already discussed previously, since it seems to be little known. I fail to find in Schoenflies . . . the sharp distinction between concept and object. In the signs a proper name (*nomen proprium*) corresponds to an object, a concept-word or concept-sign (*nomen appellativum*) to a concept."[6] Schoenflies's solution comes down to the claim that a concept cannot be admitted to a science unless it is consistent. For Frege, however, a concept can be shown to be consistent only by citing an object that falls under it, and yet "to do that one must already have the concept."[7] No one can "immediately conclude" that a concept is inconsistent, and for this reason, "mere words" can easily enter into a scientific demonstration and make it susceptible to paradoxes. As Gerhard Hessenberg notes in his *Grundbegriffe der Mengenlehre* (Basic concepts of set theory), from 1906—and this remark is so close to Frege's critique of Schoenflies that it may have dissuaded him from completing his essay—"the attempt to solve Russell's paradox [in the manner Schoenflies proposes] is . . . nothing but a displacement of the problem."[8]

A decade after his great-uncle proposed a simple solution to Russell's paradox, Benjamin did the same. His interest was less in the set-theoretical formulation of the paradox than in its linguistic-categorial counterpart. In Hessenberg's formulation, from which Benjamin probably derived his own, the paradox runs as follows:

> Every predicate a . . . can either be asserted of itself and may in this case be designated as a "predicable," or it cannot be asserted of itself, in which case

it can be called "impredicable." The predicate *thinkable* is predicable, for it is itself thinkable. The predicate *virtuous* is impredicable, for it is not itself virtuous. This disjunction between predicable (p) and impredicable (i) is complete.... Therefore, the predicate i = impredicable is either i or p. And if it is impredicable (i is i), it is thereby asserted of itself, therefore predicable (p).[9]

In response to this form of the paradox, Benjamin sketches several versions of a solution, the nucleus of which is contained in the briefest:

> *Attempt at a Solution of Russell's Paradox*
> Nothing can be predicated of a sign. The judgment in which a meaning [eine Bedeutung] is subordinated to a sign is not a predicative judgment. Russell conflates judgment of meaning and judgment of predication [Bedeutungs- und Prädikatsurteil]. (6: 11)

The solution proposed by Benjamin corresponds in a certain sense to Schoenflies's, since both nephew and uncle identify a fundamental error not so much in the argument that leads to the paradox as in the language that gives rise to the argument. When the error is recognized as such, the paradox resolves itself without an elaborate "doctrine of types." The terms through which Benjamin formulates the source of the argument, however, are closer to Frege's than to Schoenflies's. It is unlikely that Benjamin had a chance to study Frege's work, much less his posthumously published notes. Scholem, as it happens, studied with Frege while he was in Jena and even prepared a presentation for Bruno Bauch's seminar in which he argued in favor of a "concept-notation" as a major advance beyond the kind of logic Hermann Lotze propounds (S, 2: 109–11). Benjamin expressed an "urgent" wish to read Scholem's presentation (*GB*, 1: 404); but his inquiries into source of the paradox were probably undertaken long before Scholem arrived in Jena, and in any case, his understanding of the cardinal word *Bedeutung* differs significantly from Frege's. In Benjamin's case, the term is probably best translated by "meaning," the basic structure of which lies in "intentional immediacy" (6: 11): a name immediately means the thing named. Among translators of Frege's writing, *Bedeutung* is generally rendered by "reference" in order to distinguish it from "sense" (*Sinn*). A name does not immediately mean the object named; rather, it secures reference by means of its sense in the context of an assertion that, for its part, "refers" to or "means" its truth value. The "reference" or "significance" of all true or false assertions is the same, namely truth or falsehood.[10] By making it impossible for the distinction between concept and object or *nomen appellativum* and *nomen proprium* to be overlooked, Frege's "concept-notation" obviates the problems and paradoxes to which subject-predicate utterances succumb.

The distinction Benjamin draws between judgments of predication and judgments of meaning, by contrast, is meant to secure a concept of meaning in which the difference between terms that acquire their meaning by means of stipulation and those that are meaningful in the absence of any such act is insuperable. As a result, the term *judgment of meaning* is misleading: any term x that acquires meaning by means of a judgment "x means y" is not properly meaningful. Benjamin therefore replaces "judgments of meaning" by "judgments of designation" and develops his most extensive analysis of the origin of Russell's paradox in a sketch written under this title.[11] Not by chance is the first example of a judgment of designation drawn from the field of mathematics, which begins its demonstrations by assigning unambiguous meanings to arbitrary terms: "The judgment 'a designates the *BC* side of a triangle' serves as an example of a judgment of designation" (6: 9). The difference between the category of designation and all other categories lies in the fact that the former "cannot enter into logical relations with one another or with other judgments" (6: 9). In other words, nothing about the judgment "a designates the *BC* side of a triangle" compels any further judgment, even if the other sides of the triangle are designated for reasons of either convenience or convention by b and c. Benjamin's reformulation of the paradox named after Russell thus emphasizes nothing so much as *Russell's* part in its generation: "Russell *designates* a word to which one can attribute its meaning as a predicate . . . 'predicable.' He designates a word in which this is not the case 'impredicable.' . . . The subjects in [the resulting paradoxical formulations] are signs; that is, as complexes fixed in sound and writing, they do not mean anything" (6: 9–10).

Benjamin's inquiry into the source of the Cretan liar's paradox follows along similar lines of thought: just as Russell's paradox derives from Russell's judgment of designation, so the origin of the liar's paradox lies in the "I"-character of the speaker who refers to him- or herself as a liar (6: 59). Whenever meaning depends on an "I" who confers it, there is something mendacious, for the term that acquires meaning in this way is actually without meaning—or more exactly, without "authentic meaning" (*eigentliche Bedeutung*). In his *Origin of the German Mourning Play*

Benjamin specifies the character of "improper" (*uneigentliche*) time as "parasitical" (1: 313). Similarly, "improper meaning" can be called parasitical, for it depends for its existence on "proper meaning." So divergent are the two meanings of *meaning* that Benjamin proposes to replace Russell's "doctrine of types" with something like a doctrine of "disparate orders" (6: 10), a schematic version of which appears in "Judgment of Designation" as follows: "Improper meaning [uneigentliche Bedeutung], which is designation, is to be distinguished from meaning proper [der eigentlichen]. '*S* is *P*' does not designate; rather it means that *S* is *P*. *Impredicable* designates the predicate of a determinate judgment; *unapproachable*, however, means something. Whence this difference of words. In meaning there lies representation [Räpresentation], not in [designation]" (6: 10).

In terms of Russell's paradox, *meaning* is not only predicable; it is the predicable par excellence, since the "something" that it means is meaning—and so, too, every word that does not owe its meaning to a judgment of designation, including, for example, the meaning of the word *unapproachable*. Terms such as *predicable* and *impredicable*, by contrast, owe their origin to judgments of designation, which determine their meaning. Despite the schematic character of Benjamin's notes, the point of drawing a distinction between proper and improper meaning is unmistakable: *unapproachable* in contrast to *impredicable* is not so much indefinable as unapproachable by way of definition; conversely, *impredicable* can be grasped in no other way. Whenever *unapproachable* is defined, it is no longer predicable, that is, unapproachable. And this is true of all words that are properly meaningful: they mean "meaning" in a certain way. Every word is as unapproachable by way of definition as *unapproachable*, for the "something" that it "represents" can be approached only within the sphere of meaning, properly speaking—not from that of the subject who means something by means of a written or oral complex. When, in the passage above, Benjamin poses the question "whence the difference in words," he is asking not only about the difference between a word like *impredicable* and one like *unapproachable* but also about the "variety" (*Verschiedenheit*) of words in general: if the meaning of every word is "meaning," whence arises the multiplicity of words, all of which mean the same, as long as one is speaking of meaning in the proper sense of the word? In response to a question of this kind, Benjamin enlarges his reflections on Russell's paradox into a condensed theory of knowledge.

Taken to the limit, the separation of proper from improper meaning leads to the thesis of an original ordering of language, in which each word is itself a language, and all languages mean the same thing, namely meaning, properly speaking. Just as, for Frege, the rigorous application of the distinction between name and concept requires a new formal language, so, for Benjamin, the insuperable distinction between improper meaning and meaning proper requires a correspondingly definitive distinction between pragmatic versions of languages in which proper and improper meaning are inextricably intertwined and an original order of word-languages that is constitutively incapable of supporting sovereign acts of ostentation and stipulation. At the end of "The Judgment of Designation," Benjamin thus presents the circular structure of language as the point of entrance for a well-grounded theory of knowledge: "Language rests on meaning; it would be nothing if it did [not] have meaning. Here, in this double appearance of meaning in logic, there is a rudimentary and indicative reference to the linguistic nature of knowledge, which is clarified in the philosophy of language" (6: 10–11). Knowledge enters into the circular structure of language—which is based on what it supports—insofar as it is knowledge of something, that is, of an object. "Whence comes the variety of words?" In brief, the answer runs as follows: each thing speaks its own language, and knowledge of the thing consists in knowledge of its language, which simply means "meaning" in its own particular way. The relation of language to knowledge can thus be represented without reference to a world of things that stand in a causal relation to subjects who invest signs with meaning, so as to archive knowledge or communicate with one another.

"The Magic Circle of Language"

In reflecting of the technical terms through which Russell's paradox is formulated, especially *predicable* and *impredicable*, Benjamin thus arrives at a point where he can propose a theory of language in which the so-called "natural" attitude is "turned off": there is no world of things to which terms in a language refer whenever language remains true to its fundamentally circular structure. One version of the linguistic circle appears in the aforementioned passage from the notes toward the solution of Russell's paradox: "Language rests on meaning; it would be nothing if it did [not] have meaning." Another version can be found in a previously quoted draft for a *Habilitationsschrift* on Scotian speculative grammar: "if one is able to abstract from the complete correlation between signifier and signified [Bedeutendes und Bedeutete] with respect to the question of foundation, so that the circle is avoided, then the signifier aims for the signified and at the same time rests on it.—This task is to be solved by considering the domain of language. . . . The domain of language extends itself as a critical medium between the domain of the signifier and that of the signified" (6: 22).¹² And still another version explicitly appears under the rubric of the circle, specifically "the magic circle of language [magische Kreis der Sprache]" (7: 786):¹³

Initially conceived as a letter on the "infinitely difficult theme of language and mathematics," Benjamin eventually dropped the mathematics and produced, instead, a "little treatise" (*GB*, 1: 343) on language alone. The starting point for the letter and for the resulting treatise are nevertheless the same: "Every language communicates itself [Jede Sprache teilt sich selbst mit]" (7: 786). The self-communicating character of language means that it is based on, and refers to, nothing beyond itself. Benjamin arrives at this thesis by means of a reduction that is more apparent in the fragments of the letter than in the subsequent treatise: "What do languages communicate? Here it seems as though one comes up against a great divide in the realm of language: languages in which only their own speakers [Eigen-Sprechendes] communicate themselves and languages in which other things also communicate themselves. Yet the appearance of this division is deceptive, for all languages belong to the second kind, that is, there is no speaker, if one means by this term someone or something that communicates itself through these languages" (7: 785–86). "Proper" speakers of a language would be those who could say what they mean by any given term in language, for language is, after all, their property, which they obviously

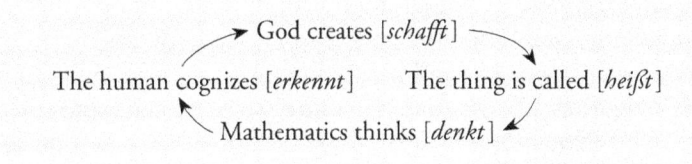

Figure 1 Benjamin probably drafted the "magic circle of language" in the fall of 1916, in response to certain conversations with Scholem in the previous summer.

have a right to use as they see fit. As Benjamin emphasizes, however, there are no such speakers, for language cannot be "one's own" and cannot be owned, in turn. It is not as though language is always on loan to speakers, who are thus indebted to its creator every time they enter into its realm; it is, rather, that the circular structure of language precludes a place for speakers to say what they mean, even if they only mean to say something as seemingly innocuous as "*a* designates the *BC* side of a triangle."[14] For this reason, mathematics, as the nadir point of the circle, does not speak: it silently "thinks."

What is meant by any language is meaning pure and simple. In the draft of the letter to Scholem, Benjamin thus concludes that every language is infinite without being altogether infinite, for it still has its own internal limit, which does not so much "define" it as give its infinitude a definitive shape: "Nothing is communicated *through* language and precisely for this reason what communicates itself in language cannot be delimited. Its essence, not its content, designates [bezeichnet] its limits" (7: 786). The essence of language does not define or otherwise determine its limits; rather, a language can be named after its essence as long as the function of naming consists in "designating" (*bezeichnen*) rather than "meaning" (*bedeuten*) something. The difference between designation and meaning points to a fundamental—although entirely negative—feature of both the letter fragment and the ensuing treatise: Benjamin rigorously avoids the term *Bedeuten*, while relying on *Bezeichnen* at crucial moments in his exposition. The task of the letter can thus be formulated as follows: develop a theory of language based on its circular structure without describing the circle in terms of the "double appearance" of meaning as both ground and content. Or, to express the same task in terms drawn from the Duns Scotus project: show that language is a "critical medium" without describing its range as extending "between the domain of the signifier [Bedeutende] and that of the signified [Bedeutete]." In place of *signifier* and *signified*, Benjamin introduces into the letter fragment two corresponding terms: "linguistic essence [sprachliches Wesen]" and "spiritual essence [geistiges Wesen]" (7: 786). As with signifier and signified, there is a rigorous "correlation" (6: 22) between linguistic and spiritual essence, yet the two must not be confused. The supposition that the terms in each of the two pairings are identical opens up the "*abyss* into which all theory of language threatens to fall" (7: 786; cf. 2: 154). But neither in the draft of the letter nor in the "little treatise" does Benjamin correlate signifier with linguistic essence and signified with spiritual essence, such that the correlation of linguistic and spiritual essence can be understood as the foundation of the correlation between signifier and signified.

Monologue and Panlogue

Only in a single extant fragment does not Benjamin bring together the pairing of spiritual and linguistic essence with that of the signifier and signified. At stake is a technical term he introduces for the sake of describing the immediacy of meaning: "word-skeleton" (*Wortskelett*).[15] The skeleton of a word lacks its "flesh and bones," so to speak, which suggests that it consists in the word in the absence of its "animation" by the "living soul," to use terms drawn from the first of Husserl's *Logical Investigations*. This, however, is precisely not what Benjamin means by the term—which is not to say that he is simply ignoring the investigation into meaning and expression that Husserl conducts at the beginning of his first groundwork. On the contrary, with the exception of his paper on "Eidos and Concept," nowhere else does he undertake a similarly "frontal assault" (*GB*, 2: 410) on the concepts of phenomenology. As Benjamin indicates at the opening of a fragment, the word-skeleton comes to light only as a consequence of a change in "attitude" (*Einstellung*). More exactly, it represents a "test case of an intentional alteration [Schulfall einer intentionalen Umstellung]" (6: 15): whenever the skeleton of a word appears, the word itself disappears. More exactly, the word-skeleton appears in place of the word whenever a word is understood to derive its meaning from a "living soul." A certain "turning off" of the attitude in which the word-skeleton emerges is required for the word in the proper sense of the word to appear: "It is strange that, with respect to the word, in many circumstances the intention toward its meaning [Intention auf seine Bedeutung] is lost in order to make a place for another intention, which one could reasonably call the 'word-skeleton' [Wortskelett]. (For its sign, one can designate the skeleton of any given word, for example, the word *tower* in the following manner: '—tower—')" (6: 15).

The change in attitude that replaces the word by its skeleton does not involve the loss of its meaning; rather, it consists in a general loss of meaning that is replaced by the representation of a living speaker who gives a skeletal word "flesh and blood" by using it for a determinate purpose. As Benjamin continues in the same fragment, "meaning" (*Bedeuten*) as opposed to "designation" (*Bezeichnen*) does not mean the same thing as "communicating" (*Mitteilen*). The paradigmatic case of designation, for Benjamin, again derives from mathematics, and designation requires a self-indexing designator, that is, an "I": "The word *tower* means something—which means nothing other than it communicates something. When I designate something, it does not communicate it; rather, I abstract in general from its communicability in order to index it in another context. When I designate the three sides of a triangle with *ABC*, these letters do not mean the three sides; that is, they do not communicate them" (6: 15–16). As with the word *inapproachable* in the discussion of Russell's paradox, Benjamin does not say what *tower* means—only that it means "something." In the word *tower* a name can be discerned with the addition of "of Babel." The word-skeleton of any given word, starting with *tower*, refers perhaps to this name, whereas the word does not. The signified in any case can be designated only by means of the word: "*tower* alone is the signified." With regard to signifier, matters are even less straightforward, and it is in this context that Benjamin begins to translate *eidos* by "spiritual essence": "The word *tower* communicates in the first instance its own communicability. As a word, it communicates that it is communicable, and this 'it' is its spiritual essence. It is something original, and it therefore communicates that a determinate, original, spiritual essence is communicable" (6: 16). Benjamin immediately adds, however, that the word "does not thereby mean" the object in question. For the word to meaning *something*, the object that it means must somehow be involved. What cannot take place, according to Benjamin's schema, is an act of ostentation, whereby speakers point toward an object as they simultaneously indicate their own presence. Such a scenario is excluded from the beginning because it transcends the sphere of language. It is little wonder, then, that Benjamin writes up a note on "Eidos and Concept" that is principally concerned with the condition under which the singularity of an object can be preserved after its factuality "succumbs to a reduction" (6: 30). A word communicates an eidos; but the communication of a "determinate and final" communicability or "spiritual essence" is not yet meaning: "[the word] does not communicate that from which it communicates communicability; it means that [dasjenige aber von dem es die Mitteilbarkeit mitteilt, teilt es selbst nicht mit, das bedeutet es]. And in order to determine the object of its meaning, it therefore needs a *virtus* in the word other than the communicating" (6: 16).

Without a supplementary *virtus*—understood as an *-ability* or *-barkeit*—the communicability of a "spiritual essence" falls short of meaning precisely that object whose essence it is. To mean an object, the word that communicates its essence must somehow include its "representation" (6: 10); and there is only one way for representation to take place: the word must derive from the object it represents. In other words, the object must speak of itself to a speaker in order for the speaker to "mean" *this particular* object and not only its eidos or "spiritual essence." In the first of the six *Logical Investigations*, Husserl turns away from the pragmatics of communication and looks, instead, toward "expression in the solitary life of the soul" (Hu, 19, 1: 41), wherein expression would presumably be free of all indicative entanglement. When I am silently speaking with myself, according to Husserl, my speech disposes of all indication, for I, of course, know exactly what I mean by what I say.[16] The premise of Husserl's initial inquiry under the rubric of phenomenology is that expression owes its origin to a living subject, who, by animating certain sensible complexes, lends them meaning; but whenever an expression enters into communication, regardless of the situation, it gets caught up in indication. Benjamin, for his part, adopts and transforms this schema to a point where it is almost unrecognizable. Any word that acquires meaning by means of a living subject is a word-skeleton, which, as Benjamin parenthetically notes, always "grimaces" (6: 15), thus indicating its origin in an "intentional alteration" that robs it of its original, paradisal condition. The phenomenon of meaning not only cannot be located in the situation of soliloquy; soliloquy has to be replaced by its opposite, namely universal loquacity. There is meaning in the proper sense of the word only when everything speaks. The theory of meaning should thus begin with panlogue rather than monologue. And so, pace Husserl, the starting point for any logico-linguistic investigation is not the elimination of the communicative character of language but its infinite expansion.

Everything must be able to speak, and the specific meaning of any given speech lies in the interval between the "spiritual essence" and the "linguistic essence" of the speaking thing: the greater the interval, the less meaningful the speech. The draft of the letter on the theme of mathematics and language begins at this point: "The linguistic essence of authentic things [eigentlichen Dinge] does not constitute their spiritual essence; their spiritual essence does not entirely get taken up in their linguistic essence. Something nonlinguistic is attached to all things, something that can appear as such only because its language cannot express some existing spiritual element. This is what we mean when we say that things are dumb [stumm]" (7: 786). The function of human beings in the theory of meaning can also be determined from this point: they are those for whom the interval between "spiritual essence" and "linguistic essence" vanishes. Here is also where the subject can enter: an "I" is that which in which the interval in question again expands. And here, finally, is where a word can be postulated that is the exact opposite of a sign that arbitrarily designates something: a word, namely, that does not simply mean the thing itself in contrast to its eidos but immediately creates it. The divine, creative word not only exhibits no interval but has no difference in the first place. A tripartite schema is thus established: creation, meaning, designation. The creative word immediately makes what it means; meaning immediately means what is meant; and designation immediately adds a "meaning," improperly speaking, to something already in existence. To the extent that magic consists in immediacy, the linguistic circle constructed from the aforementioned schema can be called "magical": the creative word makes the things that are known by their names when things express themselves to a human being; and a human being to whom things are dumb because they have themselves become things introduces designation after designation in the absence of meaning in the proper sense.

All of this accords with the "magical circle of language" that Benjamin inscribes into the draft of his letter to Scholem, with one notable exception: the nadir level of language—"Mathematics thinks"—makes no mention of designation. A mathematical demonstration cannot begin without designation; and mathematics as a whole may be unrelated to reality, such that its mode of intellection—call it "thinking"—should be distinguished from the cognition of objects; but the thesis "mathematics things" only goes so: it does not determine how the thing flows into the mathematical sign, which then flows into the human being, who knows. In describing the movement of language from creation to created thing to mathematical thinking to human cognition, Benjamin may be indebted to a neo-Kantian schema: the thing must be mathematicized in thought before it can be known. For the "little treatise" that he begins when progress on the letter stalls, however, he drops the theme of mathematics altogether. And in place of the mathematical sign the treatise proposes an idea of an utter absence of language as the nadir point of its circle.

A Transcendental Argument

It is in view of the idea of utter nonlinguisticity that the treatise "On Language as Such and on Human Language" begins to be expound a systematic idea of language in the absence of its mathematical counterpart:

> The existence of language is coextensive not only with all the areas of human spiritual expression in which language is always inherent in one sense or another, but with absolutely everything. There is no event or thing in either animate or inanimate nature that does not in some sense partake of language, for it is essential to everything that it communicate its spiritual contents. But this use of the word *language* is by no means metaphorical. For the following thought yields a full, content-laden knowledge [eine volle inhaltliche Erkenntnis]: we cannot represent to ourselves the spiritual essence of anything not communicating in expression [wir uns nichts vorstellen können, daß sein geistiges Wesen nicht im Ausdruck mitteilt]; the greater or less degree of consciousness apparently (or really) bound to such communication cannot alter the fact that we cannot represent to ourselves a total absence of language in anything [wir uns völlige Abwesenheit der Sprache in nichts vorstellen können]. An existing something entirely without relationship to language is an Idea; but this Idea cannot bear fruit even within the realm of ideas, the circumference of which designates [bezeichnet] the idea of God. (2: 140–41)

Benjamin's argument for the applicability of the term *language* beyond the conscious subject is modeled on Kant's argument for the space and time as pure intuitions. Thus, Kant writes in the opening section of the "doctrine of elements" that begins the *Critique of Pure Reason*: "Space is a necessary representation, a priori, which is the ground of all outer intuitions. One can never represent that there is no space, although one can very well think that there are no objects to be encountered in it. It is therefore to be regarded as the condition of the possibility of appearances, not as a determination dependent on them, and is an a priori representation that necessarily grounds outer appearances" (K, A 24; B 39).[17] Both arguments can be described in phenomenological terms: as demonstrations of the paradoxical primacy of fantasy for the securing of content for philosophical knowledge.[18] Through the destructive power of fantasy things are eliminated from space, for Kant; language is eliminated from things, for Benjamin. For Kant, the elimination of things from space results in the realization that space is an intuition, not a function of discursive thinking; for Benjamin, the elimination of language from things results in the realization that language is immediately communicative, not a means of communication. As an intuition, space is, for Kant, immediately related to its object; by virtue of its purity, there is no object that determines and delimits it. It can therefore be represented only as an infinite given whole. The medial character of space derives from its immediacy, and its immediacy is the basis of its infinitude. The same can be said—almost word for word—of language understood as the communicability of "spiritual essence":

> Whatever is communicable *of* a spiritual essence, *in* this it communicates itself; that is to say, all language communicates itself. Or more precisely: every language communicates itself *in* itself; it is in the purest sense the "medium" of the communication. The medial, which is the *immediacy* [*Unmittel*barkeit] of all spiritual communication, is the fundamental problem of linguistic theory, and if one wishes to call this immediacy "magical," then the primary problem of language is its magic. At the same time, the notion of the magic of language points to something else: its infinitude. It is conditioned by its immediacy. For precisely because nothing is communicated *through* language, what is communicated *in* language cannot be externally limited or measured, and therefore every language contains its own incommensurable, uniquely articulated infinitude [inkommensurable, einziggeartete Unendlichkeit]. Its linguistic essence, not its verbal content, designates its limit. (2: 142–43)

The infinitude of language does not derive from an iterative operation of synthesis; rather, every language is immediately infinite—again like space understood as a pure intuition in which objects appear and not as a concept under which they fall. As the first "Antinomy of Pure Reason" shows, there can be no answer to the question of whether the world, composed of homogeneous parts, is finite or infinite, for "the" world cannot be known as such (K, A 25; B 39). According to Kant, however, there is no question that the whole of space is an immediately given, infinite whole. The infinitude of language, for Benjamin, is similarly structured: there is nothing outside of a language that can limit it, least of all something that the language is supposed to say. But this also indicates where the two arguments are no longer compatible. Space, as Kant discusses it, is immediately available, and the "transcendental argument" consists in demonstrating the conditions of its possibility; the language of which Benjamin speaks—infinite language, which, as such, is defined by nothing exterior, neither a world to which it refers or a purpose to which it is put—may be given; but if so, its givenness is suppressed by the "natural" attitude, according to which language is defined by its referential function and determined in accordance with certain pragmatic ends. The point of Benjamin's argument is to show that the expansion of the word *language* corresponds to its concept and does not therefore require a distinction between the natural or literal usage of a term and its counternatural or metaphorical counterpart: whatever we represent to ourselves has an eidos or "spiritual essence" that can be given within certain limits and is therefore communicable in its own internally delimited language.

Once the word *language* is no longer defined by a referential function or pragmatic use, the expansion of its application is correspondingly limitless: there are an infinite variety of infinite languages, each of which—with the exception of the creative word—is internally limited by the interval between its "linguistic essence" and "spiritual essence" in general. And this, too, distinguishes Benjamin's argument about language as an immediate medium from Kant's exposition of space as a pure intuition: "If one speaks of many spaces," Kant writes, "one understands thereby only parts of one and the same unique space" (K, A 25; B 39). This was not only Kant's position on space, so to speak. In his very earliest writing, *Wahre Schätzung der lebendigen Kräfte* (True estimation of living forces), he entertains the bold conjecture that "it is really possible that God created many millions of worlds in the genuine metaphysical sense of the term" (K, 1: 22). In addition to our own three-dimensional world governed by the inverse square law of gravitation, there is a four-dimensional world would be governed by an inverse cube law, a five-dimensional word governed by an inverse quadrupled law, and so forth. Speculation of this kind shares some affinity with Benjamin's expansion of the word *language*, and so, too, does reflection on non-Euclidean geometries, each of which can be said to describe a world of its own. Around the time Benjamin wrote "On Language" he became interested in an article by Ernst Barthel, "Die geometrischen Grundbegriffe" (Basic concepts of geometry), which draws a sharp distinction between real geometry and geometries built on the basis a consistent set of axioms, regardless of their applicability to the world.[19] But Benjamin's discussion of an infinite variety of infinite language is significantly different from both speculation on higher-dimensional worlds and the construction of non-Euclidean geometry: in the latter case, there is no particular ordering of axiomatically consistent geometries, and in the former case the order is established simply by integer iteration. The infinite variety of infinite languages, by contrast, is ordered according to tripartite structure that he first describes in terms of its "magic circle." Such a structure is akin to the one Cantor discovered under the rubric of set theory, in which an order of transfinite numbers can be developed on the basis of the distinction between those that are denumerable and those that are not. Even as Benjamin put aside the "infinitely difficult theme of language and mathematics" in order to complete his "little treatise," he borrowed from the former a model for an ordering of language as such.

Equation and Translation

Cantor showed—although he famously said he did not believe what he saw—that the "size" of the linear continuum is greater than that of the set of integers.[20] To use the technical term he introduced, the "power" (*Mächtigkeit*) of the set of points on the linear continuum is higher than that of the aggregate of integers. To the latter he assigned the term *aleph null* as the lowest transfinite cardinal number; to the former he assigned the letter *c* for "continuum," without knowing—for good reason—whether it was equal to the aleph that immediately followed aleph null in a continuous order of ever-higher, transfinite cardinal numbers.[21] As for the absolute infinite, this Cantor reserved for God, who guarantees the integrity of the concept of an actual as opposed to merely a potential infinite.[22] As Schoenflies notes in one of his reports on the progress of set theory of the German mathematical association, other mathematicians respond to the crisis created by the uncertain status of the "continuum hypothesis" by proposing an ultra-continuum that would be related to the linear continuum as the latter is related to the set of algebraic numbers.[23] A tripartite schema thus arises, in which the set of algebraic numbers stands below that of the linear continuum, which, in turn, stands below an ultra-continuous set—or perhaps below the infinite infinity of God. Correspondingly, the schema of infinite language Benjamin adopts for this "little treatise" places the language of things at the lowest point, that of human being a grade higher, and the creative word transcending them both.[24]

Despite the affinity between the order of transfinite numbers and that of infinite languages, there is one insurmountable difference: the relation among mathematical terms is expressed by an equation (*Gleichsetzung*), the relation among languages by translation (*Übersetzung*). Equation and translation are not, however, unrelated to each other: the latter is the basis of the former. Infinite languages are translatable into one another because an "equation" (2: 146) holds good for language as such: "linguistic essence" = "spiritual essence." Translation thus assumes the function otherwise attributed to predication: that of expressing identity. In the notes on Russell's paradox Benjamin distinguishes between predication and designation in terms of a more exacting distinction between proper and improper meaning: whereas the latter derives from definition, the former is fundamentally indefinable. The treatise on language takes the same line of argument one further step. A word like *unapproachable*—to the extent that it is a word and not its skeleton, which grimaces at whomever would use it—is indefinable because language as such is so. Translation, in turn, consists in the equalization of the "spiritual essence" of a thing, which designates its inner limit, with its corresponding "linguistic essence." In this way, the interval that defines the meaning of any given thing converges with the original language of the human being, which knows of no such interval. At this point an infinite language is raised to a higher power of infinitude, and the human being speaks the name of the thing. Since things speak in their own languages, the names they receive are precisely their names, not designations of what they are supposed to mean. The panlogue Benjamin sketches in response to Husserl's description of monologue shows that meaning derives from things, not from "proper speakers." Once it can be shown that the language of a thing is translated into that of a higher language, which consists in names, Benjamin returns to the initial paradox of linguistic theory and makes an otherwise abyssal equation into the thesis on which the theory henceforth rests:

> If spiritual essence is identical with linguistic essence, then a thing, by virtue of its spiritual essence, is a medium of communication, and what is communicated in it is—in accordance with its medial relation—precisely this medium (language) itself. Language is thus the spiritual essence of things. Spiritual being is therefore posited [gesetzt] at the outset as communicable; or rather, it is posited *in* communicability, and the thesis [Thesis] that the linguistic being of things is identical with the spiritual, insofar as the latter is communicable, becomes in its "insofar" a tautology. *There is no such thing as a content of language; as communication, language communicates a spiritual essence—a communicability pure and simple.* The differences among languages are those of media that are distinguished, as it were, by their density [gleichsam nach

ihrer Dichte], therefore gradually; and this with regard to the density both of the communicating (naming) and of the communicable (name) in the communication. (2: 145–46)

Whereas Benjamin argues that his use of the world *language* is nonmetaphorical, he emphasizes that his use of the word *density* is figural: languages are not material media, even if they are not exactly spiritual media either. In proposing the metaphor of density Benjamin is following a mathematical precedent: a set of points is dense if another point can be constructed between any two given points. The point set of integers is not dense, because one cannot discover or construct new integers between any two of them. By contrast the set of rational numbers is dense, but it is not continuous, because it is missing all of the irrational numbers. The density of infinite language is constructed accordingly. Languages of things are dense yet discontinuous, each in its own way; by contrast, the language of names is continuous and thus functions as the matrix of translatability:

> By the fact that, mentioned earlier, languages relate to one another as media of varying density, the translatability of languages into one another is given. Translation is the conveying of one language into another through a continuum of transformations [Übersetzung ist die Überführung der einen Sprache in die andere durch ein Kontinuum von Verwandlungen]. / The translation of the language of things into human language is not only a translation of the mute into the sonic; it is also the translation of the nameless into the name. It is therefore the translation of an imperfect language into a more perfect one, and cannot but add something to it, namely knowledge. (2: 151)

Because the name of a thing derives from its language—not from the will of a speaker—it contains its "spiritual essence," and so knowledge of the name is equivalent to knowledge of the thing. "But," Benjamin immediately adds, "the objectivity of this translation is guaranteed by God" (2: 151). Without the premise of a word that creates the thing, there can be no assurance that the translation of its language into the language of names yields precisely its name and only its name, which is, from the beginning, distinct from all others. In two respects, Benjamin's argument corresponds to Kant's. On the one hand, his theory of language forges a path beyond the opposition of "bourgeois" to "mystic" theories of language, just as Kant develops a critical path that overcomes the fruitless debates between dogmatism and skepticism. On the other hand, the divine language that Benjamin postulates in terms of the creative word is related to the human language of names as *intuitus originarius* is related to *intuitus derivativus*, according to Kant. Whereas the creative word is spontaneously creative, the language of names represents a unity of spontaneity and receptivity, which does not create things but, rather, only cognizes them: "The infinity of human language always remains limited and analytic in nature, in comparison to the absolutely unlimited and creative infinity of the divine word" (2: 149). The four-part "magic circle of language" thus turns into a new semantic triangle: the divine word creates the thing, which is called by the name that the human being knows; and this knowledge is wholly objective because it is addressed to the creator of the word that has been translated into its "proper" name. As for the position that was originally assigned to mathematics, it turns into a convoluted space below the base line of the triangle that is in a certain sense above it as well. And as for the direction of Benjamin's own investigation, only this much is certain: he replaces the theme of "language and mathematics" with the exegesis of the Book of Genesis.

More than One Human Language

In his studies of color and fantasy "something spiritual" designates the objective counterpart to pure reception: just as the colors of fantasy are seen by the dreamer, so unclouded colors are received by the child. A similar schema underlies the reflections on language: the original human being—call him "Adam," whose name recalls the "redness" (*adom*) of both "earth" (*adamah*) and "blood" (*dam*)—remains untouched by the things whose language he translates into the ethereal medium of sound. In Hölderlin's terms, Adam is "holy-sober"—"sober" because the things he hears generates no excitement and "holy" because he is unaffected by whatever approaches him. In the same vein, the infinite variety of the colors seen by both the dreamer and the child corresponds to the infinitude that is originally predicated of the language of things: differences among these languages, like those among pure colors, consist only in "nuances." There is, nevertheless, a sharp distinction between Benjamin's studies of color and fantasy, on the one hand, and the logico-linguistic studies, on the other: none of the former discusses the transition from the original way of viewing color to the derived way, whereas the exegesis of the Book of Genesis proposed in "On Language" is concerned with just such a transition. In this context Benjamin makes only a single claim about the event in question: language somehow gets excited. As a result, the languages of things as well as the language of the human being lose their inviolability and become subject to affections of all kinds. The so-called fall arises from the excitation of a particular form of language, which rises above things and allows the language user to speak "about" or "over" (*über*) them as a result:

> The knowledge of things resides in the name, whereas that of good and evil is, in the profound sense in which Kierkegaard uses the word, "chatter" [Geschwätz], and knows only one purification and elevation, to which the chattering one, the sinner, was therefore submitted: judgment. This judging word expels the first human beings from paradise; they themselves have excited [exzitiert] it in accordance with the eternal law by which this judging word punishes—and expects—its own awakening as the sole, the deepest guilt." (2: 153)

The "judging word" occupies the position in the treatise on language that Benjamin ascribes to "judgments of designation" in his attempt to solve Russell's paradox. In both cases something new is added: a valuation in one case, a designation in the other. And just as designation singles out the designator—this is why *Russell* is responsible for the paradox named after him—evaluation evaluates the evaluator, who is always found guilty of evaluating, that is, adding valuations to creation. The proximity of Benjamin's exegesis of Genesis to his logico-linguistic studies appears most clearly in the suggestion, derived from "Eidos and Concept," that the "judging word" be understood as the source of abstraction (2: 154). The treatise adds something, nevertheless, to the logico-linguistic exercises: a brief exposition of the temporal character of the novelty thus introduced. The judging word not only punishes those who "awaken" it; punishment also stretches out over time and so, too, does the awakening which means that there is none—only a confusion of dreaming with awakening. In other words, the punishment is also always "expected." Unlike the language of names, the "judging word" is anticipatory, and what those who are condemned anticipate, above all, is further punishment and thus deepening guilt.[25]

All of which would suggest that prelapsarian language, by contrast, knows no division; but this cannot be the case as long as the Book of Genesis is the guide, for Adam does not simply name the animals, he also names another one of his kind—which means that the name is already divided between names in the proper sense of the word and proper names, which do not derive from the language of the thing named. The "judgment of designation" can also be found in paradisal language insofar as the naming of human beings is similar to the construction of a sign system for geometrical demonstrations: "The theory of proper names is the theory of the limit [Grenze] of finite language in relation to infinite language" (2: 149). In calling something "woman" or "Eve"—Benjamin registers the discrepancy between the two versions of creation in Genesis—Adam traverses the limit separating infinite from finite language in advance of the collapse of the infinite language of names and the accompanying fall under the regime of judgment. And how could it be otherwise? Giving someone a name is suspiciously akin to the act of designating one thing as something else. The name Eve does not come from Eve. Benjamin's theory of the proper name silently seeks to maintain an ever-dissolving distinction between an utterance such as "I call you Eve" and a designation such as "*a* is the *BC* side of the triangle":

> Of all beings, the human being is the only one who names his own kind, as he is therefore the only one whom God did not name. It is perhaps bold, but scarcely impossible, to name the second part of Genesis 2:20 in this context: that the human being named all beings, "*but* for man there was not found a helper fit for him." Accordingly Adam names his wife as soon as he receives her (woman [Männin] in chapter 2, Eve in chapter 3). By giving names, parents dedicate their children to God; the names they give do not correspond—in a metaphysical rather than etymological sense—to any knowledge, for they name newborn children. In a strict sense, no name should (in its etymological meaning) correspond to anyone, for the proper name is the word of God in human sounds. By it every human being is vouchsafed his creation by God, and in this sense he is himself creative, as expressed by mythological wisdom in the intuition (which doubtless not infrequently comes true) that one's name is one's destiny. The proper name is the communion of the human being with the *creative* word of God. (2: 149–50)[26]

Benjamin discovers a supplementary commandment in the early chapters of Genesis, which could be formulated in the following terms: the name bestowed on a human being shall not derive from any discerned correspondence with the one so named. The commandment is either tautological or dubious: tautological if the one named, as a newborn, is deemed to have no qualities of its own; dubious if it is applied to practices of name giving, especially those recounted in Genesis, as Scholem especially would have known. God, for example, names the newly created man in the proscribed manner: "the LORD God formed man ['adam] from the earth ['adamah]" (Gen. 2:7). Benjamin implicitly recognizes the violation of the commandment in the case of Eve: "this one," Adam says, "shall be called Woman ['ishshah], for from man ['ish] was she taken" (Gen. 2:23). Or, alternatively: "The man named his wife Eve [hawwah], because she was the mother of all the living [hay]" (Gen. 3:20). Nothing changes in this respect after the expulsion from paradise: Eve, for example, names her first-born, Cain, in recognition of her "gain [qanithi]" (Gen. 4:1). Once names no longer mean anything beyond the act of name giving—as, for instance, the name *Agesilaus Santander* (6: 520–23)—then the commandment can be fulfilled, for the name then simply "means" the event of its bestowal. In the beginning, however, this is not the case. Or it is not so if the Book of Genesis gives guidance. Benjamin proposes a commandment God, Adam, Eve violate: whereas the names of things derive from their languages, the names of human beings are parodic versions of the purely creative word, which in its nonparodic form is denied to human beings.[27] In this way, the theory of the proper name that Benjamin proposes in the "little treatise" assumes the systematic function assigned to the mathematical sign in the draft of the letter to Scholem: the latter does not participate in the creative word but, rather, communes with the cosmos and can be absolved of arbitrariness only for this reason: "Mathematics speaks in signs. The language of mathematics is doctrine [Lehre] / Its writing is the sign. The signs of mathematics are also recovered, so to speak, in the sky-heavens [Himmel]: there signs are read, whereas in mathematics they are written" (7: 788).

Not only, then, does the exegesis of Genesis replace the theme of mathematics; the theory of the proper names, as the replacement for a theory of the mathematical sign, runs counter to the account of proper names that can be found in the Hebrew Bible as a whole and in Genesis in particular. As soon as there is more than a single human being, human language, unlike its lower and higher counterparts, is split in two: an infinite language of names and a finite language of proper names, which are comparable to mathematical signs. The triangle of language, like the "magic circle," circulates under the condition of vertical inequality: lower-order languages are created by the highest-order one, which, in turn, creates a creature whose middle-order language consists in the translation of lower-order ones. Nothing of this schema applies to the giving of proper names: Adam names Eve, and parents name their children, but neither

Adam nor parents are superior in linguistic kind to those whom they name. As Benjamin develops his treatise beyond its exegesis of Genesis, the double character of human language begins to characterize language as such. The names Adam "gives" the animals turn out to be improper, after all: "Things have no proper name except in God. For in the creative word God of course calls them forth with their proper names" (2: 155). Every created thing, then, finds itself in a situation similar to that of Eve and her children: they are named by someone who did not create them in first speaking their names. To use terms drawn from Benjamin's attempt to solve Russell's paradox, every name other than those that are identical with the creative word includes a degree of designation. By "giving" something a name of its own, the subject improperly raises *itself* over the object—and both fall as a result into the sphere of "improper meaning." The equation "spiritual essence = linguistic essence" is, above all, meant to obviate this movement: instead of "giving" names to things, Adamic language consists solely in the translation of their "own" words into the immaterial medium of sound. Nothing of this scenario holds for the naming of one's own kind, however, and as it turns out, the scenario cannot hold for the naming of things either. The names that Adam gives them are only "his" names, for his language alone is one of names. And the condition of being named finds expression in mourning for the absence of one's own name: "To be named, even when the namer is godlike and blissful, perhaps always remains an intimation of mourning" (2: 155).

What Kierkegaard finds in anxiety, Benjamin locates in mourning: a mood so pervasive that it traverses the bliss of paradise. For the author of *The Concept of Anxiety*, prelapsarian uneasiness is the very temptation to which Adam and Eve succumbed when they sought to eliminate it from their otherwise blissful condition.[28] Benjamin declines to follow Kierkegaard in this regard: the fall has nothing to do with the hint of uneasiness in the prelapsarian condition. Things become "overnamed" only after the infinite language of names has broken into many finite languages. And guilt accrues to Adam and Eve for "exciting" the judging word by exercising judgment, that is, by predicating abstract qualities such as good and evil of things that are good by virtue of being created in the divine word. This theory of the origin of the guilt that expels Adam and Eve from paradise implicitly—and ironically—absolves Adam of culpability for creating a general atmosphere of sadness by giving names to everything and everyone around him: things are improperly named because they own names that are contained only in the word that created them; and Eve is improperly named, not only because she is overnamed, but also because both of her names violate the commandment that proper names not correspond to anything yet in the world, not even the "life" she is.

Language at Last Disentangled from Designation

All of this can be summarized in a single sentence: the attempt to identify the condition in which expression is completely free of indication by replacing monologue with panlogue cannot in the end be carried out, for even in its paradisal condition human language is "entangled," to use Husserl's term, with judgments of designation, the paradigmatic instance of which can be found in Adam naming Eve.[29] Benjamin probably pursued the theme of mathematics and thinking for precisely this reason: mathematical theorems begin with judgments of designation, but their objective character is not thereby damaged. The turn to the Book of Genesis is decisive in this regard, for the narrative requires that Adam name both things unlike himself and one who is his equal; in addition, it suggests that there was only one human language until the collapse of the Tower of Babel. As the pseudonymous author of *The Concept of Anxiety* says with characteristic directness, the narrative suffers from a certain "imperfection."[30] Benjamin makes no similar claim, and his treatise is divided against itself as a result. Specifically, it provides two opposing answers to the question: was there a basis for the plurality of human languages in the paradisal condition? At a transitional point in his argument Benjamin says "no": "After the fall into sin—which, by making language into a means, laid the ground for its plurality—it took only one more step for there to be linguistic confusion. . . . *Signs* must confuse where things are entangled" (2: 154). Earlier in the treatise, however, Benjamin proposes the opposite thesis: "Since the unspoken word in the existence of things falls infinitely short of the naming word in human knowledge, and since the latter in turn must fall short of the creative word of God, there is reason [Grund] for the plurality of human languages" (2: 152).

The tension between the two passages is far from inconsequential, for it calls into the question the unity of the "little treatise." On the one hand, it develops a triangular schema of infinite languages, each of which is distinct from the others; on the other—and as a replacement for a treatment of the mathematical sign—it takes direction from the Book of Genesis. The latter requires that the plurality of languages be a postlapsarian phenomenon; the former presents the plurality as an essential element of language as such. In one of the notes for a *Habilitationsschrift*, Benjamin goes even further in this direction, as he implicitly repudiates his own earlier position: "The plurality of language is . . . a plurality of essences [Wesensvielheit]. The doctrine of the mystics concerning the decay [Verfall] of the true language cannot, therefore, truthfully yield a doctrine of dissolution into a plurality that would somehow contradict the original and God-willed unity, for the plurality of languages is not the product of decay any more than is the plurality of peoples, and is indeed so far removed from any such decay that precisely this plurality alone expresses their essential character" (6: 24). Once it is conceded that the plurality of human languages is essential to their linguisticity, the condition in which designation is eliminated can no longer be identified with "the" original human language. Nor can translation be understood primarily in terms of lower languages giving way to higher ones. If the plurality of human language is original, none is higher than any other, and the equality of all human languages means that in the end they all mean the same thing: meaning in the proper sense of the word. Or, as Benjamin explains in the preface to his translation of Baudelaire's *Tableaux parisiens* that he published under the title "The Task of the Translator," regardless of what any speaker intends to say, every medium of communication, hence every human language "as a whole [im ganzen]," in its "complementary [ergänzende]" relation to every other language, "wants to say [will sagen]" the very same nonthing: namely, "pure language [reine Sprache]" (4: 13).

The translation of one human language into another makes possible an always partial reduction of designation. Even in the absence of any elaborate theory of meaning, it is readily apparent that the act of translation runs counter to the judgment of designation. If the basic formula of designation runs "I designate x by y," that of translation proceeds in almost the reverse direction: "x means y, regardless of what I say." The only space for the subject lies in the meaning of the original, to the extent that it derives from the intention of its author. The task of the translator, then, consists in reducing this interval as much as possible. If the task sounds as though it were a mathematical one, this should come as no surprise. Something of Benjamin's conversation with Scholem in August 1916 about the differentiability of certain functions and the direction of the corresponding curves may underlie his summary exposition of the task that translators are required to solve: "Just as a tangent touches [berührt] a circle fleetingly and at only one point; and just as this touch, not the point, prescribes the law according to which it is to continue on its straight path to infinity, a translation touches the original lightly and only at the infinitely small point of sense [Punkte des Sinnes] in order to pursue its own course according to the law of fidelity in the freedom of linguistic movement" (4: 19–20). By presenting the task of the translator in term of a mathematical image, Benjamin retrieves the theme he had originally put aside. Without entering into a lengthy discussion of mathematics and thinking, he presents the task of the translator in terms of two infinities: the infinitely small point touched by an infinitely long line, which represents the interval Δ—that is, the text—under consideration.

The end point of translation thus lies in a reduction that no translator can accomplish without immediately ceasing to be a translator: the reduction, that is, of the last remaining residue of designation, which takes shape as an infinitesimal "point of sense." A translator who did not touch the original with a law-bound tangent line would no longer be a translator but would, instead, fall silent—or worse, become "creative." In the final paragraph of "The Task of the Translator," while drawing on both the original and the mathematical sense of *Sinn* as "direction," made even more emphatic by its association with the French word *sens* (sense, direction), Benjamin describes the situation of late Hölderlin in precisely these terms: "The Sophocles translations were Hölderlin's last work. In them, sense [der Sinn] plunges from abyss to abyss, until it threatens to lose itself in the bottomless depths of language" (4: 21). In terms of the geometric figure that describes the solution to the task of the translator, the point at which the translation touches the original turns so sharply—in the language of mathematical, the curve would be called "pathological"—that no tangent line can be drawn. The translations of Sophocles have no regular interval Δ, hence no direction, and therefore verge on senselessness. As Benjamin indicates in conversations with Scholem, Hölderlin and the Bible are similarly "doctrinal." At the end of "The Task of the Translator," however, he distinguishes between the two and abstains from undertaking an exegesis of either one. Whereas, for Benjamin, Hölderlin's translation of *Oedipus Rex* and *Antigone* provides the "prototype" of the form, its "archetype" can be found in certain translations of the Bible, which can put a "halt" (*Halten*) to the abyssal movement in which translators at once find and lose themselves whenever the point of sense upon which they touch is so sharp that it gives them no direction for the solution of their task: "No text vouches for [this halt] other than the holy one, in which sense has ceased to be the watershed for the flow of language and the flow of revelation" (4: 21).

It is not as though the Bible, for Benjamin, is without any points of sense; on the contrary, there are an infinite number: every word—indeed, every phonetic and graphic element, no matter how small—is such a point. It is for this reason that interlinear translation of the Bible is the "ideal or archetype of translation" (4: 21). The continuity of the line does not consist in a smooth flow of language; rather, every element takes a sharp turn, which diverges from its predecessor and successor, yet the line of translation continues all the same. The Bible thus assumes in reverse the function Benjamin had first attributed to Adam's language: just as every lower language is translated into the latter, so is the former translatable into every equally human language. This inversion indicates the decisive difference between the two: the original language of the human being is supposed to be absolved of all designation, whereas the "interlinear" translation of the Bible only "vouches" for this absolution. In the language of *Logical Investigations*, interlinear translation indicates a sphere of expression beyond indication, even its own. So understood, biblical language performs the same function as what Heidegger later in the same decade will call "the call of conscience." Unlike the "call of conscience," however, interlinear translation, as Benjamin briefly describes it at the end of his preface, does not awaken guilt; rather, it guarantees that, even as translators fail to solve their task—or occasionally fall silent in approaching their goal—the task has always already been solved: there is, after all, meaning in the proper sense of the word, or, in other words, language disentangled from designation.

§6 Pure Knowledge and the Continuity of Experience

"On the Program of the Coming Philosophy" and Its Supplements

The Kantian Typic

Among those who reject philosophy in its traditional forms but do not want to abandon the idea of philosophy altogether, there is a relatively simple solution: propose a program for the philosophy of the future. The point of the program does not consist so much in outlining a new system of philosophy as in disclosing something that contemporaneous philosophical research, caught in a narrow nexus of problems, fails to recognize. And this "something" is generally called "experience." Feuerbach's *Grundsätze der Philosophie der Zukunft* (Principles for the philosophy of the future) is instructive in this regard: the Hegelian system, for Feuerbach, has shown how very little is accomplished by abstract philosophizing; but in recognition of this nullity, a new form of philosophy can develop that is oriented toward "empiricity."[1] Nietzsche's *Jenseits des Gut und Böse* (Beyond good and evil)—which describes itself as a "Prelude to the Philosophy of the Future"—is considerably more complicated than Feuerbach's *Principles*, but it moves in the same direction: toward a concept of experience that cannot be cast in terms of subject and object, regardless of how fully they are dialectically developed in reciprocal relation to each other.[2] The brief treatise that Benjamin wrote in late 1917 and early 1918 under the title "On the Program for the Coming Philosophy" follows Feuerbach and Nietzsche in this regard. Instead of propounding a theory of experience based on the opposition of subject and object, it seeks out almost the direct opposite: "a sphere of total neutrality in relation to the concepts of object and subject" (2: 153).

Neither Feuerbach nor Nietzsche is mentioned in Benjamin's "On the Program of the Coming Philosophy." Instead of placing his program in relation to the opponents of German idealism, Benjamin associates it with its primary precursor, namely Kant, whose *Prolegomena zur einer jeden künftigen Metaphysik, die als Wissenschaft wird auftreten können* (Prolegomena to any future metaphysics that can come forward as science) functions as a model for future plans for philosophies of the future. Just as Feuerbach and Nietzsche declare that metaphysical inquiries have failed to grasp the nature of experience, so, too, does Kant. And Benjamin follows suit. The form of this "following" is determined by what Benjamin enigmatically calls the "Kantian typic," which provides a framework for a system that otherwise departs from the *Critiques* in every other conceivable way: "It is a matter of obtaining a prolegomenon to a future metaphysics on the basis of the Kantian typic," Benjamin writes near the beginning of "On the Program of the Coming Philosophy," as he specifies the mode of experience under consideration, "and of fixing in view this future metaphysics, this higher experience" (2: 160). Kant had introduced the technical term *typic* in the second *Critique* as the practical counterpart to a technical term introduced in the first, namely, *schema*, which is itself defined as the sensible counterpart to concepts and charged with the function of making concepts applicable to appearances (K, A 140; B 179). As for the typic, it is the near opposite: the "typus" of moral judgment consists in the concept of a universal law of nature without any specification of the law itself, and "it is therefore permitted to use the nature of the sensible world as typus of an intelligible nature, so long as I do not translate intuitions, whatever is dependent on them, onto the latter" (K, 5: 70). Benjamin, for his part, shows little interest in "the typic of pure practical judgment," except perhaps insofar as it represents a translation of the biblical prohibition on graven images into philosophical vocabulary. Instead, he is specifically attracted to the idea of the typic as such, which mediates between Idea and concept, just as the schema mediates between sensibility and conceptuality.

The typic is thus a pure concept that, unlike a category, can be applied only reflexively, on the one hand, and in relation to an action, on the other. The typus of "intelligible nature" is applicable in the transition from the awareness of the moral law to the action it dictates—or, for Benjamin, who takes it outside of moral philosophy and makes it into a term for a theoretical program, only in the transition from philosophy to "doctrine." As he tells Scholem in a letter from October 1917, with apologies about the inexactness of his formulations, the transition from philosophy to doctrine can take place only in the context of the "Kantian typic" (*GB*, 1: 389); and as he further elaborates in a letter from December of the same year, the "typical" character of Kant's *Critiques*, which corresponds in the sphere of philosophy to the "canon" in the sphere of art, can be attributed to a singular phenomenon: Kant alone—with the possible exception of Plato—somehow was able to introduce technical terminology into philosophy without tainting the new terms with his own subjectivity: "Precisely the study of Kantian terminology, indeed the only one in philosophy that in its entirety has not only emerged but was also created, leads to the knowledge of its extraordinary potency" (*GB*, 1: 389). Even the choice of the word *doctrine* as the *terminus ad quem* of theoretical philosophy is grounded in the study of Kant's singularly objective and therefore "potent" terminology: according to Kant, all previous versions of metaphysics derive from a "dogmatic" theory of knowledge, whereas the metaphysics of the future consists in the "doctrinal" principles of nature and morals that result from the critique of theoretical and practical reason respectively; similarly, according to Benjamin, all previous forms of metaphysics derive from a "mythological" epistemology, whereas the metaphysics of the future takes shape as "doctrine."

Despite all of this, however, there is a striking difference between Kant's *Prolegomena* and Benjamin's: unlike the former, the latter does not take its point of departure from a major philosophical treatise for which it serves as a commentary. Something similar can be said of Feuerbach's and Nietzsche's programs for the coming philosophy: Feuerbach's represents the philosophical foundation of *Das Wesen des Christentums* (The essence of Christianity), and Nietzsche's consists in an implicit exposition of what he had accomplished in *Thus Spoke Zarathustra*. Benjamin, by contrast, had written nothing similarly systematic—with the ambiguous exception of "On Language as Such and on Human Language." The exception is ambiguous for two reasons: the "little treatise" (*GB*, 1: 343) remains incomplete, especially as regards the theme of mathematics; and it is also traversed by a structural uncertainty with regard to the status of "human language." The systematic exposition of the three modes of infinite language requires that human language originally be plural, whereas the exegesis of the Book of Genesis demands that the plurality of human languages be treated as a consequence of the fall. Instead of revising "On

Language as Such and on Human Language" in response to this uncertainty, Benjamin writes another little treatise beginning with *on*.

The "Kantian typic" is already evident in "On Language as Such and on Human Language," even though it makes no mention of Kant's work. What Benjamin says about the difference between the creative word of God and the cognizing name of human language reproduces the distinction through which Kant captures the specificity of human cognition: divine intuition for Kant, like divine language for Benjamin, is spontaneously creative, whereas human cognition, like the original human language, consists in a certain unity of spontaneity and receptivity. The primary term treated in Benjamin's first "little treatise," namely "language as such" (*Sprache überhaupt*), suggests its Kantian counterpart: "consciousness in general" (*Bewußtsein überhaupt*), which is the term through which the *Prologemena* seeks to capture the idea of a "transcendental unity of apperception" in relatively nontechnical terms (K, 4: 300). A certain dimension of the coming philosophy can be determined accordingly: toward a theory of knowledge in which "language as such" replaces "consciousness in general" as the name for what grounds experience. As he notes near the end of the essay, specifically referring to the *Prolegomena*, the idea of language may function as a "corrective to the one-sided concept of knowledge, which is mathematically-mechanically oriented" (2: 168), but two points are worth emphasizing from the start: there is no certainty that "language in general" will take over the systematic position of "consciousness in general" in the system of theoretical philosophy, and there is no similar suggestion at the beginning of the essay. Following a long tradition of commentary, Benjamin distinguishes between the underlying aims of the *Prolegomena* and those of the *Critique of Pure Reason*: the former derives its concept of experience from "the sciences and especially mathematical physics," whereas the latter does not immediately suppose that experience is "identical with the object-world of science" (2: 158). In relation to Kant's texts, then, the program of analysis is clear: leave aside Kant's *Prolegomena* and discover the program of the coming philosophy that remains latent in the first *Critique*.

Contra *Erlebnis*

About the concept of experience that surpasses the one developed in Kant's *Prolegomena*, this much is certain: it is indeed a concept of experience, that is, of *Erfahrung*, not a concept of "lived experience" (*Erlebnis*).

Neither Feuerbach nor Nietzsche made much of *Erlebnis* in their own philosophical programs; but their writings laid the groundwork for the *Erlebnis*-discourse that Wilhelm Dilthey and Martin Buber, among others, developed and promoted in response to the threat of a one-sided, scientific *Erfahrung*, which no one could claim as his or her own. So little does Benjamin want the concept of experience to be associated with "lived experience" that he avoids all mention of the term. In this regard, Benjamin's procedure in drafting his program gains a high degree of tension: he borrows, as it were, the pathos of contemporaneous *Erlebnis*-discourse, which summarily repudiates the "mechanical-mathematical" experience of the physical sciences; but he does not then expend this pathos in evocations of the higher life that awaits whomever has enough courage to break out of the narrow confines of Kantian critique in particular and Western rationalism in general. "On the Program of the Coming Philosophy" evinces none of the intoxicated talk of an eruptive "breakthrough" that Thomas Mann memorably recounts in the wartime discussions among students of theology at the University of Munich who had read their Feuerbach and their Nietzsche.[3] But Benjamin also does not go in the opposite direction and defend the concept of experience derived from the mathematical sciences as the bulwark against irrationalism in general and *Schwärmerei* or "fanaticism" in particular.

The tension in the concept of experience that Benjamin seeks to develop can be localized in one of its defining attributes: "transience" (*Vergänglichkeit*). For the proponents of *Erlebnis*, the advantage of "lived experience" over scientific experience consists in its singular, momentary, and unrepeatable quality, which disappears from view whenever attention is turned only to the recording of regularities and the corresponding discovery of laws. The distinction Dilthey draws between the sciences of nature and those of spirit derives from the attempt to recognize the value of transience.[4] So, too, does the distinction Buber draws between rule-bound religion and theophanically inclined religiosity.[5] When Benjamin takes Kant to task for overlooking the transient character of experience in formulating its concept, he would appear to place his program squarely in the same argumentative field as Dilthey and Buber; because, however, he indicates that it is only in experience that transience acquires "dignity"—implicitly opposed to "value"—he unequivocally departs from their precedent, while recasting the opposition between *Erfahrung* and *Erlebnis* in terms of the relation between knowledge and experience:

"The problem of the Kantian theory of knowledge, like that of every great epistemology, has two sides, and he was able to give a valid explanation to only one of those sides. It was in the first case the question of the certainty of knowledge that is remaining [verbleibend]; and in the second case the question of the dignity of an experience that is transient [vergänglich]" (2: 158).

For Kant, the concepts of knowledge and experience are structurally equivalent: any object that can be known is also and at the same time an object of possible experience. To develop a concept of experience that is fundamentally distinct from that of knowledge without thereby becoming a concept of *Erlebnis* is, in short, the task that Benjamin assigns himself in the opening paragraphs of "On the Program of the Coming Philosophy." Although the terms in which the task is cast are his own, its point of departure lies in a negative epistemological principle that his program shares with others of the period: any concept of knowledge that derives from the image of subject-object causation, such that the source of knowledge lies in the effect that the object exercises on the subject, is intrinsically erroneous. One version of this image—which can be attributed to Dilthey's writings—appears under the rubric *psychologism*. Another version of the image—which applies to Buber's writings—appears in the guise of mysticism, which, as Benjamin notes in 1913, "commits the mortal sin of naturalizing spirit, taking it as self-evident that it is causally conditioned" (2: 32). Whatever else may be said of the concept of experience that Benjamin proposes, its corresponding concept of knowledge is of a piece with at least three philosophical movements of the period that seek to secure their scientific character, each in its own way: the Marburg school, to which Benjamin explicitly refers; phenomenology, which he briefly mentions; and the "logistical" programs advanced by Russell and Frege, to which he alludes in its final pages. These philosophical schools—combined with Einstein's discoveries, about which Benjamin was clearly aware, if only in outline—make the concept of knowledge proposed in Kant's *Prolegomena* a thing of the past. In response to this situation, Benjamin does not so much propose a program of his own as set out the lineaments of a program for a program; hence, he writes an essay "*On* the Program of the Coming Philosophy," which, if it manages to distinguish the concept of experience from that of knowledge without falling into either mysticism or psychologism, will lay the groundwork for the program of the coming philosophy.

Cohen and the Concept of "Pure Knowledge"

The separation of the concept of knowledge from that of experience is the starting point for the coming philosophy—not a total separation, of course, still less a supersession into a higher term, but rather an ordered disambiguation in which experience is shown to derive from knowledge: "It is in this that philosophy rests: experience lies in the structure of knowledge and is to be developed from it" (2: 163). This is the nucleus of the program under construction: experience derives from neither the interaction between the empirical subject and empirical objects, as proponents of *Erlebnis* suppose, nor from the synthesis of concept and intuition, as Kant proposes; instead, it "develops" out of knowledge that, for its part, does not owe its origin to any sensible intuition and can therefore be called "pure." The basis of the concept of "higher experience" is thus the "purification of the theory of knowledge" (2: 163), which is not so much a desideratum as an accomplishment that can be attributed to Hermann Cohen, whose *Logic of Pure Knowledge* appeared in its second and definitive edition in 1914 as the first volume of his "system of philosophy." To a certain extent, then, the Marburg school already lays the groundwork for the program of the coming philosophy.

And Cohen may have done more. Although he often denigrates post-Kantian idealism, his *Logic of Pure Knowledge* is far closer in spirit to Hegel's *Wissenschaft der Logik* (Science of logic) and even to his *Phänomenologie des Geistes* (Phenomenology of spirit) than to contemporaneous revisions of formal logic such as those proposed by Frege and Russell. Just as Hegel's *Phenomenology* shows the experience of spirit, as it appears to itself, so Cohen's *Logic of Pure Knowledge* describes the journey undertaken by pure thinking as it—on its own, without any assistance from sensibility—generates the basic modes of knowledge.[6] Unlike Hegel, Cohen does not specifically describe the journey of thinking in terms of experience; but he comes very close to doing so, especially in the initial sections of his *Logic*, in which the origin of knowledge in the "judgment of origin" comes under discussion. To the extent that all experience is perilous—as the *per* in *experience*, like the *fahr* in *Erfahrung*, suggests[7]—the "adventure" that Cohen recounts at the beginning of his major epistemological treatise is eminently experiential: "Judgment does not shy away from an adventuresome detour [abenteuerlichen Umweg], if it wants to trace a 'something' to its origin [in seinem Ursprung das Etwas aufspüren will]. This adventure of thinking presents nothingness. Through the detour of nothingness judgment presents the origin" (C, 84). Thus begins the perilous journey of thinking in the absence of any self-conscious thinker to whom its incidents could be described as accidents of a substance. A "something" is generated from the nothingness that pure thinking presents precisely because it is "pure," that is, absolved of all relation to sensibility, even the forms of sensibility that Kant identifies as space and time. The adventure is so perilous that it would end with a mere "something," if it were not for an auxiliary concept that gives an otherwise erratic journey a sense of direction: "The adventuresome path requires a compass for the discovery of the origin. Such a compass offers itself in the concept of continuity" (C, 90). And further: "Unburdened and confidently continuity may undertake its journey in the lands of nothingness. It is not a land, though; it would be one. Only according to the problems are the regions staked out. But there is always in all questions one problem: the origin. Only thinking generated from its origin is valid as knowledge. The journey is led by a secure star, by continuity as the law of thought" (C, 117–18).

It is in relation to the "compass" of continuity that Benjamin establishes the relation of his own thought to the Marburg school. This dimension of Benjamin's "critical altercation [Auseinandersetzung] with Kant and Cohen" (*GB*, 1: 441) is less apparent in "On the Program of the Coming Philosophy" than in a preliminary sketch he wrote under the title "On Perception." Its argument can be summarized as follows: the *Critique of Pure Reason* demotes the Leibnizian *lex continua* from a constitutive to a regulative principle of experience; space and time are continuous, to be sure, but since the laws under which spatio-temporal things fall are contingent and cannot therefore be derived from the theory or "concept" of knowledge, the continuity of experience can be posited only as a methodological principle that guides empirical research. According to Benjamin, with this doctrine Kant not only breaks away from his predecessors but also introduces a discontinuity in the course of philosophy, which had previously understood experience as derivable from knowledge and the concept of the former as therefore continuous with that of the latter. By doing away with the Kantian "doctrine of elements" and thus absorbing space and time into the categories of pure thinking, Cohen restores the systematic function of continuity in the theory of knowledge, for the object generated by any "factual" science is constitutively continuous. Because, however, this very principle—which expresses itself in the famous "fact of science" (C, 57)—requires that philosophy abstain from dictating a priori concepts to the sciences, Cohen introduces an equally intractable discontinuity into the theory of experience: every science methodologically generates its own object and thus lays out a field of "experience" that belongs to it alone.

Having established this tense situation, in which continuity in knowledge excludes the continuity of experience, "On Perception" seeks a single and immediate solution by proposing a distinction between "experience" and "knowledge of experience" (6: 36). Cohen's sublation of the distinction between forms of intuition and categories turns into a distinction internal to experience: "experience" assumes the role assigned to the forms of intuition, whereas "knowledge of experience" takes over the function of the categories. What remains especially problematic, however, is the precise manner in which "experience" is related to the "experience" to which the term *knowledge of experience* refers. Benjamin begins to formulate this question at the point where "On Perception" begins to break off: "in what does the identity of experience in both cases consist" (6: 37)? In response, Benjamin offers an image—or, more exactly, an image to the second power, since the "image" (*Bild*) he proposes consists in a comparison of "knowledge of experience" to the image of a landscape, in which the landscape itself functions as "experience" in the absence of its corresponding knowledge. But it is clear that there is something misleading about the comparison: a spectator who occupies a transcendent position can compare the image of the landscape and the landscape under the assumption that the former is "about" the latter. Such is not the case for the fundamentally different cases of "experience." As if in response to this objection, Benjamin formulates a term that resolves the problem by absolving experience of all relation to knowledge. Such is "absolute experience" (6: 37): it cannot be experienced directly but must be deduced from a "symbolic complex," which assumes the function of the image of the landscape insofar as the latter does not represent but rather symbolizes its "artistic connection [künstlerischen Zusammenhangs]" (6: 36). The degree to which Benjamin was unsatisfied by this proposed solution can be discerned from the numerous revisions to which he submitted the subsequent definition of philosophy, which so abruptly concludes "On Perception" that it never comes to discuss perception per se: "Philosophy is absolute experience deduced in a systematically symbolic complex as language" (6: 37 and 6: 657, for variants).

None of this makes its way into "On the Program of the Coming Philosophy." Instead of defining philosophy in terms of "absolute experience," the programmatic essay proposes only a concept of *higher* experience—which suggests that the absolutely highest experience exceeds the program of the coming philosophy and perhaps the coming philosophy as well. The assessment of Cohen's achievement in the *Logic of Pure Knowledge* alters accordingly: according to "On Perception," it fails to restore the continuity that the *Critique of Pure Reason* had decisively interrupted; in "On the Program," by contrast, Cohen's treatise, as a "purification" of the theory of knowledge, is the basis on which the concept of higher experience can be developed. Benjamin gave Scholem a copy of "On the Program of the Coming Philosophy" in celebration of his twentieth birthday; and they soon thereafter undertook a joint reading of Cohen's treatise *Kants Theorie der Erfahrung* (Kant's theory of experience) with the expectation that it contains an intimation of the concept of experience that should arise out of a purified theory of knowledge. The results of their reading, however, were highly unfavorable to Cohen. Long before they finished the book, they broke off their discussions with a damning indictment: "It is goyish to the umpteenth power; as Walter correctly says, the genius [Genuis] is missing, in the full sense that Walter associates with this word" (S, 2: 275). Even more damning, though, is a comment a month earlier that probably derives from Benjamin as well: "Cohen wastes the power of the *coming* philosophy for the sake of a dogma that makes Kant profound [tiefsinnig]" (S, 2: 257). In recording this comment, Scholem seems not to have noticed that the term *profound* alludes to the program of the coming philosophy that Husserl published in the first volume of *Logos* under the title "Philosophy as Rigorous Science."

Husserl and the "Sphere of Total Neutrality"

Near the conclusion of "Philosophy as Rigorous Science" Husserl identifies a purely negative characteristic of the coming philosophy: whatever it may become, it will certainly not be "profound." This follows from the newfound scientific character of philosophy, which phenomenology represents. Whereas alchemists, according to Husserl, gave off airs of profundity, chemists do not. The same transformation will take hold of philosophy: "Only when the decisive separation between one philosophy and another has made its way into the consciousness of the time is it also conceivable that philosophy would assume the form and language of a genuine science and recognize as an incompleteness what often elicits praise and is even imitated: profundity. . . . The essential process of the new constitution of rigorous science lies in recasting the intimations of profundity [Ahnungen des Tiefsinns] into univocal rational formations" (Hu, 25: 59). To this memorable assertion, which gives rise a lively discussion between Lev Shestov and Jean Héring in the 1920s,[8] Benjamin responds in the opening paragraph of "On the Program of the Coming Philosophy," with an implicit distinction between "profundity" (*Tiefsinn*) and "depth" (*Tiefe*):

> It is the central task of the coming philosophy to allow the deepest intimations [Ahnungen] that it draws from the time and the presentiment of a great future to become knowledge by relating them to the Kantian system. The historical continuity that is guaranteed by the connection to the Kantian system is at the same time the only such continuity of decisive systematic importance. For Kant is the most recent of those philosophers for whom the primary concern is not immediately the extent or depth of knowledge but first and foremost its justification and, next to Plato, he is also really the only one. These two philosophers have in common the confidence that the knowledge of which we can render the purest account will at the same time be the deepest. They have not banished the demand for depth [Tiefe] from philosophy; rather, by identifying depth with the demand for justification, they do it justice in a singular manner. (2: 157–58)

Benjamin's rejoinder to the concluding paragraphs of "Philosophy as Rigorous Science" could scarcely be clearer. Kant and Plato did not intend to deepen knowledge, much less to present themselves as profound; but they did not therefore issue an edict against depth per se, for they understood—without exactly knowing—the following thesis: as knowledge becomes more fully justified, it thereby becomes deeper. This thesis accords remarkably well with the very final remarks of "Philosophy as Rigorous Science," where Husserl, citing a phrase from Empedocles, demands of the coming philosophy that it once again conceive of itself as the "science of the true beginnings, of origins, of the *rizōmata pantōn*" (Hu, 25: 61). By going to its roots, philosophy achieves a kind of depth that has nothing to do with profundity; but the imperative itself is not a matter of knowledge; rather, it is, as Benjamin emphasizes, a matter of "confidence." In the case of Kant—and Kant alone is of consequence here, since Benjamin leaves Plato aside for the remainder of the treatise—the *rizōmata pantōn*, understood as the root of everything knowable, lies in the "transcendental unity of apperception." To the extent that a deeper ground of knowledge can be found, the concept of correspondingly higher experience can be developed. And to the extent that the project of transcendental phenomenology that Husserl launches in *Ideas* represents an unexpected departure from *Logical Investigations*, it accords with the claim with which Benjamin concludes the opening paragraph of "On the Program of the Coming Philosophy": "The more unforeseeably and the more boldly the development of the coming philosophy announces itself, the more deeply must it struggle for certainty, whose criterion is systematic unity or truth" (2: 158).

Certainty does not consist in the self-evidence of phenomenon but, rather, in a guarantee of truth that takes the form of systematic unity. At first glance it seems as though the coming philosophy would be more closely associated with a project like Cohen's than with one like Husserl's: the systematic intention of Cohen's "system of philosophy" is unmistakable, whereas *Ideas* eschews systematicity precisely because its criterion of certainty lies in the givenness of the phenomenon. But the struggle for certainty is as independent of its criterion as the attempt to justify knowledge is unconcerned with its depth. Certainty is attained when knowledge is systematically unified—but not *because* the philosopher intends to construct an unshakable system. For Husserl, the struggle for certainty requires ever more thoroughgoing and definitive reductions of the so-called natural attitude. In this way, a foundation of knowledge deeper than that of the "transcendental unity of apperception" can be discovered. There is a danger, however, that the discovery will be misconstrued as a form of psychological investigation. And it is precisely in this context that Benjamin aligns his own line of inquiry with phenomenology. If a single thesis can be identified as the point of departure for the coming philosophy, it is doubtless the following: "All genuine experience rests on pure epistemo-theoretical (transcendental) consciousness, if this term is still applicable under the condition that it is disrobed of everything subjective" (2: 162–63). For Cohen, the appeal to the putative "fact of science" obviates the need to rely on the term *consciousness* in matters of epistemology; but the "fact of science" is, as it were, a shallow foundation for the concept of knowledge, for it requires that experience be as discontinuous as the factual sciences. In order to elucidate how *consciousness* can be retained in a nonpsychologistic concept of knowledge, Benjamin enlists assistance from Husserl's school: "Pure transcendental consciousness is fundamentally different from every empirical consciousness, and it is therefore questionable whether the application of the term is acceptable. How the psychological concept of consciousness is related to the concept of a pure knowledge remains a principal problem of philosophy that is perhaps to be restored only with recourse to the time of the scholastics. Here is the logical place of many problems that phenomenology has recently renewed" (2: 163).

By describing the place for phenomenological research, Benjamin not only presents its program as an auxiliary to his own but also distinguishes it from the program proposed by his former instructor Heinrich Rickert, whose essay "Vom Begriff der Philosophie" (On the concept of philosophy) launches the first volume of *Logos*, a journal he founded.[9] According to Rickert, philosophy secures a future for its own "concept" only insofar as it avoids the twin errors of historical relativism and Bergsonian vitalism, on the one hand, and scientific objectivism, on the other; and it can avoid these errors only by identifying a "domain that lies beyond subject and object."[10] This domain, for Rickert, consists in "values," which are realized in "sense" rather than in reality. To this, Benjamin responds with his own program for a theory of value, which is confined to the strata of empirical consciousness: "The cognizing human being, cognizing empirical consciousness is a mode of insane [wahnsinnigen] consciousness. By this, nothing more is said than that there are only gradual differences within the different modes of empirical consciousness. These differences are only those of value, whose criterion cannot consist in the correctness of knowledge, which is never a concern of the empirical, psychological sphere; one of the highest tasks of the coming philosophy will be to establish the true criterion of the difference in value of different modes of consciousness" (2: 162). Which means, in short, that the "domain" that Rickert seeks to identify cannot be described in terms of value; rather, the "domain beyond subject and object" is the sphere of knowledge, in which there is no trace of the "natural" attitude. Higher, then, than the task of value-discrimination is the task of discovering a sober sphere that has no "value" from the perspective of "insane" consciousness: "It is the task of the coming theory of knowledge to find for knowledge the sphere of total neutrality in relation to the concepts of subject and object—in other words, to isolate the autonomous, primordially proper [ureigne] sphere of knowledge in which this concept no longer designates the relation of two metaphysical entities" (2: 163).

Just as the "domain beyond subject and object" cannot be ascribed to value, it cannot be represented in terms of the putative "fact of science." Rather, for Benjamin, as for Husserl, the domain in question is "primordially proper" and thus solicits a term like *consciousness*, despite the ineradicable ambiguity of this term. "Pure transcendental consciousness" is altogether different from empirical consciousness, and the determination of this difference requires two independent tasks: one is associated with the project of scholastic philosophy, with its examination of the objective strata of *intentio*, as renewed among members of Husserl's school; the other with a critical project for which there is no corresponding equivalent in contemporaneous phenomenology. The critique in this case is directed, not so much against the "insanity" of empirical consciousness, which is irreparable, as against the "epistemo-mythology [Erkenntnismythologie]" (2: 161) from which it derives. In this way, "On the Program of the Coming Philosophy" diverges from *Ideas*: instead of being content with a blanket description of the "natural" attitude, the philosophical program for which Benjamin's serves as preparation identifies historically different ways in which subjects conceive of themselves as affected by objects. The project of "turning off" the "natural" attitude thus turns into a critique of epistemo-mythology, which, in turn, transforms epistemo-critique into a historical yet nonhistoricist enterprise.

Toward a Critique of Epistemo-Mythology

The presumption that objects affect subjects who then seek to dominate them is not natural but, rather, mythological, and the elements of any theory of knowledge that has not been purified of the image of subject-object interaction are at the same time the elements of myth. This, means, of course, that modernity, and specifically the Enlightenment, has not escaped from myth, but has, on the contrary, introduced its own mythology, in which subjects and objects are starkly opposed to each other and the "meaning" of experience accordingly tends toward zero. It is in this context that Benjamin locates the historical significance of Kant. Comparable to a tyrant who encounters no resistance, he measures out the sphere of experience and violently disposes of an unruly element, namely the thing-in-itself:

> That Kant was able to undertake his enormous work under the constellation of the Enlightenment indicates that it was carried out, as it were, on the null-point, that is, experience reduced to the minimum of meaning. Indeed, one may say that the greatness of his attempt, his own radicalism presupposes such an experience, whose proper value approached zero and which could have attained a (we may say: sad) meaning only in its certainty. None of the pre-Kantian philosophers saw themselves confronted with the epistemological task in this sense; none, in any case, had such a free hand, since an experience, whose quintessence, whose best side was a certain Newtonian physics could be treated roughly and tyrannically without suffering. (2: 159)

Once the meaning of experience is reduced to a minimum, nothing hinders a metaphysical theorem of the following type: there is no object of experience, for objects mean nothing. In other words, they are incommunicable and thus "sad." Such a "rough and tyrannical" treatment of the object, which takes shape as Kantian critique, would have met resistance if experience had any meaning—as, for instance, if a divine place or time could be experienced. With the Enlightenment pegged as a near null-point, a gradation of *Weltanschauungen* can then be developed: the greater the meaning of experience, the more resistant a *Weltanschauung* is to the recognition of the mythic character of its corresponding epistemo-mythology. Benjamin thus outlines a theory of historical "experiences or intuitions of the world [Erfahrungen oder Anschauungen von der Welt]" (2: 159) that remains immune to the critique of Dilthey's *Weltanschauungsphilosophie*, with which "Philosophy as Rigorous Science" concludes (Hu, 25: 41–57). In contrast to Kant, who affirms that "there is only one experience" (K, A 110), Benjamin asserts that every historical epoch consists in "an experience [einer Erfahrung]," which is, for this reason, always "transient" (2: 158). Because the concept of experience has been distinguished from that of knowledge, the transient character of the former does not "relativize" the validity of the latter. The coming philosophy is in this sense as nonhistoricist as the "rigorous science" for which Husserl argues.[11] Experience lasts only as long as its object persists. With the elimination of the object, it tends toward its limit, namely the "singularly temporal." Any experience that is purported to last longer than its time is no longer a "genuine experience" but, rather, an experience in the derivative sense, the "experience" in quotation marks that characterizes everything mythic.

This is, therefore, what philosophy has hitherto missed: the singular-temporal character of experience. For the sake of such experience Benjamin proposes a program of the coming philosophy. Contemporaneous philosophical programs derived from the writings of Feuerbach and Nietzsche misrecognize what is missing when they replace *Erfahrung* with *Erlebnis*. The term "singularly temporal" can doubtless be predicated of *Erlebnis* but only because of its subjective and therefore epistemo-mythological character. Benjamin's verdict is categorical: "Philosophers have not been conscious of this experience in its entire structure as a singularly temporal one, and Kant was also not conscious of it" (2: 158). But—and this is crucial—this is no fault of the philosophers, least of all a fault that can be ascribed to Kant, for the experience in question cannot be ascribed to anyone. As soon as someone holds onto a "singularly temporal" experience, it is recast as *Erlebnis*. The two strands of Benjamin's argument—one drawn from an inquiry into the concept of knowledge, the other from an analysis of "experiences or intuitions of the world"—join at the following point: higher experience is structured in such a way that it tends toward the singularly temporal, but at the same time the concept of experience that rests on the pure concept of knowledge is altogether continuous. If there is to be a coming philosophy; if, in other words, *philosophia* is not always to be *perennis*, then the two irreducible yet seemingly irreconcilable characteristics of experience—continuity and temporal singularity—must somehow be synthesized.

"Primordial Concepts" and the Problem of Identity

Benjamin declines to follow Husserl in replacing the concept of *Erfahrung* with that of the primordial *Erlebnisstrom* (stream of consciousness), and he declines to follow Cohen in replacing transcendental consciousness with the "fact of science" as the ultimate ground for the justification of knowledge. The sole basis for experience is "pure epistemo-theoretical (transcendental) consciousness" (2: 162), which no one can claim as his or her own. Here, if anywhere, there is good reason to introduce a technical term as a replacement for *consciousness*. In this case, as in almost all others, Benjamin declines to do so. Instead, he poses a question as to whether the "logistical" programs of Frege and Russell may result in the requisite formulation. The guide for Benjamin's reflections is, once again, the Kantian typic, and the question he poses is whether any of the "logistical" programs forms an adequate basis for a transformation of the Kantian table of categories that would accord with Cohen's transformation of space and time into "logical" elements:

> Just as the Marburg school has already begun with the sublation of the distinction between the transcendental logic and the transcendental aesthetic (even if it is also questionable whether an analogue to this distinction must return at a higher level), so the table of categories is to be fully revised, as is now universally demanded. Precisely in this instance, then, the transformation of the concept of knowledge announces itself in the acquisition of a new concept of experience, since the Aristotelian categories, on the one hand, were set up arbitrarily and, on the other, were exploited by Kant in an entirely one-sided manner with regard to a mechanical experience. Above all, it must be considered whether the table of categories must remain in its current isolation and lack of mediation, or whether it could not take a place among other members in a doctrine of orders [Lehre von der Ordnungen] or itself be built up into such an order, grounded on logically primordial concepts [Urbegriffe] or connected with them. What Kant discussed in the "Transcendental Aesthetic" would also then belong in such a universal doctrine of orders. (2: 166)

Perhaps because the lament over the haphazard state of the Kantian table of categories is almost as old as the table itself—it was not until Klaus Reich's dissertation in 1932 that its rigorous principle of ordering was duly recognized[12]—Benjamin goes no further in this direction; he begins, instead, to formulate a series of ever more schematic suggestions for further research, each of which is marked by a *furthermore*, until he inserts a dash, and makes his most far-reaching proposals. One of these is of particular importance, for it identifies one of the "primordial concepts" from which a new table of categories can be developed: "The fixing of the concept of identity, which was unknown in Kant" (2: 167). Of course, it is not as though Kant was unaware of the concept of identity; but in some of his earliest work, such as the *Nova dilucidatio* (K, 1: 288–91), as well as in the *Critique of Pure Reason* (A 150–53; B 189–93), he treats the principle of identity as a derivate version of the principle of noncontradiction. In this regard, representatives of "logistics," especially Frege, are more incisive. Thus does the opening paragraph of his essay "Über Sinn und Bedeutung" (On sense and reference) challenge Kant's conception of analytic judgments by asking whether equality, understood as identity, is a relation among objects or between an object and a sign, and in a now-famous appendix to his *Grundgesetze der Arithmetik* (Fundamental laws of arithmetic)—which Benjamin may have consulted in conjunction with his attempt to solve Russell's paradox—he laments the "misfortune" that has befallen the "foundations of his edifice" as a result of the paradox and refutes one of the proposed solutions: one cannot associate different kinds of identity-relations with different kinds of objects for the simple reason that "identity is a relation given to us in such a specific form that it is inconceivable that various kinds of it should occur."[13] To an assertion of this kind, if not to this very claim, Benjamin responds in his "Theses on the Problem of Identity" (6: 27–28), which was written around the same time as "On the Program." The problem under consideration in Benjamin's terse document consists in uncovering and then "fixing" at least two modes of identity, only one of which expresses itself in tautological formulations.[14]

The point of departure for Benjamin's line of argument is the categorial moment of infinity, which takes the following form in the *Critique of Pure Reason*: "x is not y" (K, A 72–73; B 97–98). To the extent that a categorical judgment identifies something as something else—for instance, x as y—the first of the eleven theses Benjamin proposes is valid: "Everything nonidentical is infinite, but this does not mean that everything identical is finite" (6: 27). Drawing on the traditional Aristotelian distinction between potential and actual infinity, Benjamin then proposes that there are two ways for the "nonidentical infinite" to be "nonidentical": "(a) potentially identical" and "(b) actually nonidentical." Before breaking off this line of inquiry, he points toward an investigation that he is, as yet, unable to carry out: "which mode of mathematical infinity falls under (a) and which under (b)" (6: 27). The two modes of mathematical infinity to which he refers are presumably the ones upon which Cantor founded transfinite set theory: denumerable point sets such as the aggregate of rational numbers and nondenumerable point sets such as the set of all points on the linear continuum.

Benjamin then turns to the place where the specific "problem" of identity arises, that is, in "the identity-relation," which appears to be expressible only in the form of tautology. The point of seeking to "fix" the problem of the identity-relation is to arrive at a point where he can directly counter a thesis like Frege's and identify at least two different modes of identity, only one of which expresses itself, either in the traditional form of tautology such as "a is a" or in its self-reflective transformation: "I = I."

As the eleven theses become progressively less "thetic"—thus ending up in parentheses—Benjamin poses the question that generates his project as a whole: what is this other mode of identity that solves the problem of identity by disclosing its relational character without recourse to tautological formulations? An intimation of an answer can be found in the idea of translation, to the extent that translation instantiates the identity-relation in a nontautological manner. Benjamin, however, responds to the question that implicitly begins his project by emphasizing its open-ended character: "How identity with itself—for the principle of identity attests to its possibility, and it is its authentic content—is to be distinguished from another kind of identity, and whether this other kind is something like a purely formal-logical identity of the thought [Identität des Gedachten] in contrast to that of the object with itself, this must remain an open question" (6: 29).

In a letter to Scholem from December 1917 Benjamin proceeds a little further in the same direction. After indicating that he has abandoned his initial plan to write a dissertation on Kant's theory of history, turning instead to the neo-Kantian idea of "the infinite task," Benjamin seeks to capture, however provisionally, the ultimate goal of his "Theses on the Problem of Identity": "fixing" an identity-relation of thinking that cannot be captured in any thought, but is not therefore a mystical *Erlebnis*, which would somehow be post- or prerational. Instead of simply distinguishing the identity-relation from tautology, he presents it in the letter to Scholem as its internal limit. So fully tautological is the identity of thinking that it cannot be expressed *as* tautology: "The affirmation of the identity of thinking would be the *absolute tautology*. The illusion of a 'thinking' emerges only through tautologies. Truth *is* just as little thought as it thinks. '*A* is *a*' designates, in my view, the identity of the thought"— to which he adds a footnote, as if the letter were a little treatise: "*better* said (*alone* correctly): of the truth itself" (*GB*, 1: 409).[15] In trying to determine the identity-relation that transcends the identity of the thought and correspondingly characterizes that of truth, Benjamin finds himself at a loss for words. So difficult is the project of "fixing" the concept of identity that he even violates his stylistic imperative and invents a technical term: "Again and again this is how matters appear to me: I would deny that there is an identity of thinking as a particular case, neither of an 'object' nor of a 'thought,' because I doubt that 'thinking' is a correlate of truth. Truth is 'thi(n)ck [denkicht]' (if I must create this word because none stands at my disposal)" (*GB*, 1: 409).

Truth is so thick with thought that no act of thinking—which is always the thinking of *a* thought, identical to itself—can capture it. Truth cannot therefore be understood as its correlate. Which is to say, it belongs to a sphere apart. But once again, this is not because the identity of truth is somehow beyond or before thinking, as a proponent of irrational *Erlebnis* would suppose. Rather, truth is so oversaturated with thinking that that it escapes any *one* thought, even if this one thought is systematically ordered. A higher identity of unity is required to characterize truth than the identity of any particular thought or even the identity of thought in general, which presents itself as a system of philosophy. All of this is closely connected with Benjamin's decision to write a dissertation on "the infinite task," but for the purposes of his dissertation, he will pursue the same problem in terms of another "primordial concept": not identity but, rather, unity, which is less troubling than the concept of identity for a number of reasons. Above all, the concept of unity, unlike that of identity, is recognized by Kant in two forms: on the one hand, it is a category; on the other, it is the transcategorial unity of consciousness in general.[16] Benjamin thus has at his disposal a Kantian framework for the "fixing" of the concept. Following his own suggestion in the third of the "Theses on the Problem of Identity," he sketches the problem of "the infinite task" by aligning the two kinds of unity with the two modes of transfinite cardinals. Just as Cantor shows that the "set" or "complex" (*Menge* or *Inbebriff*) of rational numbers is of a lower "power" than the point set or complex of the linear continuum, so Benjamin proposes that the "unity of science rests on the fact that its complex [Inbegriff] is of a higher power [Mächtigkeit] than the complex of all of the infinite number of finite, that is, given questions which can be posed. This means that the unity of science consists in the fact that it is an infinite task" (6: 51). The unity of science as a whole, in contrast to the unity of any particular program for the acquisition of knowledge, is comparable to the identity of truth in contrast to that of thinking: it is a higher unity that is not, however, the unity of transcendental consciousness. So high is this unity that any judgment about its infinitude must itself be infinite: "the unity of science itself is neither finite nor infinite" (6: 52).

Two Dissertation Projects and the Division of Domains

Although Benjamin ultimately decides against writing a thesis on "the infinite task," he retrieves the line of argument he outlined for its development for the final section of his major essay on Goethe's *Elective Affinities*, where he distinguishes between the unity that characterizes the work of art and the unity that makes philosophy systematic. One of the drafts for this section, which Benjamin succinctly entitled "Theory of Art Critique," shows the set-theoretical character of the relevant distinction even more emphatically than either the published essay or the brief sketch of "The Infinite Task": "The unity of philosophy, its system, is an answer of a power [Mächtigkeit] higher than the infinite number of finite questions that can be posed. It is of a higher kind and power than the complex [Inbegriff] of questions that can be posed, because the *unity* of the answer cannot be questioned. It is therefore also of a higher power than any individual philosophical question, as a problem, which can be posed" (1: 833; cf. 1: 172). Even the choice of Goethe's novel can be seen to derive from the underlying project of "fixing" the concept of identity and solving its corresponding problem, inasmuch as affinity is a mode of that epistemo-theoretical unity which Kant had explored in the first version of the Transcendental Deduction, under the title of "transcendental affinity" (K, A 113). This is what the "doctrine of orders" aims to identify: a "sphere of total neutrality," the elements of which are so affine with respect to one another that their unity is of a higher power than the unity of the "I." The "sphere" thus replaces the "transcendental unity of apperception" as the most thoroughly justified and therefore the "deepest" concept of knowledge. Characterized by the category of identity, the neutral sphere combines the two desiderata of the program for the purification of knowledge that Benjamin develops in consultation, as it were, with Cohen and Husserl: although nonsubjective, it is nevertheless "primordially proper," insofar as identity is the "essence" of propriety: "[the concept of identity] presumably constitutes the highest transcendental-logical concept and is perhaps truly appropriate for the grounding of the sphere of knowledge beyond subject-object terminology" (2: 167). Benjamin concludes "On the Program of the Coming Philosophy" with a summary account of the relation between experience and knowledge once it is understood that the former rests on the latter and does not, therefore, owe its origin to either a metaphysical entity or a methodologically generated object: "Experience is the unified and continuous multiplicity of knowledge" (2: 168).

The problem that Benjamin pursues in his dissertation *The Concept of Art Criticism in German Romanticism* derives almost directly from the definition of experience with which "On the Program of the Coming Philosophy" concludes. Art criticism, for Friedrich Schlegel and Novalis, consists in the "unified and continuous" multiplication of whatever is knowable in any given work. The "sphere of total neutrality" thus takes shape as a medium in which subject and object are constitutively identical to each other—in other words, as a "medium of reflection" (1: 36) in accordance with the model of self-positing first proposed in Fichte's *Über den Begriff der Wissenschaftslehre* (On the concept of the *Wissenschaftslehre*). In search of a term that would characterize the unity of knowledge, Benjamin turns away from the early Romantics themselves and directs attention, instead, to a phrase that can be found in the first of Hölderlin's Pindar commentaries, which appears under the appropriate title of "Das Unendliche" (The infinite): "infinitely (precisely) connecting [unendlich (genau) zusammenhängen]" (1: 26). Similarly, in some highly condensed notes, Benjamin associates the imperceptible "plane" of perception, which makes the interpretation of appearances possible, with "the absolute connection [absolute Zusammenhang]" (6: 34).[17] Such is the unity of science or knowledge, which is of a higher "power" than the infinite complex of cognitions, perceptions, or reflections. The connection, coherence, or continuity of knowledge is so "dense"—this is the mathematical term Benjamin borrows in "On Language as Such and on Human Language" (2: 151)—that it solicits the literary form of the fragment and, more importantly, escapes the opposition through which any lower power of unity can be characterized, beginning with the traditional opposition between finite and infinite.

Despite the confidence with which Benjamin punctuates the conclusion of "On the Program" with a succinct definition of philosophy, which he then explicates in his doctoral dissertation, the final full paragraph of the essay is marked by a palpable degree of tentativeness. In place of a bold statement that would ultimately determine the precise program of the coming philosophy—something like this can be found at the end of "On Perception"—Benjamin simply emphasizes a point where Kant was right and where Cohen went wrong. Drawing attention to Kant's programmatic "prize essay" on the difference between metaphysical and mathematical methods (K, 2: 275–301), Benjamin notes that the future author of the *Critiques* recognizes that there is only one fact that philosophy must acknowledge in its inception—not the putative fact of science but, rather, the fact of language, which expresses itself, first of all, in the requirement that philosophy be conducted in linguistic rather than mathematical terms:

> The fact that, for Kant, all philosophical knowledge has its unique expression in language and not in formula and numbers has completely retreated behind the consciousness that philosophical knowledge is absolutely certain and a priori, [that is] behind the consciousness of this side of philosophy in which it is the equal of mathematics. In the final analysis, however, this fact must assert itself as decisive, and it is because of this fact that the systematic supremecy of philosophy over all the sciences as well as over mathematics will ultimately be asserted. (2: 168)

Benjamin makes the same point in a brief fragment, "On Transcendental Method," in which he asserts that Cohen would have been more loyal to the *Critique of Pure Reason* if he had based his system of philosophy on the "fact of language" rather than the "fact of science" (6: 52). Once the "fact of language" is recognized, the proper relation between the theory of knowledge and the concept of experience can emerge: "A concept of [such philosophical knowledge], acquired in reflection on its linguistic essence, will create a corresponding concept of experience, which will encompass regions whose truly systematic ordering Kant did not reach" (2: 168). Among the regions that exceeds Kant's grasp, although perhaps not Cohen's,[18] that of "religion" is of particularly significance, for it is both the "highest" (2: 168) region and the one Kant expressly excludes from the domain of possible knowledge.

Few utterances in the *Critique of Pure Reason* are as well known as a passage in the preface to its second edition where Kant explains the motivation for the entire critical project: he wishes to "make room for faith" (B xxx) by—in Benjamin's topology—making sure that the "sphere of knowledge" only abuts upon the "domain of religion." The program Benjamin ends up proposing moves in almost the opposite direction. It should be emphasized, however, that there is as little place for knowledge of a metaphysical entity called "God" in the domain of religion as there is for the knowledge of an object in the domains of the individual sciences. And it should be further emphasized that the revision of Kant's attempt to "make room for faith" follows from the specifically *critical* program proposed in "On the Program of the Coming Philosophy." The line of argument is in this case relatively simple: the critique of "epistemo-mythology" requires the integration of the "domain of religion" into the "sphere of knowledge" because *religion* means nothing other than: "the absence of mythology." As the place in which there is no room for epistemo-mythology of any kind, religion is the name of the purest, hence the "highest" domain within the sphere of knowledge, on the basis of which a concept of higher experience arises.

"Transition"

In this way, however, the "sphere of knowledge" appears to collapse into the "domain of religion"—so much so that they might even be two names for the very same place. Insofar as the sphere and the domain decisively break with epistemo-mythology, they can even be said to be identical. A critique akin to the one Hegel launched against Schelling's "philosophy of identity"—as the "night in which all cows are black"[19]—could be applied to the line of argument with which "On the Program of the Coming Philosophy" comes to a close. And the critique could extend even further: to the extent that every science excludes "epistemo-mythology," every one is identical to every other. The "total neutrality" of the "sphere of knowledge" means that it cannot be divided into discrete domains on the basis of its putative objects. And because of this neutrality, the sphere cannot be articulated in terms of the intentional structures of consciousness either. Nothing, it seems, can be said of this sphere other than that it breaks free of epistemo-mythology and therefore appears to be indistinguishable from its domains, beginning with the highest: "In the interest of clarifying the relation of philosophy to religion the content of the former is to be repeated insofar as it concerns the systematic schema of philosophy" (2: 168).

Thus begins the "Nachtrag" (Supplement) that Benjamin posts to "On the Program of the Coming Philosophy." The "repetition" results in a topological alteration: "the philosophical" is no longer described in terms of a "sphere" but as a "domain" into which theoretical philosophy as a whole has been absorbed. The schema to which Benjamin appeals is drawn again from Kant—but not from the outline of the *Critiques*. Rather, Benjamin turns to the Kant of the *Opus postumum*, which supplements the critical system by means of a "science of transition" (K, 21: 642). In a letter to Christian Garve, whom Benjamin briefly mentions in "On the Program" (2: 159), the author of the *Critiques* describes the reason that he feels compelled to develop a supplement to his work: having found a "gap in the system of critical philosophy," he now has no choice but to develop the "transition from the metaphysical first principles of natural science to physics" (K, 12: 254). For the late Kant, the tantalizing idea of "transition" from critique, through "doctrine," to experience is the precarious basis on which the future of philosophy stands or falls.[20] Something similar is true for early Benjamin, whose "Supplement" both generalizes and simplifies the schema by which the postcritical Kant proposes something other than post-Kantian philosophy: it generalizes the schema by making physics into one dimension of experience among others; and it simplifies the schema by placing both the "metaphysical first principles" and its corresponding "transition" under the rubric of transition: "Where the critical dimension stops and the dogmatic begins does not perhaps show up exactly, because the concept of the dogmatic should simply mark out the transition from critique to doctrine, from the more general to the particular fundamental concepts. All of philosophy, therefore, is the theory of knowledge, precisely only theory, the critical as well as the dogmatic part of all knowledge" (2: 169).

Benjamin thus relinquishes, in part, the schema through which "On the Program of the Coming Philosophy" develops: the theory of knowledge remains free of all metaphysical elements, to be sure; but metaphysics does not then find a place of refuge in "genuine experience." Rather, metaphysics has no other place than in the transition from critique to doctrine. The "Supplement" thus moves in the direction of a philosophical kinematics: experience is not so much "transient" (*vergänglich*) as "transitional" (*übergänglich*). The putative "objects" of knowledge are neither realities nor potentialities; they are, rather, virtualities that can be known as such—as "something" that exists in a medium that it instantiates—only at the limit point where epistemo-critique makes a transition to metaphysics: "The meaning of the term *metaphysical*, as it is earlier introduced, consists now only in the fact that this limit [between philosophy and individual sciences] is not explained as something already in existence [vorhanden], and the renaming of 'experience' as 'the metaphysical' means that so-called experience is virtually contained in the metaphysical or dogmatic part of philosophy, into which the highest epistemo-theoretical, that is, critical part make a transition" (2: 169). Benjamin probably borrows the idea of virtuality from Leibniz, whose own program for the coming philosophy, "Discours de métaphysique" (Discourse on metaphysics), describes "substantial forms" in terms of their complete concepts, which "virtually" contain all of their predicates and thus include everything they seemingly "experience."[21] The structure of any given science is, for Benjamin, similarly monadic: its "so-called experiences" are contained in its hypotheses, and its apparent objects are actually virtualities, which are interior to the theory that recognizes them. In line with the criticism of Kant's *Prolegomena*, which relies on "a certain Newtonian physics" (2: 159) for the exposition of his theory of experience, Benjamin adds a second supplement in which he implicitly takes issue with one of Newton's most famous pronouncements: "*hypotheses non fingo* [I do not feign hypotheses]."[22] In place of Newton, there is now Einstein, who does indeed make hypotheses.

Thought Experience

By way of coincidence, Benjamin proposed his prolegomena to a future metaphysics in the same Swiss city where Einstein had begun to develop a program for the coming physics by reflecting on the "electrodynamics of moving bodies," to cite the title of the famous paper he published while he was living in Bern. In 1915 Einstein generalized the original theory of relativity, so that it would include all cases of motion, not simply the special cases of constant motion in a nongravitational environment. Benjamin would not have known the theory of relativity in detail, of course; but there is also no question that he was familiar with its outlines. Among the many places where Benjamin would have acquired some familiarity with the theory, an essay Heidegger published in 1916 is worth mentioning. After briefly discussing a few remarks drawn from Einstein's original paper, Heidegger quotes a striking passage from a series of lectures on contemporary theoretical physics delivered by Max Planck: "The conception of time that develops out of the theory of relativity 'outdoes in boldness everything that has yet been accomplished by speculative research into nature, indeed by the philosophical theory of knowledge'" (He, 424).[23] Not considering himself competent enough to judge this aspect of Heidegger's essay—a reticence that stands in obvious contrast to its author, who considers himself in a position to dismiss Planck's claim with a few words of "clarification"—Benjamin asks Scholem for an assessment of Heidegger's account of time in contemporary physics; but he also expresses a suspicion that Heidegger is as misguided in his discussion of Einstein as in his analysis of the historical concept of time (*GB*, 1: 344).

And in a postscript to a letter he wrote to Scholem in the winter of 1918, he briefly discusses a comment made by Erich Gutkind about the idea of "filled space" (*erfüllten Raum*), causally adding: "I don't know whether he puts it in relation to the theory of relativity" (*GB*, 1: 435)—which indicates at the very least that he associates the idea of "filled space" with the notion of interpenetrated space-time that Einstein develops in the general theory of relativity.[24]

In "On Perception" Benjamin makes the following suggestion in response to the question of why Kant, breaking with the philosophical tradition, had no desire to deduce experience from knowledge: "the supposition is perhaps allowed that in a time in which experience had in fact sunk into an extraordinary flatness and godlessness, philosophy—if it was honest—could no longer have any interest in the saving of this experience for the concept of knowledge" (6: 36). So much for the era of Newton. As for the time of Einstein: the experience of a "relativistic" cosmos may be equally "godless," but it is decidedly not "flat." On the contrary, it is "curved" in such a way that "space-time"—a term Benjamin uses relatively often, including in "On Perception" (6: 33)—is characterized by a certain plasticity. And in the second supplement he adds to "On the Program of the Coming Philosophy," under the title "Versuch eines Beweises, daß die wissenschaftliche Beschreibung eines Vorganges dessen Erklärung voraussetzt" (Attempt at a proof that the scientific description of a process presupposes its explanation), Benjamin presents the "experience" of modern mathematical physics as so saturated by thought that the experience of thinking can verify its hypotheses: "In the experiment the question concerning the systematic dignity (that is, mathematicity) of a hypothesis receives a response in the form of an answer to the question concerning its validity for our experience. In principle, however, a physics without experiment is possible, for it remains certain that the experiment is simply a methodological means of verifying the relation of the hypothesis to mathematics, which in principle must also be able to be located in thinking" (6: 42).[25]

While implicitly presenting Einsteinian electrodynamics as an epistemo-critical advance over Newtonian mechanics, Benjamin makes an appeal to Plato that is otherwise lacking in the program for the coming philosophy. To the extent that Kantian critique is bound up with Newtonian science, the Kantian typic gives way to its Platonic counterpart, in which science "saves" phenomena by demonstrating their mathematicity, that is, their general "thinkability." The minimal thesis Benjamin proposes at the bottom of the "magic circle of language"—"mathematics thinks" (7: 786)—thus finds an application in contemporary science: "The logical origin of the hypothesis does not lie in experience but, rather, in the task: to rescue phenomena under the presupposition that there are phenomena in general; that is, to grasp and maintain a moment of necessity, a mathematical moment, in them. In this presupposition there lies a moment of contingency, and this is indeed because we are able absolutely to think necessity only in the sphere of mathematics. For this reason, experience is contingent insofar as we are unable immediately to think the necessity (mathematicity) of appearances. Therefore, it is always a matter of the Platonic problem: if we want to think the world, we must *ta phenomena sozein* [save the phenomena]. This is what the hypothesis does" (6: 41). Not only, then, can the hypotheses of mathematical physics be verified by means of so-called "thought experiments," they must be verifiable in this way, for hypotheses do not so much explain the behavior of phenomena as reveal why there is a world of phenomena in the first place—so that they can be "saved" by thinking. Mathematical physics is therefore a dimension of doctrine in which space, time, and intellect so thoroughly permeate one another that thinking alone is itself observation. The experience of this physics is "virtually contained" in its hypotheses, and its so-called objects are in fact virtualities, which are "there" only in the medium of their exposition.

Ultra-Continuity

Just as the putative objects studied in mathematical physics are actually virtualities, so is the thinking subject an illusion generated by a false analogy between acting on something in the world and thinking about worldly things. In this context, Benjamin adds still another supplement to his program in which he takes philosophical kinematics one step further. Replacing the phenomenological phrase *pure epistemo-theoretical (transcendental) consciousness* with its epistemo-critical counterpart, Benjamin does not thereby follow Cohen in presenting the latter as altogether distinguished from transcendental subjectivity; rather, he seeks to show that consciousness in general misconstrues "thinking" when it represents it as intentional, which is to say, directed at an object of thought: "But can the human being as an empirical being think in general? Is thinking in general in this sense an activity like hammering, sowing, or is it not an activity toward something [Tätigkeit auf etwas hin] but, rather, a transcendental *intransitivum*, as walking [gehen] is an empirical one" (6: 43). On the basis of a schema of systematic philosophy drawn from Kant's *Opus postumum*, the "Supplement" to "On the Program of the Coming Philosophy" prepares the way for this question and suggests an answer as well: wherever the transcendental "walking" (*Gehen*) that goes by the name of "thinking" immediately "makes a transition" (*übergeht*), there is the "doctrine of religion" (*Lehre von der Religion*)—which is to say, the doctrine that comes from religion as much as the doctrine that is about it.

An *immediate* transition, however, is in a certain sense a contradiction in terms. If something immediately turns into something else, the turning-point is less a transition than a transfiguration. From an empirical perspective, however, the immediacy of a transition can be easily imagined—as a bump. A moving object suddenly comes to a halt, and all of its momentum is transferred to whatever blocks its passage. The "something" that brings the "transcendental *intransitivum*" to a standstill cannot, of course, take the form of an object, for thinking is neutral with regard to the metaphysical entities called "subject" and "object." Benjamin therefore has no choice but to assign a technical term to the "something" against which thinking bumps, and he calls it *Dasein* (existence). In choosing this term, Benjamin clearly takes his point of departure from the so-called "pre-critical Kant," who elucidates *Dasein* as the "position" in judgment in the context of developing "the one possible ground of proof for the demonstration of God's existence" (K, 2: 73; cf. *Critique of Pure Reason*, A 598; B 626). And just as clearly, Benjamin departs from the line of thinking developed under the direction of the Marburg school, for, according to Cohen's "infinitesimal method," the category of reality is correlated with the mathematical operation of differentiation, such that the real in any given complex is equivalent to the tangent of the continuous function that represents the complex in question.[26] According to the fundamental theorem of the infinitesimal calculus, discovered independently by Newton and Leibniz, the inverse of differentiation is integration, and Benjamin elucidates *Dasein* accordingly: "Yes, it must be said: philosophy in general, in its posing of questions, can never bump up against the unity of existence [Daseinseinheit] but always only encounters unities of lawfulness whose integral is 'existence'" (2: 170). The integration of all the "domains" of the "sphere of knowledge" yields *Dasein*—but not its "unity," which is to say, not something in existence against which the progress of philosophy would bump.

Philosophy can thus proceed along its own autonomous path, without knowledge of the higher power of unity to which the integration of its domain refers. Wherever there is an *immediate* transition from the "domain of the philosophical" to its corresponding doctrine, however, the former bumps up against that which it is constitutively incapable of recognizing—not simply *Dasein* as the sum of its parts but *Dasein* as a unity in its own right. There is obviously no process by which this transition can be achieved and no formula for its accomplishment. It can only happen and is therefore at once transient and transitional, *vergänglich* and *übergänglich*. "The formal dialectic of the post-Kantians" (2: 166) is of no consequence in this case, even if the term *bump* (*Stoß*) owes a debt to the motif with which Fichte begins post-Kantian thought: "a mere check [Anstoss], actually there, without any action of the positing I, gives the I the task of limiting itself."[27] For Benjamin, however, the "domain of the philosophical" cannot be represented in terms of the I's self-activity, especially if, as in the case of Fichte's original conception of philosophy as "Wissenschaftslehre" or "doctrine of science," the reason for proposing a "check" lies in securing a thought-thing against which the I in its theoretical capacity can transitively act. Not only is the bump that emerges in the "Supplement" neither a thing in the world nor an object of thought; it is not even the forever receding starting point that establishes the "infinite task" of the empirical sciences.

Philosophy, understood as the thoroughgoing critique of epistemo-mythology, proceeds as though *Dasein* were *only* the sum of its parts—and not somehow also "there" in its own right, "unified" in a manner that cannot be captured by the identity sign in the following equation: $\int$ "the theory of knowledge" = "existence." It is in this context that Benjamin once again, and for the last time, draws on the concept of experience in contrast to that of knowledge in accordance with the following proposition, which corresponds to the brief sketch of a dissertation on the "infinite task": "There is a unity of experience that can in no way be understood as a sum of experiences" (2: 170). The argument proceeds along the following lines: philosophy itself has nothing to do with experience, for it is solely the sphere of knowledge; but its "highest region" serves as the basis for the concept of higher experience, which is characterized by a higher "power" of unity. More precisely, the unity of experience consists in "concrete totality" (2: 171), which means that every "part" of the unity is identical to the unity as a whole. Translated into the language of experience, this means: every experience is experience pure and simple—which reproduces, in a sense, Kant's dictum that "there is only one experience" but makes it possible for this "one" *of* experience to be dense enough, as it were, to be "there" in a systematically and historically diverse complex of so-called experiences. Whenever the domain of the philosophical makes an immediate transition to the doctrine of religion, there is *Dasein*, which is to say, experience as a whole in one of its systematic or historical instances. The "Supplement" thus culminates in something like a shock to the system: "Now, however, the source of existence [Dasein] lies in the totality of experience, and only in doctrine does philosophy bump up against an absolute, as existence, and therefore up against that continuity in the essence of experience whose neglect is presumably the lack of neo-Kantianism" (2: 170).[28]

The formulation here is precise, and so, too, is the critique of the Marburg school. Benjamin does not maintain that philosophy bumps up against *the* absolute. In making an immediate transition to doctrine, philosophy encounters *an* absolute that is *an* experience—one of many experiences, all of which are irreducibly diverse or transient but which are not therefore relativized by either their diversity or their transience. The continuous course of thinking that Cohen describes in *Logic of Pure Knowledge* thus comes up against a more powerful continuity: an "ultra-continuity" that certain mathematicians of the period proposed in conjunction with the enigma of Cantor's so-called "continuum problem."[29] In Hölderlin's terms, drawn from "The Infinite," experience can be called "infinitely (precisely) connecting" (1: 26).[30] In Benjamin's terms, experience is "thi(n)ck"—so thick with thinking that it cannot be captured by any one thought, nor even by the system of philosophy as a whole. For this reason, the concept of experience rests on that of knowledge but is not reducible to it.

The Platonic Counterpart to the Kantian Typic

Beyond the various sketches and fragments Benjamin prepared around 1918 in conjunction with the drafting of "On the Program of the Coming Philosophy," he wrote a further supplement to the program that was published a decade later as the preface to the *Origin of the German Mourn-*

ing Play. From the perspective of the earlier versions, there is something decidedly odd about the preface: nowhere is the name of Kant mentioned. And only in two places does the preface gesture in the direction of neo-Kantianism: in the first word of its title, "epistemo-critical," and then again, in a brief reference to Cohen's *Logic of Pure Knowledge* that elucidates the first word of the work as a whole, namely *origin*. Otherwise, the "Epistemo-Critical Preface" is so free of Kant and neo-Kantianism that it could scarcely be considered a continuation of, much less a supplement to, the early program, which entrusts Kant with the task of guaranteeing the continuity of philosophy. A similar assertion, however, could in reverse apply to "On the Program of the Coming Philosophy," in that it almost completely declines to develop one of the claims with which it opens: "For Kant is the most recent of those philosophers for whom the primary concern is not immediately the extent or depth of knowledge but first and foremost its justification and, next to Plato, he is also really the only one" (2: 157). The accent doubtless falls on Kant; but the almost total absence of Plato from the program-related writings is as striking as the total absence of Kant from the "Epistemo-Critical Preface." It is as if Benjamin silently follows a rule of the following kind: a philosophical program can be developed from either the Kantian or the Platonic typic, but not from both at the same time.

With the replacement of the Kantian typic by its Platonic counterpart, a number of terms change; but the basic structure of divergent strata remains largely untouched. Instead of empirical consciousness, there are the phenomena, which must be "saved" (1: 214). And instead of analyzing the relation between the individual sciences and science itself, there is the relation between knowledge and truth. Rather than drawing on the set-theoretical term *power*, Benjamin retains the terminology of the "Supplement" and distinguishes between lower and higher modes of unity on the basis of terms drawn from the two modes of the infinitesimal calculus:

> Knowledge is questionable but not truth. . . . The unity of knowledge—if it were even to exist—would only be a mediated one, that is, on the basis of individual cognitions and, to a certain extent, through their equalization, a complex that can be produced, whereas the unity in the essence of truth is thoroughly unmediated and a direct determination. If, then, the integral unity in the essence of truth were questionable, then the question would have to be: to what extent is the answer already given in every conceivable answer that truth would answer to questions? And confronted with the answer to this

question, the same would again repeat itself, with the result that the unity of truth would escape from every question-posing. As unity in being and not as unity in concept, truth is beyond all questioning. (1: 209–10)

In the very next sentence of the "Epistemo-Critical Preface," the fundamental disadvantage of the Kantian typic in comparison to its Platonic counterpart becomes apparent. The first part of the sentence refers to former, the second to the latter: "Whereas the concept proceeds from the spontaneity of the understanding, ideas are given to contemplation" (2: 210). Nowhere in either the "letter" of the *Critiques* or in the "spirit" of epistemo-critique in general is there a place for what Paul Linke calls "the nonempirical given."[31] In the Platonic counterpart to the Kantian typic, by contrast, a place is reserved for the mode of pure receptivity called "contemplation." In the same vein, the distinction that Héring draws between "essence" (*Wesen*) and "essentiality" (*Wesenheit*) allows for a transformation of Husserlian *Wesensschau* (intuition of essence) into a "primordial hearing [Urvernehmen]" (1: 206) of essentialities, which are by nature isolated from one another and which enter into objective—not arbitrarily constructed—relations with one other in the form of ideas: "An essentiality that would have to determine itself as in harmony with others would thereby cease to be one" (1: 928). As for the response of Kant or Cohen to talk of nonempirical *Vernehmen* (hearing, perception), which retrieves the Platonic idea of *dianoia*, there is no question that both would insist on the discursive character of "reason" (*Vernunft*), which has nothing to do with *Vernehmen* beyond the fact that the latter term descends from the former.

Given the absence of a place for the "nonempirical given" within the Kantian typic, its replacement by a Platonic one must be strict and, as it were, unforgiving. Benjamin goes so far as to skip over Kant, as he briefly lists the "great philosophers," each of whose work retains its validity despite the fact that the scientific concepts from which it derives its categories have long since become obsolete: "Plato with the doctrine of ideas, Leibniz with the monadology, Hegel with the dialectic" (1: 212)—but not Kant with the critical system. In the draft of the preface, soon after the passage that corresponds to the previous quotation, Benjamin refers to Cohen's introduction to *Kant's Theory of Experience*, in conjunction with the following remark: "The more intensively the thinkers sought to project the image of reality, the more richly they were bound to develop a conceptual order that must appear to later interpreters as though it were at bottom meant as an original description of the world of ideas" (1: 930–31; cf. 1: 212). Not only does Benjamin delete this reference to *Kant's Theory of Experience* from the published work; he erases all reference to Kant's theory of experience and indeed to the very concept of experience. One suspects that Benjamin would also have wished to erase the last remaining reference to Cohen in the preface but ultimately decided that it had to be retained, for at this point—and this point alone—the Kantian typic must be invoked in the form of its negation. After distinguishing between the category of origin and the empirical concept of "emergence" (*Entstehung*), Benjamin briefly discusses Cohen's *Logic of Pure Knowledge* in order to forestall an inference authorized by the principles of epistemo-critique: if a concept is not empirical in origin, and if, in addition, it claims a cognitive function, it must be "logical." Benjamin responds by denying the premise of the inference and distancing his work from Cohen's: "the category of the origin is therefore not, as Cohen claims, a purely logical one but is, rather, historical" (1: 226).

As neither an empirical concept nor a *purely* logical category—which means that its historical character is logical as well—origin cannot be ascribed to a place within any system of epistemo-criticism. Yet in this instance, probably the only one in the entire "Epistemo-Critical Preface," Benjamin does not then seek to substitute a Platonic typic for a Kantian one by singling out a word from within the Platonic lexicon that would correspond to *origin*—and this despite the fact that such a word is readily identifiable: *khōra* (spacing, receptacle, matrix), which Plato famously introduces into *The Timaeus* (51d–52d) in order to identify a "third kind" (*titros genos*) that mediates between the eternal beings and transient phenomena. In the absence of a reference to the Platonic idea of logico-historical origin, the retention of a reference to Cohen stands out as the sole representative of the Kantian typic. Despite its brevity, then, the reference to the *Logic of Pure Knowledge* in the "Epistemo-Critical Preface" points in the same direction as the slightly more expansive critique of neo-Kantianism that takes shape in the previously quoted passage from the end of the "Supplement": "only in doctrine does philosophy bump up against an absolute, as existence, and therefore up against that continuity in the essence of experience whose neglect is presumably the lack of neo-Kantianism" (2: 170). In other words, origin is higher experience. In a more nuanced form, the ultra-continuity that characterizes a mode of experience whose *essence* is continuous can be ascribed to origin as well. Both higher experience and origin are, to use Benjamin's neologism once again, "thi(n)ck"—so saturated with thinking that they escape every thought, regardless of how systematically it may be ordered. Little wonder, then, that in the context of the origin Benjamin is drawn to the baroque *Trauerspiel*, whose "law of bombast [Gesetz des Schwulsts]"—formulated by Johann Christian Hallmann in terms of certain "Asiatic words" (1: 384)—makes it especially thick with thinking. It is not as though higher experience and origin are unthinkable, much less "ineffable." Rather, they "bump" thinking off its smoothly running course. The ultra-continuity of experience corresponds to the supersaturation of the origin, which Benjamin represents through the image of the "whirlpool" or "vortex" (*Strudel*): "The origin stands in the flux of becoming as a vortex and rips into its rhythm the material of emergence" (1: 226).

The vortex is not so much the interruption of a flux as its intensification: whereas the flow of a current is continuous, a vortex in the current represents a moment of ultra-continuity and appears, in turn, as an interruption of the current. The vortex thus acquires its historical character: the course of time, from past to future, gains independence from all "judgments of designation" (6: 10). Instead of deriving from the stipulations of meaning-bestowing subjects, *before* and *after* owe their origin to the thing itself, that is, to the origin, which is nonderivable and thus "original" for precisely this reason: "the origin does not lift itself beyond factual findings; rather, it pertains to their pre- and posthistory" (1: 226). To use the terms that Benjamin developed around 1916 in order to capture "the problem of historical time": to the extent that chronology begins with the beginning, the years are properly reckoned and can be called "historical." The starting point for the reckoning is not, as Heidegger proposes, an "historically significant event" (He, 432) but, rather, an origin, in other words, a higher experience, which generates its own pre- and posthistory, regardless of whether or not some regime makes the distinction into a convention with designations such as *BC* and an *AD*.

§7 The Political Counterpart to Pure Practical Reason

From Kant's "Doctrine of Right" to Benjamin's Category of Justice

A Borrowed Notebook

In the middle of a series of notes that Benjamin wrote in conjunction with his decision, later rescinded, to write his doctoral dissertation on Kant's concept of history, there is a passage in which he proposes a major revision of *Die Metaphysik der Sitten* (The metaphysics of morals). As it stands, Kant's last systematic treatise is divided into two parts, the first of which contains the *Metaphysische Anfängsgründe zur Rechtslehre* (Metaphysical starting grounds for the doctrine of right), while the second contains the corresponding the *Metaphysische Anfängsgründe zur Tugendlehre* (Metaphysical starting grounds for the doctrine of virtue). The *Doctrine of Right* is concerned with the virtue of justice under the condition that the term *justice* be applied only to circumstances of "right" or "law" (*Recht*), which, as he argues, is equivalent to the authorization to use coercive force. Of course, Kant is neither the first nor the last philosopher to treat justice as a special kind of virtue that in its restricted form as right or law must be treated apart from the others. In his revision of the *Kritik der praktischen Vernunf* (Critique of practical reason) under the title *Ethik des reinen Willens* (Ethics of pure will) Hermann Cohen directs his attention to the concept of justice only after having treated the other traditional virtues, such truthfulness, humility, bravery, and loyalty; but instead of presenting the virtue of justice as fundamentally distinct from the others, he identifies humanity as the outstanding virtue.[1] By contrast, Kant unambiguously distinguishes the concept of *Recht* from that of virtue: whereas right or law "concerns only the *formal* conditions of outer freedom," virtue has to do with the "*ends* of pure practical reason" (K, 6: 380). The revision of the *Metaphysics of Morals* that Benjamin proposes in a highly concentrated form calls for the replacement of a "doctrine of right" with the category of justice—which immediately prompts the question of whether there are any formal conditions under which coercive or "legal" force can be authorized:

> The above thoughts lead to the supposition that justice is not a virtue among other virtues (humility, love of one's neighbor, loyalty, bravery); rather, it grounds a new ethical category, which one cannot perhaps even call a category of virtue. Rather, it must be called a category that is of the same order as the other virtues. Justice does not seem to refer to the good will of a subject but, rather, constitutes a state [Zustand] of the world. Justice designates the ethical category of the existent [des Existenten], virtue the ethical category of the demanded [des Geforderten]. Virtue can be demanded; justice in the end can only exist—as a state of the world or as a state of God. (S, 1: 401)

This remark cannot be found among the seven volumes of Benjamin's collected writings; it first appeared in Gershom Scholem's *Diaries* under the following title: "From a Notebook Walter Benjamin Lent Me, 'Notes Toward a Work on the Category of Justice.'"[2] For some reason, Scholem did not detach these "Notes" from his *Diaries* when he gathered together his copies of Benjamin's papers for inclusion among both the two-volume collection of Benjamin's writings he co-edited with Adorno and the seven-volume collection edited by Tiedemann and Schweppenhäuser. Nor did Scholem mention the "Notes" in his *Walter Benjamin: The Story of a Friendship*. The reason cannot be that he simply jotted them down in October 1916 and then thought no more about them. At the conclusion to the "Notes"—perhaps as Scholem's addition—there are some striking remarks on "the problem of historical time," including one that began their afternoon-long conversation in August 1916, which Scholem discusses at length in his *Story of a Friendship*.[3] He also produced a roughly contemporaneous response to the "Notes" in which he arranged his remarks into two columns: under A he commented on Benjamin's remarks, while under B he proposed his own theory of justice.[4] Among the possible reasons for this puzzling situation, the following is probably the more plausible: despite the fact that Benjamin specifically distinguishes his theory of justice from those of socialists and communists, his "Notes" can be seen as a gesture in their direction. From the perspective of the elder Scholem, such a perception damages the image of Benjamin that he was in the process of constructing, according to which his conversion from "metaphysics" to "Marxism" in the 1920s was a disaster that befell him as a consequence of his decision to remain in Europe rather than emigrate to Palestine. If, in 1916, Benjamin was always intensely involved in a theory of justice that resonates with socialism and communism, his attraction to Marxist terminology a decade later can be seen as a development rather than a repudiation of his early "metaphysics."

Regardless of the reason that Benjamin's "Notes Toward a Work on the Category of Justice" did not enter into the editions of his writings overseen by Scholem, this much is clear: the opening note, which generates the "thoughts" quoted above, is a direct response to the opening paragraphs of the *Doctrine of Right* in which Kant claims that it is contrary to right to institute a law that would permanently bar a useful thing from being used. A law to this effect, according to Kant, amounts to the "practical annihilation" (K, 6: 250) of whatever is declared unusable. The exclusion of such a law is the first step in the "deduction"—itself a term drawn from German doctrines of property law—of the concept of legal possession, which then underlies the demand that everyone who enters into relation with others must enter into a civil state based on public law. Without concerning himself with the latter demand, Benjamin begins his notes by rescinding the postulate with which the *Doctrine of Right* begins: "To every good, limited as it is in the temporal and spatial order, there accrues a possession-character [Besitzcharakter] as an expression of its transience. But the possession, caught as it is in the same finitude, is always unjust. Therefore, no order of possession, however it may be articulated, leads to justice" (S, 1: 401).

In very broad terms, the difference between Kant and Benjamin can be stated as follows: in a bewildering series of arguments, the *Doctrine of Right* claims that useful things—that is, "goods" in the broadest sense of the term—are so susceptible to possession that they can be made into possessions even if there is no definitive "right" to do so; Benjamin, by contrast, declines to take this step. To paraphrase a saying of Kafka that Max Brod recorded for posterity and Benjamin quotes in the essay he wrote on the tenth anniversary of Kafka's death: there are possessions—"but not for us" (2: 414). Which means that a certain violence characterizes every actual possession. To the extent that all justifications of violence are at bottom mythic, the *Doctrine of Right* can be the starting point for a consideration of the relation of myth to law. Scholem inadvertently indicates as much in his *Diaries*. Immediately after his conversation with Benjamin in August 1916 about "the problem of historical time," he places the following note in his diary, which he then cites verbatim in his *Story of a Friendship*: "Benjamin's spirit revolves, and will for a long time continue to revolve around [the theme of] myth, which he regards from the greatest variety of angles. From the perspective of history, where he proceeds from romanticism; from that of poetry, where he proceeds from Hölderlin; from that of religion, where he proceeds from Judaism, and from law [Recht]" (S, 1: 391). In the last clause Scholem fails to complete his thought and identify the point of departure for Benjamin's inquiry into the concept of law or right. From the evidence of the notebook that Scholem borrowed, his remark about Benjamin's "spirit" should be supplemented as follows: "and from law, where he proceeds from Kant's *Doctrine of Right*."

Kant and the Metaphysical Problem of Possession

The purpose of the *Doctrine of Right* lies in developing the political counterpart to pure practical reason, in which "pure practical reason" is simply the technical term for what is generally called "morality." Whereas the *Critique of Practical Reason* draws a sharp line between rational actions that are grounded entirely in reason and those that are determined by nonrational factors such as pleasure and pain, a corresponding "doctrine of right" establishes the basic structure of relations among finite rational beings within which they can use themselves as well as things without thereby degrading themselves into the things they use. The "political" character of the doctrine emerges from the equivalence that distinguishes right from virtue: the former is, once again, equivalent to the authorization to use coercive force, whereas the latter has only the "internal" force that derives from the demand that pure reason imposes on rational agents by virtue of their rationality. In a small book Kant published in 1795 under the title of *Zum ewigen Frieden* (Toward eternal peace), he describes this "politics" as "true": "True politics [wahre Politik] cannot take a step without having already honored morality" (K, 8: 380).

Nevertheless, the critique of aggressive warfare that Kant undertakes in *Toward Eternal Peace* is solely concerned with the "ending grounds," as it were, of the "doctrine of right," namely the principles of international law.

Two years later, after many false starts and much delay, Kant finally published—with a number of lacunae and in an oddly mangled form—the "starting grounds" of the doctrine that represents the crowning achievement of his critical system. The order of publication is no accident, for it reflects the degree to which Kant was confident in the force of his argument. With respect to the inception of the doctrine in the "axiom" of right or the "basic principle" of externally enforceable law, he had no doubts, even if he could not quite determine its precise relation to the categorical imperative, whose form it reflects: "act outwardly in such a way that the free use of your elective will can cohere with the freedom of everyone according to universal laws" (K, 6: 231). And Kant likewise had no doubts that a "doctrine of right" should end with the principles of international law as they are outlined in *Toward Eternal Peace*. Furthermore, he knew that "public law" should take the form of the republican State, in which political powers are appropriately separated from one another; but he was altogether unclear as to how the "truly" political counterpart to pure practical reason could get started in the absence of already established civil constitutions. More exactly, he could not figure out how the axiom of right could be originally expanded beyond the sphere of its immediate applicability—a sphere that is exceedingly small and fundamentally discontinuous. Without a supplementary principle, the axiom of right only prohibits every rational being from inhibiting every other rational being's range of action without consent, except in cases where the inhibition directly counters a current violation of the axiom. The sphere in question can be described as the region of the human body with a plus and a minus: plus whatever the body happens to hold; minus its "sexual properties," which, for Kant, are not one's own to use as one chooses, insofar as the use of these properties tends to degrade the personhood of their user.[5]

Unless the sphere can be expanded beyond the limits of the body, roughly speaking, not only is there no basis for the State, there is no right to move from one place to another, since in doing so one uses space that is not one's own to use. In the absence of a supplementary principle of right that would justify the expansion of its sphere, there thus arises the threat of juridical paralysis, in which everyone is free to do anything that does not interfere with everyone else; but since almost everything anyone does potentially interferes with someone else, everyone is actually permitted to do almost nothing. For Kant, who for some enigmatic reason confined himself to the space around Königsberg for his entire life, the paralytic situation is intolerable. The juridical body has to be able to expand beyond its physical shape, and the form of this expansion lies in the concept of possession—but not, of course, in that of merely physical possession but in its metaphysical counterpart. In the dozens and dozens of drafts of a solution that have been collected into volume 23 of his complete writings, Kant tested a series of adjectives that would distinguish physical possession from its juridical counterpart, including *intentional* (K, 23: 282), *virtual* (K, 23: 287), and even *potestival* (K, 23: 311). In the absence of a "starting ground" beyond that of the original axiom of right, the only power will be physical might—which not only means that there is no justice in the world but also that things and persons are ultimately peers: the moment a thing is no longer held in physical possession, it returns to its original juridical condition as *res nullius*. The expansion of the person beyond the limits of its body in the concept of nonphysical possession thus becomes the guarantee of "personal" integrity; similarly, the expansion of its time beyond that of its life, such that its will extends beyond its "elective will." To use a lexicon closer to the one that Benjamin adopts for "Notes Toward a Work on the Category of Justice," a possession-character accrues to things because of their spatio-temporal finitude, whereas those who would possess them correspondingly confirm their nonfinite character.

The impasse that Kant encountered in expanding the concept of right beyond the sphere of its immediate applicability can be described in a few brief steps. The juridical body must be stretched, so that the use of the term *mine* as opposed to *yours* is no longer determined by the contingent shape of the body along with whatever it happens to hold in its prehensile apparatuses. It is clear, though, that no one can "see" the enlarged body, for it is not a phenomenon in the Kantian sense. Similarly, but from another perspective, there would be no impasse on the way to the crowning principles of international law if there were no history of reason; that is, if everything in the world were always already divided into "mine" and "yours," with the result that "mine" can become "yours" by means of a bilateral agreement. The impasse thus materializes only in the beginning—when things necessarily lie in the contingent condition of being *res nullius*. At this point a principle must be established so that, without an explicit agreement of any kind, usable things can be used in such a way their use is not limited by physical power. But there simply is no such principle. Of course, there are nonprincipals: rules that can be applied to given situations based on pragmatic reasoning; but there is no universally valid principle that can be deduced from the axiom of right. After seeking for years to discover a principle, Kant gives up and begins the *Doctrine of Right* by turning the missing principle into a demand, that is, into a "postulate of practical reason related to right," which in its published form reads as follows: "It is possible for me to have *any* external object of my elective will as mine; that is, a maxim which states that an object of the elective will would *in itself* (objectively) have to be without a lord (*res nullius*) is contrary to right, if it were to become law" (K, 6: 246).

In some of the drafts of the *Doctrine of Right* Kant emphasizes that the missing principle must take the form of an "apagogic" proof, that is, a so-called *reductio ad absurdum*. The explication of the postulate that Kant appends to the published version is doubtless apagogic; but he does not actually use this term, fearing perhaps that it would compromise its necessity. A *reductio ad absurdum* is valid only if—to continue with the Latin terminology—*tertium non datur*. In other words, if it can be shown that *p* is "absurd," then one can draw the conclusion "not-*p*" only if *p* and not-*p* are the only alternatives. In the case under consideration, "not-*p*" would be a situation in which the law says of some good that it can never be used and is therefore "practically annihilated" (K, 6: 250). The absurdity of this situation, for Kant, implies "*p*": everything is potentially a matter of possession, in which the right to use something knows no prescribed bounds. If, however, there is a third term, the *reductio* no longer leads to an absurdity: there is an alternative to "*p*" and "not-*p*" that would consist in a condition where things can be legally possessed, but it would be unjust for anyone to possess anything in particular. An historically sanctioned third term is "the sacred." And this is precisely what divine law in the Roman legal tradition does, as Kant would doubtless have known: it takes "commercial" things out of commerce, including, for instance, any plot of land under which a legally buried corpse lies.[6] The postulate with which the *Doctrine of Right* begins says, in effect, that divine law runs counter to right. Without alluding to the idea of divine law, Kant's attempt to demonstrate that the basis of the postulate nevertheless indicates the precise point at which *reductio ad absurdum* fails:

> An object of my elective will is that which I have the physical capacity [Vermögen] to use as I please, that whose use lies in my power [Macht] (*potentia*); this must be distinguished from having the same object under my

control [in meiner Gewalt] (*in potestam mean redactum*), which presupposes not simply a capacity but also an *act* of the elective will. But in order to *think* of something simply as an object of my elective will it is sufficient for me to be conscious of having it within my power. It is therefore a representation a priori of practical reason to regard and treat any object of my elective will as an objectively possible mine or yours. (K, 6: 246)

Something remarkable happens in the course of Kant's justification of the postulate through which the sphere of right begins to expand: any object in the world can be treated as mine—or yours. Which means that there is no basis on which it can be treated as mine alone or yours alone. But this is precisely what the postulate is charged with doing: not so much providing the actual principle for the division of things as showing that the division can be principled and does not ultimately amount to the "law" that anyone with sufficient power gets whatever he or she wants in the absence of actual resistance. The weight of the argument then falls on a term that is ambiguous in the extreme: *Gewalt*, which Kant translates into Latin as *potestas* but elsewhere in the treatise translates as *violentia* (K, 6: 307). If ever a fourth critique were needed, it would be here, for only a critique of *Gewalt* can distinguish between legitimate *potestas* and illegitimate *violentia*. Perhaps because Kant recognized that pure *Gewalt* could not be identified with *potestas*, he refrains from undertaking a critique of *Gewalt* along these lines and barely even acknowledges the ambiguity of the term. Instead, he seeks to uncover a degree of minimally justified "control" (*Gewalt*, *potestas*) within the phenomenon of "might" (*Macht*, *potentia*) by arguing along the following lines: something can be mine only if it is first in my physical power and thus contingent on my might; whenever something comes under my "control," however, it remains mine even if it is no longer in my physical possession, since "control," unlike "might," includes a rational element, namely an "act of the elective will." The distinction between *Gewalt* and *Macht* allows Kant to show how a physical condition turns into a metaphysical one: might makes right whenever might is minimally rational. In other words, insofar as might is rational, it is always somehow right. However the argument is framed, its point is clear: unless might is originally deemed to be right, there can be no expansion of right beyond the confines of the body, roughly speaking.

Thus, in the very next paragraph of the *Doctrine of Right*—as an out-of-place remark that was soon recognized as such—Kant almost admits to failure.[7] He does not announce in so many words that no one can make something his or her own simply as a consequence of a unilateral "act of the elective will." Rather, he takes up an ambiguous term he had introduced into the final section of *Toward Eternal Peace*, namely "permissive law," which suggests that the expansion of the sphere of right is contrary to justice, even if it is the foundation of all later law:

> One can name [nennen] this postulate a permissive principle (*lex permissiva*) of practical reason, which gives us authorization that could not be obtained from the mere concept of right as such, namely [nämlich] authorization to put all others under an obligation, which they would not otherwise have, to refrain from using certain objects of our elective will because we have *first* taken [genommen] them into possession. Reason wills that this holds as a principle, and it does this as *practical* reason, which extends itself a priori by this postulate of reason. (K, 6: 247)

From *nennen* (name) to *nämlich* (namely) to *in unseren Besitz genommen* (taken in possession)—this is the nominal path through which *Vernunft* (reason) begins to expand the legal sphere across the surface of the globe. The naming of the postulate makes it into a law that permits the "first taker" to make the thing taken into his or her own possession. Nevertheless, the naming is as wrong as the newly named law is unjust.

The potential paralysis that Kant encountered while drafting the *Doctrine of Right* would never have materialized if he had not rejected a supplementary principle of right that he had earlier accepted as valid: the so-called labor theory of property, often associated with John Locke, according to which the "mixture" of my labor with an unclaimed thing in a state of nature gives me title to it.[8] Not only does the *Doctrine of Right* repudiate this principle, it identifies its origin in the trope of personification:

> The first working, enclosing, or, in general, transforming of a piece of land can furnish no title of acquisition to it. . . . This is so clear in itself that it is hard to assign any other cause for this opinion, which is so old and still widely dominant than this: there is a secret deception that consists in personifying things and of thinking of a right to things as being a right immediately against them, as if someone could, by the work he expends on them, put things under an obligation to serve him and no one else; for otherwise one would not have passed with such a light foot over the natural question . . . "how is a right to a thing possible?" (K, 6: 268–69)

The mythic character of the personification is undeniable: those who wish to acquire an unclaimed thing imagine that it possesses a "guardian spirit [bewahrenden Genius]" (K, 6: 260), with which a bilateral agreement can be negotiated. All of this would be fine, so to speak, if Kant did not make use of the trope he so vehemently rejects—and thus proves by way of his inconsistency that *Gewalt* replaces labor as the supplement to the axiom of right. In response to the question as to how far one is allowed to "take possession of a parcel of land," Kant answers: "As far as the capacity to place it in one's charge [Gewalt], that is, as far as whoever wants to appropriate it can defend it; as if the piece of land itself were to say, 'if you cannot protect me, you cannot command me'" (K, 6: 265). Each plot of usable land, it seems, does have a genius loci with which one can enter into a feudal agreement. The contradiction between the two passages cannot be ascribed to Kant's absent-mindedness or supposed senility; it silently attests to the wrong with which the *Doctrine of Right* begins. As it takes over the function of labor in the justification of original acquisition, *Gewalt* acquires a mythic character: things are given voices as vassals, who always say, "I am yours as long as you protect me from powers of your own kind."

What Heidegger says of Kant with regard to the difference between the first and second versions of the *Critique of Pure Reason*, especially in light of its doctrine of schematism, can be applied, with less violence, to the opening sections of the *Doctrine of Right*: Kant shrank from the abyss uncovered by his own inquiry into the starting grounds of metaphysics.[9] In the case of the *Doctrine of Right*, the abyss is identified with the legal vacuum of "anarchy" or "lawlessness" (*Rechtlosigkeit*). A consistent, unambiguous doctrine would not replace labor with *Gewalt* as the source of an originally expanding sphere of right. Nor would it convert the absence of an additional principle of right into a demand that there be one. Still less would it allow such a postulate to be interpreted as a presumptive "right" to take whatever one comes under one's *Gewalt*, if only the thing is not already under someone else's. Instead, it would assert what it discovers: useable things can be possessed, but no one can possess them. This is precisely what Benjamin sets down as the first of his "Notes Toward a Work on the Category of Justice": "To every good, limited as it is in the temporal and spatial order, there accrues a possession-character as an expression of its transience." With remarkable concision he adds: "But the possession, caught as it is in the same finitude, is always unjust." Kant, by contrast, proposes a *lex permissiva* that permits a wrong to be right for a time. It is only because Kant holds to the conviction, as expressed in his writings on history, that an original injustice can be gradually alleviated by means of ever-increasing conformity of enforcable law to the idea of right, that he fails to draw the same conclusion as Benjamin: "Therefore, no order of possession, however it may be articulated, leads to justice" (S, 1: 401).

"The Good-Right of Goods"

In some of his most extensive drafts for the opening paragraphs of the *Doctrine of Right* Kant seeks a way out of the impasse by invoking the doctrine of schematism, as it was first developed in the *Critique of Pure Reason*. The schema of a concept is the rule by which something can be constructed in intuition. If the basic concept of right could function as a category of the understanding, then the supplementary principle for the application of the axiom of right beyond the sphere of the body could be determined on the basis of the following analogy: just as the categories apply to intuitions through their corresponding schemata, so would the axiom of right apply to phenomena by virtue of its schematization of "mine" and "yours" into a distinguishable object of the elective will. Eventually, however, Kant drops the argument. Just as there is only a "typic" of pure practical reason, not a schema, so there is no schematization of the axiom of right: "The concept of right as a rational concept is entirely incapable of a schema" (K, 23: 325). At the end of "Notes Toward a Work on the Category of Justice," Benjamin draws a simple conclusion from the nonschematizable character of both right and morality: in the case of the moral law the absence of a schema protects its purity; in the apparently corresponding case of right, by contrast, the absence of a schema indicates that it is shielded from justice: "The empirical individual act is related to the moral law somehow as a (nondeductible) fulfillment of the formal schema. In reverse, right or law relates to justice as the schema to the fulfillment. Other languages have designated the enormous gulf that divides the essence of right from that of justice: *ius, fas; themis, dikē; mishpat, tsedek*" (S, 1: 402).[10] As recorded by Scholem, the "Notes for the Work on the Category of Justice" then segue into the remarks on chronology with which their conversation in August 1916 began, and soon thereafter Scholem's own reflections resumes.

Despite the fragmentary character of Benjamin's "Notes," they outline a systematic contribution to the critique of reason. Just as Kant at the beginning of the introduction to the *Critique of Judgment* discusses the "immense gulf" (K, 5: 175) that separates the domain of knowledge from that of freedom, so Benjamin presents his work on the category of justice as a reflection on the "enormous gulf" separating the sphere of law from the state of justice. And whereas the *Critique of Judgment* attempts to bridge the gulf in the form of pure aesthetic judgment, the "Notes" go in the opposite direction, accepting the unbridgeable character of the impasse thus uncovered. The way to justice is not entirely blocked, however: "Rather," the third of Benjamin's "Notes" reads, "this [the passage from right to justice] lies in the condition of a good that cannot be a possession. This is alone the good through which goods become possessionless [besitzlos]" (S, 1: 401). Benjamin says nothing further about the enigmatic good to which no possession-character accrues. He may have initially developed the idea of writing his dissertation on Kant's concept of history under the assumption that it would be intimated in this context. His expression of disappointment with the relevant essays indicates as much: "In Kant['s writing on history], it is less a matter of history than certain historical constellations of ethical interest. And in addition, it is precisely the ethical side of history that is set up as inaccessible for specific consideration, and the postulate of a scientific mode of observation and method is posited (introduction to the 'Idea of History')" (*GB*, 1: 408). From this brief remark—which does an injustice to the complexity of Kant's writings on history, beginning with his "Idea for a Universal History from a Cosmopolitan Perspective-Intention [Absicht]"—at least one source of Benjamin's disappointment can be easily discerned: instead of making his reflections on history into a site where the metaphysics of right gives way to another metaphysics, he proposes a mechanistic "postulate" that is designed to disengage the idea of history from that of justice. The category of justice that Benjamin begins to outline in response comes down to a deceptively simple question: what is the name of the enigmatic good that situates itself in the "enormous gulf" that separates the sphere of law from the state of justice and thus opens a path to the latter?

Instead of directly answering this question, Benjamin identifies a basic error that would prevent it from being posed: "In the concept of society one tries to assign a possessor to the good that cancels its possession-character" (S, 1: 401). Any theory of justice based on the proposition that society takes possession of whatever individuals are barred from appropriating fails to pose the question of whether there is, after all, a good to which no possession-character accrues: "Every socialist or communist theory," Benjamin continues, "misses its goal for the following reason: because the claim of the individual extends to every good. If individual A has a need z that can be satisfied with good x, and one therefore believes that good y, which is like x, may and should and be given, for the purpose of justice, to individual B in order to assuage the same need, then one errs. For there is the entirely abstract claim of the subject, in principle, to every good—a claim that in no way refers back to needs but, rather, refers to justice, the ultimate direction of which probably does not tend toward a possession-right of the person but, rather, toward a good-right of the good" (S, 1: 401). To the abstractness of the postulate Kant proposes at the beginning of the *Doctrine of Right*, Benjamin responds with similar abstractness: everyone has a claim on everything, which means at the very least that there can be no "permissive law" of practical reason granting a right to the "first-taker" that would reciprocally obligate everyone else to desist from using the good in question. And the absence of the so-called "right of the first-taker"—to say nothing of the absence of the equally venerable "right" of the laborer to the fruits of his or her labor—forms the matrix of Benjamin's attempt to determine the category of justice.

Every individual has a claim on everything that he or she deems good for whatever reason. There are neither spatial nor temporal qualifications on the rightfulness of this claim: the goods of the future are as claimable as those of the present, even if, as Benjamin will write many years later, human beings demonstrate a remarkable lack of "envy" with respect to future generations (1: 693). For this reason, however, it is immediately evident that the demands of justice not only cannot be fulfilled; the category of justice cannot be expressed in terms of demands at all. The corresponding demands of morality, as Kant emphasizes, may never have been fulfilled; but the absence of any actual example of genuinely moral action says nothing against the categorical character of the demand. By contrast, the demand-character of justice cannot even be formulated as an imperative. It is not simply that there is a scarcity of goods, which makes it impossible to distribute them on the basis of a neutral principle; rather, there is a fundamental mismatch between the number of claims and the number of goods. The former is of a higher power than the latter, for even if there were an infinite number of goods—as if paradise were a matter of material plenitude—the claims would "outnumber" goods, since no particular claim carries greater weight than any other, such that it would tilt the scale in favor of one potential possessor over another. In discussing "need," Benjamin specifically considers one mode of weighing claims. To say of a good that it derives from a need instead of a desire means that it can be satisfied by any arbitrary member of a class of objects—food or water, for instance. In this way, the concept of need not only does not contribute to the construction of the category of justice; it makes it impossible to pose the question as to the good that cannot under any condition be possessed. All of this leads Benjamin to a philosophical-juridical *novum*, which directs attention away from the character of the subject and toward that of object made into a good by virtue of a claim made upon it: justice consists in the "good-right of the good" as opposed to the "possession-right of the person" (S, 1: 401).

Apropos the "possession-right of the person," the *Doctrine of Right* is uncharacteristically clear: the term *possession-right* means nothing other than nonphysical possession. The fact that a person has a right to possess something simply means that the person possesses it, regardless of who happens to hold physical power over it. The idea of a "good-right of the good" can be understood accordingly: it is the right of the good to be what it is, namely desirable, regardless of who has it. As with every other right, this one corresponds to an obligation: goods must not be turned into something else—for example, into evils, or, in reverse, into "sacreds," if a neologism can be introduced here. The "good-right of the good" demands that nothing be holy and nothing impure. In terms of Jewish law, things are to be neither kosher nor *treyf* (unkosher). None of this fits very well with the canons of either ancient or rabbinical forms of Judaism, of course, especially when it is brought into relation with the quotation with which Scholem had concluded his attempt to summarize the "spirit" of his friend. At the point where he should have determined the point of departure for Benjamin's reflections on law, he enters the following quotation into his diary, which he retrieves for his *Story of a Friendship*: "'When I finally have my philosophy,' Benjamin says, [according to Scholem] 'it will somehow be a philosophy of Judaism'" (S, 1: 391). When, however, Benjamin cites the Bible in his "Notes Toward a Work on the Category of Justice," the text he chooses does not come from the Hebrew Bible but, rather, from the Sermon on the Mount. Under the heading of "the Lord's Prayer," Benjamin writes the following: "'lead us not into temptation but deliver us from evil' (*several words unreadable*) is the prayer for justice, for the just state of the world" (S, 1: 402; cf. Matt. 6:13). The degree to which this note departs from the spirit of Scholem's *Diaries* can be measured by the fact that its editors, who are otherwise highly meticulous in tracking down every reference and allusion, fail to identify the passage Benjamin discusses, as if the source of the Lord's Prayer somehow exceeded the resources of historical philology. Equally strange is the inclusion of the editorial remark that apparently stems from Scholem, "(*several words unreadable*)," suggesting that the same thing happens to Benjamin's "Notes" that occurred in the case of Kant's *Doctrine of Right*: a breakdown of communication for which no single agent can be held responsible. Fortunately, however, in the case of Benjamin's "Notes," the collapse makes little difference, for the exegesis of the Lord's Prayer is clear enough: it asserts the "good-right of goods" by asking of the Lord that goods not be converted into evils.

In thus sketching out the path to justice, Benjamin formulates a practical tautology: goods are to be good. In the language of rights, despite its unfitness for matters of justice, the tautology becomes the good-right of the good, which every order of possession violates by establishing the conditions under which goods become evils: whenever someone makes a good into his or her possession, it turns into an evil for everyone else. This is obviously true in the case of capitalism, about which, perhaps for this reason, Benjamin says nothing in the "Notes." Socialism and communism, by contrast, may undo the mechanisms by which goods are converted into evils under capitalist conditions of production and distribution; but they are prone to violate the good-right of the good as well: any good that does not satisfy a need is in danger of becoming a temptation that everyone is under a social or communal obligation to renounce. Let goods be goods: this, in short, is the postulate that Benjamin proposes as a replacement for the one with which the *Doctrine of Right* begins. But it is a very peculiar postulate, for goods are already goods—which suggests that justice is a manner of being and should thus be understood as an ontological category. Far from upholding the status quo as satisfying the demands of justice, the suggestion runs counter to any stance or statute that would determine the point where goods are evils. And precisely "now" is the time in which this is done, since justice does not consist in demands that are yet to be fulfilled. Even now, no claim on any good, whatever its provenance, is to be renounced; no order of possession, resting as it does on the renunciation of claims, is to be conceded; no compromise is to be made for the purpose of postponing conflicts over goods. But—and here again is the impasse—none of these "no's" can take the form of a law: for instance, a law that would enforce the right of goods to be goods and establish a communal kingdom of God on earth. This goes for the postulate "let the good be good" as well: it cannot even be formulated as a postulate. Rather, the "no" must take the form of a thing, specifically the good to which no possession-character accrues and which makes all other goods possessionless as a result.

Around the Tree of Knowledge

Scholem copied "Notes Toward a Work on the Category of Justice" into his *Diaries* at the very same time as Benjamin was completing "On Language as Such and on Human Language" as a consequence of finding himself unable to complete a long letter on the "infinitely difficult theme of language and mathematics" (*GB*, 1: 343). With the exception of the final remarks on terms for law and justice in German, Latin, Greek, and Hebrew, the "Notes" make no reference to language. Conversely, with the exception of a brief paragraph at a transitional point of its exposition—as the exegesis of the Book of Genesis comes to a conclusion and an independent reflection on the sadness of being "overnamed" begins—the treatise makes no allusion to the problem of possession. Yet Benjamin may have been drawn to the biblical text for precisely this reason: in the story of the fall from paradise the two lines of a systematic work on language and justice intersect. And the point of intersection can be identified with precision: the Tree of the Knowledge of Good and Evil. If ever there was a *res nullius* in what Kant calls its "objective sense," it lies in the fruit of this tree, which no one is permitted to use under any condition. The conflict between the beginning of the Hebrew Bible and the opening of the *Doctrine of Right* could not be sharper: the former begins with a divine law that "practically annihilates" the fruit of a tree, whereas the other declares that precisely this kind of law is contrary to right. It is little wonder in this context that Kant chooses the apple as his example of a graspable good (K, 6: 250). If Kant takes the side of law, the Book of Genesis is presumably on the side of justice, and the task that Benjamin establishes for himself in the fall of 1916 can be formulated accordingly: determine what happens when the good to which no possession-character accrues nevertheless acquires a possessor.

According to Benjamin's exegesis of the relevant biblical passage, the violation of the divine law against eating from the Tree of Knowledge gives rise to three forms of fallenness: the judging word is "excited" (2: 154), human language falls into an "abyss of chatter," and things enter into a condition of servitude. The first two consequences of the fall are directly correlated with the origin of "law" (*Recht*) in myth: "The Tree of Knowledge does not stand in God's garden because of the information about good and evil that it could deliver but, rather, as an emblem of the judgment about the questioner. This enormous irony is the hallmark of the mythic origin of the law" (2: 154). Since everything that has been created is good, whoever raises a question as to whether something happens to be good or not must be evil. The order of law thus arises as the condemnation of whomever, by evaluating a thing, incites evaluation. Yet Adam and Eve do not simply seek insight into the difference between good and evil; they are also in the paradigmatic condition of the "first takers." In discussing the third consequence of the fall, in which the theory of language converges with the theory of justice, Benjamin opens a new and uncharacteristically brief paragraph that superimposes the story of the Tower of Babel onto that of the expulsion from the Garden of Eden, perhaps with the expectation that an account of the servile status into which things fall will eventually emerge. As builder of the Tower, Nimrod is traditionally represented as the originator of property law, since the storming of the heavens is predicated on a division of the earth: "The enslavement of things [Verknechtung der Dinge] in foolishness follows the enslavement of language in chatter almost as an unavoidable consequence. In this turning away [from the original intuition] of things, which was the enslavement, there arose the plan for the Tower [of Babel] and, with it, the confusion of tongues" (2: 154).[11]

Nothing is perhaps more telling in Benjamin's formulation of the origin of property than the "almost" with which he qualifies "unavoidable consequence." It is as though the story of the expulsion and that of the Tower are *almost* combined into one. In this way, Benjamin could remain in a state of uncertainty about two questions: whether or not the original language of human beings was singular or plural, on the one hand, and whether or not the "enslavement of things into foolishness" could be described in linguistic terms, on the other. Things presumably become foolish because, like all fools, they permit themselves to be used as others see fit. Just as language succumbs to chatter when it allows itself to be used, so do things fall prey to foolishness. Instead of inquiring further into the character of thingly foolishness, however, Benjamin cuts short his reflections on the Tower and begins the treatise anew with some reflections on the all-pervasive sadness of things. Only the "fallen" character of "law" (*Recht*) is thus clarified. The restitution of justice to its original stance, by contrast, remains opaque.

Justice as Stance, Striving, and Power

In November 1916, around the time Scholem was completing the first draft of his "mathematical theory of truth," Kafka traveled to Munich to participate in a literary event where he read parts of his story "In der Stafkolonie" (In the penal colony). At the culmination of the story, the officer who explains, administers, and enforces justice in the penal colony adjusts the apparatus of justice, so that it will inscribe the imperative "be just" into his own flesh, whereupon the apparatus goes berserk and "murders" him.[12] As it happens, Benjamin was in Munich at the time of the reading but apparently missed the chance to encounter Kafka—perhaps because he was in the process of extending the following argument, which Scholem copied into his diary during the previous month:

> Justice designates the ethical category of the existent, virtue the ethical category of the demanded. Virtue can be demanded; justice in the final analysis can only be as a state [Zustand] of the world or as a stance [Zustand] of God. In God all virtues have the form of justice; the epithet *omni* in omnigracious [all-gütig], omniscient and so forth points in this direction. Only a fulfillment of what is demanded can be virtuous; only a guarantee [Gewährleistung] of the existent (*no longer perhaps* determining through demands, nevertheless in any case not any arbitrary one) can be just. (S, 1: 401–2)

In response to the question as to whether any moral action has ever been accomplished, Kant has a ready-made answer: it does not matter, for the fulfillment of the moral law is demanded in any case. Because justice cannot be made into an imperative without thereby collapsing into law, a similar response cannot be formulated. A similarly tautological answer can, however, emerge: justice adds nothing to the existent beyond the guarantee of its existence, which does not protect the existent from other existents, as a legal order would claim to do; rather, as an "instance" or higher "court" (*Instanz*), it guarantees that the existent will be what it is—"ex-istent"—and thus that which stands outside of itself: "Responsibility for the world we have is protected from the instance of justice" (S, 1: 402).

This amounts to an ethic of irresponsibility only if it is assumed that the world we "have" is already, without striving, the highest good. In the absence of this assumption, the direction of responsibility is misplaced. Instead of responsibility toward the world, there is striving to make the world, more exactly, "the striving to make the world into the highest good"—which Benjamin then defines as justice (S, 1: 401). According to Kant, any discussion of the highest good, which he represents as an exact proportionality between happiness and virtue, gives rise to an antinomy: it is necessary yet impossible. Unless the antinomy can be resolved, however, the moral law "must be fantastic and directed toward made-up ends and must therefore in itself be false" (K, 5: 114). "True politics"—to say nothing of the critical system as a whole—requires that the moral law not be false. Kant's solution to the antinomy lies in the concept of "self-contentment" (*Selbstzufriedenheit*), which is to say, in the state of being at "peace" (*Frieden*) solely with oneself: "Now it can be understood: how consciousness of this capacity of a pure practical reason through an act (virtue) can produce a consciousness of a superior power [Obermacht] over one's inclinations, thus an independence from them, consequently also a contentment that always accompanies them, and therefore a negative delight with one's own state [Zustand], that is, contentment, which in its quality is contentment with one's person" (K, 5: 119). Among the many questions raised by the proposed resolution of the antinomy, the most pressing is perhaps the following: under what conditions can a distinction be made between the phenomenon of self-contentment and that of self-complaisance, from which Kant later infers the radically evil character of humanity at large.[13] Without explicitly posing a question of this kind, Benjamin replaces self-contentment with world-striving. Only the striving to make the world into the highest good is striving in the proper sense of the word: every other so-called striving separates the goal from the act and, at best, expresses the separation in terms of the demand-fulfillment structure of virtue. Striving, by contrast, makes the world into the highest by virtue of itself: whenever there is striving, the world is ipso facto the highest good.

In a diary entry from November 1916, Scholem makes the following comment on Benjamin's concept of striving: "The essence of the Jewish notion of justice as the 'striving to make the world into the highest good,' as Benjamin writes, reveals itself very deeply in the *very* untranslatable words of the sages, which [Immanuel] Hirsch cites: *tzaddikim yachinu shechinah ba-aretz*, and which he beautifully translates, so that a *part* of the meaning, a nuance is reproduced, and it sounds as though Benjamin was familiar with it: 'Justice prepares the earth to be a site of the divine.' That is justice—to make the earth into the seat of the Shechinah, to draw *downward* the Shechinah" (S, 1: 419). It seems as though Scholem makes a similar suggestion to Benjamin, who then responds in the letter in which he announces the near completion of his "little treatise" on language and criticizes Heidegger's "Concept of Time in Historical Scholarship": "What does Shechinah mean?" (*GB*, 1: 344). The fact that Benjamin would ask such a question indicates the degree to which his point of departure for a work on justice differs from Scholem's. And in Scholem's point-by-point response to Benjamin's "Notes" the difference becomes visible. Under column B Scholem develops his own theory of justice as "the historical annihilation of divine judgment," whereas under column A he comments on Benjamin's "Notes"—but only to a certain extent.[14] In the last of the five theses under column A, Scholem criticizes Benjamin's definition of the highest good: "In this sense, from the side of ethics, defining justice as the comportment [Verhalten] that makes the world into the highest good is *thoroughly* unsatisfactory."[15] Benjamin thus receives a failing grade, perhaps because he did not know what *Shechinah* means—or perhaps because the concept of striving has been construed as a certain "comportment." But in any case, the insufficiency of the definition of justice, which Scholem associates with the untranslatable words of the sages, gives him a chance to begin anew under column B with the declaration that "the concept of justice gains its true context in the philosophy of religion."[16]

Immediately before Scholem deems Benjamin's definition of justice insufficient from the side of ethics, he comments on the most enigmatic of the notes he copies into his *Diaries*: "Justice is the ethical side of the battle; justice is the power of virtue [Tugend der Macht] and the virtue of power. Responsibility for the world we have is protected from the instance of justice" (S, 1: 402). In Scholem's corresponding note, the term *Macht* (power) is replaced by *Gewalt*, and Benjamin's assertion now appears as a response to a question that emerges in conjunction with the basic thesis of the "Notes," namely that justice, unlike the other virtues, does not consist in a demand. What happens, then, if justice is made into a demand? Kafka's "In the Penal Colony" answers this question with a single word: "murder." The answer that appears in the last of the theses Scholem places under column A is *Gewalt*:

> Justice as demand is the virtue of *Gewalt*. It is the most revolutionary and most catastrophic of all demands. For virtue has an individual bearer; the humble one stands in an unambiguous, uncomplicated relation to humility. The bearer of *Gewalt*, by contrast—which is a much deeper phenomenon than virtue—is an individual only symbolically, since the actual, nonsymbolic holder of *Gewalt* is anonymous: society. The demand that is necessarily directed toward the holder of *Gewalt* therefore requires in justice a virtue that has no bearer in the sense of ethics. No virtue accrues to symbolic figures. Thus, this demand—and it is a sublime irony—can be implemented only in a fundamental catastrophe of *Gewalt*. Revolutionary politics determines itself in it, not in the nonironic context of religion.[17]

It is impossible to say whether Scholem's thesis is a commentary on the note in which Benjamin calls justice the "virtue of power," whether it seeks to capture an independent remark made elsewhere, or whether it is Scholem's interpolation, which allows for a smooth transition to column B. Given Benjamin's previous remarks about the unfitness of "society" as the possessor in the last instance, it is unlikely that he would have proposed that it is the actual "holder of *Gewalt*." Nevertheless, Scholem's thesis suggests a lacuna in Benjamin's attempt to capture the category of justice: the absence of its critical counterpart. To say of justice that it is the "virtue of power" is to give the impression that any self-proclaimed power is, by definition, virtuous; and to say of justice that it is "power of virtue" is to suggest that all of the virtues are impotent without a supplementary power—which may be true and even accords, to a certain extent, with Kant's conception of virtue, but it scarcely goes without saying. The ease with which the formula "virtue of power" can be reversed lends credence to the suspicion that an essential element of the category of justice is missing. Whether or not the fifth of the theses that Scholem places under column A corresponds to a further remark on Benjamin's part, it nevertheless locates the missing element in *Gewalt* (power, force, violence), which, for the purpose of disambiguation, requires a critique comparable to that of the "faculty" or "capacity" (*Vermögen*) of reason and the "force" or "power" (*Kraft*) of judgment.

Another Critique, Another Antinomy

The beginning of an essay that Benjamin published in 1921 compensates for the apparent lacuna in the notebook Scholem borrowed five years earlier: "The task of a critique of *Gewalt* can be circumscribed as the exposition of its relation to law and justice [Recht und Gerechtigkeit]" (2: 179). "Toward a Critique of *Gewalt*" is unlike any other text that Benjamin published in terms of the directness with which it confronts the category of justice; but the essay belongs to a larger project he pursued in conjunction with his reflections on an appropriate topic for a *Habilitationsschrift*, a project he generally described as his "Politics." Several completed essays and a number of drafts were probably intended as part of the project, including a review essay of Ernst Bloch's *Geist der Utopie* (Spirit of utopia), which he apparently submitted to *Kant-Studien* but which has since been lost, and a reflection on Paul Scheerbart's "asteroid-novel" *Lesabéndio* that he probably wrote around 1917 or 1918 (2: 618–20). Other texts of the period that may have contributed to his larger plans include a series of brief inquiries into the sociological theories of Max Weber and Werner Sombert that he placed under the title "Capitalism as Religion" (6: 100–103), and an essay entitled "Leben und Gewalt" (Life and *Gewalt*) (6: 106), of which a brief fragment remains (7: 791). The principal concern of the fragment lies in the idea of an "original *Gewalt*," whose sphere of application overlaps with that of the axiom of right in Kant: a nonderivable and thus "original" *Gewalt*, according to Benjamin, asserts itself in any situation where the sphere of the living body, roughly speaking, is immediately threatened. In a fragment of a letter that Benjamin wrote in October 1920 to a potential publisher of his "Politics," he describes his work-in-progress as "a series of political essays" in which "anarchism" would be a principal point of reference (*GB*, 2: 101); but it would be a mistake to conclude from this remark that the volume was supposed to be an unsystematic collection of reflections on recent politics and related matters.

As with "Notes Toward a Work on the Category of Justice," the systematic intention of Benjamin's "Politics" expresses itself in Kantian terms. Near the end of the letter to Scholem in which he describes the weakness of Heidegger's *Habilitationsschrift*, he lays out a plan for the completion of his "Politics." With a single exception, all of the proposed titles derive directly from Kant: the third and final section will "probably" take the form of a "philosophical critique of [Scheerbart's] *Lesabéndio*," for which he has already written its "prolegomena" (*GB*, 2: 54); the title of the middle section, "The True Politics," comes from the aforementioned passage in "Toward Eternal Peace," where Kant distinguishes the "moral politician" from the "political moralist" (K, 8: 346–47); and the title of the first section, "The True Politician," represents a middle term that presumably takes both sides of the Kantian dichotomy to their respective extremes. In addition, the chapter entitled "True Politics" is itself divided into two subsections, the second of which is entitled "Teleology Without Final Purpose [Teleologie ohne Endzweck]" and thus represents a bisection of the *Critique of Judgment*, the first part of which revolves around two famous formulas for beauty, "lawfulness without law" and "purposiveness without purpose" (K, 5: 236), while the second part, as a critique of teleological judgment, culminates in a reflection on the "final purpose of the existence of the world, that is, of creation itself" (K, 5: 434). The only title not drawn directly from Kant is that of the first subsection of "The True Politics," which was to be called "Die Abbauung der Gewalt" (The dismantling of violence; or: The deconstruction of power). This subsection, however, would probably consist in a revision of "Toward the Critique of *Gewalt*," which expresses the Kantian character of the entire enterprise more effectively than any of the others.[18]

Benjamin, like Kant, presents the basic path of critique as an alternative to the potentially interminable conflict between two antithetical positions, both of which are generated from an objective error or "transcendental illusion." In the case of the first *Critique*, the conflict revolves around the dispute between dogmatism and skepticism: whereas the former claims to know things in themselves, the latter denies that there is any such knowledge. The common illusion is that the objects of knowledge must be things in themselves. "Toward the Critique of *Gewalt*" begins by identifying a similar conflict: legal dogmatism gives rise to the doctrine of natural law, which asserts that the rightful foundations of law are accessible to reason alone; skepticism with respect to right expresses itself in the school of legal positivism, which denies that the State has any foundation beyond its own norms. As for the illusion that unites the two schools of law and prevents them from undertaking a critique of *Gewalt* themselves: "Just ends can be attained by justified means, justified means used for just ends" (2: 17). In other words, the use of coercive force is either just in itself or leads to justice as long as it is placed under a general rule—either an established law, adjudicated by jurists for the school of legal positivism, or a universal principle, accessible to all rational beings, for the school of natural law.

In response to a similar antinomy, Hermann Cohen formulates a solution in his *Ethics of Pure Will*, which can briefly be summarized as follows: Kant sought to base his *Critique of Practical Reason* on the "fact of reason" (K, 5: 31), understood as moral self-consciousness; but since a pure object can never be grounded in the psychological sphere defined by the term *consciousness*, the relevant foundation for ethics should be sought, instead, in the "fact of legal science."[19] Such a formulation gives no solace, however, to the school of positive law, for the science of ethics methodically generates its object, which in this case is the "pure will" in its totality. Once the pure will is generated and applied to all fields of humanity—but this process, like the generation of the object of "pure knowledge," involves an infinite task—the messianic age comes into existence.[20] The conflict between the two schools of law can thus be resolved by representing legal science as a transitional sphere in which positive law functions as the indispensable *terminus a quo*, while natural law functions as the ever receding *terminus ad quem*.[21] Benjamin absorbs a passage from Cohen's *Ethics of Pure Will* into his critique in conjunction with his discussion of the ancient conception of "destiny" (2: 199); but he does not accept his solution to the originating antinomy. Instead, he replaces Cohen's idea of "pure will" with his own counterpart to pure practical reason: "pure *Gewalt*."

Benjamin does not begin his critique, however, by simply advancing the concept of pure *Gewalt*. In this way, he parts ways not only with Cohen but also with Kant, who begins the first *Critique* with the "metaphysical exposition" of space and time as "pure intuitions" and begins the second *Critique* with the exposition of the pure mode of practical reason in the form of the moral law. Instead of proceeding directly to a discussion of pure *Gewalt*, Benjamin provisionally sides with the school of positive law. As with Cohen, a certain "fact of legal science" is his point of departure, and the source of Benjamin's methodological decision is similar to Cohen's: both reject any suggestion that a philosophical dilemma can be solved simply by appealing to a rational datum, which all of us can immediately recognize. Benjamin, moreover, is explicitly skeptical of the categorical imperative. The sole substantial footnote to "Toward the Critique of *Gewalt*" subjects one of its formulations—"Act in such a way that you use the humanity in your person as well as the humanity in another person always at the same time as an end, never merely as a means" (2: 187;

cf. K, 4: 439)—to a kind of hyperbolic doubt: "One may doubt whether this famous demand does not contain too little, namely, whether it is permissible to use, or allow to be used, oneself or another as a means in any respect whatsoever. Very good grounds could be adduced for this doubt" (2: 187). Even though Benjamin does not then specify any of the grounds, certain consequences can be drawn from his expression of doubt about the viability of the imperative, so formulated: every conceivable form of "employment," including self-employment, is morally suspect. The skepticism goes so far as to make it questionable whether anyone is entitled to treat any part of his or her own body as a means to an end, thus, for example, employing a hand to consume an apple or using one's legs to get from one place to another. If, for Kant, the "sexual properties" of the body are defined as those parts of the organism that cannot simply be used as one sees fit, then the hyperbolic doubts to which Benjamin subjects the categorical imperative point toward a sexualization of the body as a whole and a corresponding moral paralysis: under current conditions no part can be employed as a means at all. Since marriage, for Kant, is the juridical form within which "sexual properties" can finally be used, Benjamin's doubts about the categorical imperative are of a piece with the critique of marriage that traverses his analysis of Goethe's *Elective Affinities*.

If every formulation of the moral law is suspect, so, too, is every form of natural law that derives from it. Provisionally, therefore, Benjamin has no choice but to align his analysis of the relation between right and justice with the school of legal positivism. To some extent, this decision corresponds not only with Cohen's reflections on the "fact of legal science" but also with Kant's own version of natural law. For Kant, law or right is distinguished from virtue because of its purely formal character; that is, law or right proposes no ends and therefore exists solely for the purpose of protecting the "elective will" of the person, which is presumably—but this is by no means obvious—an indispensable dimension of moral personality. Just as the "ethical formalism" of the second *Critique* found numerous opponents among post-Kantian philosophers, so did the "legal formalism" of the *Doctrine of Right* prompt prominent jurists in the nineteenth-century to conceive of "law as a means to an end" (*der Zweck in Recht*), to cite the title of a major treatise by Rudolf von Jhering, which Benjamin describes as "important" (6: 61). In the debate between Kant and Jhering, Benjamin ambiguously sides with former, who at least intimates that the sole "purpose" of law is its own self-assertion. Benjamin's legal skepticism goes so far as to deny that reason can propose *any* end for law that would not ultimately express itself in a tautological formulation that recalls the ontological proof of God's existence: just as the necessity of law lies in its possibility, so the end of law consists in its existence. Law thus eclipses God as the self-actualizing sum of all perfections. In the following passage, the term *fateful* assumes the function that Kant assigns to the term *pathological* in his moral philosophy, for, according to Benjamin, who follows Cohen in this regard, "fate" or "destiny" (*Schicksal*) derives from a prohibition that generates its own trespass and is, to this extent, irreducibly irrational, which is to say, in Kantian terms, pathological:

> Reason, however, never decides on the justification of means and the justice of ends; rather, fateful *Gewalt* on the former, God on the latter. An insight that is strange only because of the domination of an obstinate habit: conceiving of every just end as an end of a possible law, that is, not only universally valid (which follows analytically from the characteristic of justice) but also capable of being universalized, which, as can be shown, contradicts this characteristic. For ends that are just in one situation [Situation], capable of being universally recognized, are just for no other situation [Lage], regardless of how similar in other respects. (2: 196)

Benjamin demonstrates his point in formulating it, for the circumstance that finds expression in the word *Situation* is different from one that expresses itself in the word *Lage*, even though the two words are almost synonymous. This "almost" cannot be forgotten without a certain violation of the situation or *Lage* under consideration. Benjamin's exposition of just ends has nothing to do with "situational ethics." Rather, the very idea of a situation demands that it be given in its own terms and be determined by no others. The analysis of spatiality that Benjamin proposed in "Two Poems of Friedrich Hölderlin" is exemplary in this regard: "the poetized" generated from "Blödigkeit" (Infirmity) temporalizes space in the form of "opportunity" (*Gelegenheit*). Time thus permeates every place, which means that there is no empty space through which one travels by setting and achieving goals. Such is the situation: without a destination, hence nonfateful and without borders determined from without. The situation stands in contrast to the nontemporalized space of "politics," which solicits a pathological form of *Gewalt* to establish and enforce its borders for the purpose of luring trespassers whom it can thus punish. What Benjamin says of the situation in the essay from 1921 thus reflects what he says of goods in the notes from 1916: the subject of right is not the person; rather, the good has a "right" to be a good and the situation to be a situation. And once again, just as a good is already a good, so the situation is already a situation, which means that it requires no external apparatus for the enforcement of its status. The force internal to the situation takes shape as "pure *Gewalt*."

Just as, for Kant, practical reason is impure to the extent that the end of any given action, its outer purpose, takes precedence over its inner form, so, for Benjamin, *Gewalt* is impure whenever it is governed by an end in whose service it is enacted. The assumption shared by the school of natural law and its positivistic antagonist can be reformulated accordingly: both schools take it for granted that *Gewalt* is justified under the condition that it serve a certain purpose: the well-being of rational beings, in the case of natural law; the maintenance of particular States, in the case of legal positivism. The identification of the underlying assumption is not, however, equivalent to the dissolution of the corresponding illusion, which could be formulated as the generally unspoken thesis that *Gewalt* is always somehow "right."[22] In other words, whenever something or someone is placed under someone's "control," certain rights are established and certain laws made, even if only in provisional form. Kant was confident that the transcendental illusions that reason generates by virtue of its finitude could be dissolved by means of reason alone. For Benjamin, by contrast, who follows Marx in this regard, objective illusions cannot be so dissolved, and they can be analyzed in terms of "myth" because they protect themselves from "logical" dismantling. The mythic character of *Gewalt* most clearly manifests itself in the *Doctrine of Right* when Kant, having denounced the labor theory of property, nevertheless gives voice to certain portions of unclaimed space, which welcomes their putative possessors so long as they show sufficient *Gewalt*. For Benjamin, the mythic character of *Gewalt* announces itself with similar clarity in a polar-opposite circumstance: not when there is land under no legal title but, rather, when everything is owned by a very few, and those who possess nothing but their labor power categorically abstain from work. In the situation of a "proletarian general strike," in other words, the exercise of a right that has been granted to the workers threatens the very State that grants the right to strike. The threat stems from a mode of *Gewalt* that, unlike extortion, is wholly legal. A display of legally sanctioned force, once taken to an unforeseeable extreme and thus hyperbolized, shows that "violence" is legal in some fundamental and irreducible sense: "In a strike the State fears more than anything else that function of *Gewalt*, the identification of which this investigation sets forth as the only secure foundation of its critique.... The strike shows that ... *Gewalt* is in a position to found and modify legal relations, however offended the feeling of justice may thereby find itself" (2: 185).

The requisite criterion for the critique of *Gewalt* as a means to an end can thus be found in a struggle over so-called distributive justice. At stake in the struggle is the division of the social product. A general abstention from work reiterates, in effect, the opening thesis of Benjamin's "Notes Toward a Work on the Category of Justice": "no order of possession, however it may be articulated, leads to justice" (S, 1: 401). Benjamin's contribution to a critique of *Gewalt* ultimately complements the earlier notes. One takes its point of departure from the school of natural law, declaring at the outset that there is no principle under which the division in question can be "justified." The other begins from the perspective of legal positivism, discovering a circumstance in which a strange "feeling of justice" emerges among legal positivists, for whom moral sentiment is anathema. Just as the "feeling of justice" is without justification from the perspective of those whom it affects, so, too, is the threat from the perspective of the strikers whom Benjamin describes; for whatever else may said about "proletarian general strike," it is not a means to a predetermined end, much less the expression of a demand for a greater share of the total social product on the basis of unfulfilled needs. By no longer allowing themselves to be employed under any condition, even one in which they would also be respected as ends-in-themselves, the strikers express a maximal version of the categorical imperative. If there is any end in view, it is not as telos but only as demise—in this case, as the demise of any order of persons and things that is founded and maintained by pathological *Gewalt*.

On the Program of the Coming "Moral Philosophy"

In describing the "proletarian general strike," Benjamin identifies a phenomenon that the school of positive law recognizes only at its peril: an instance of *Gewalt* that abstains from making law despite the fact that it is in a legal and physical position to do so. In the same stroke he identifies the political counterpart to pure practical reason as pure *Gewalt*. The two lines of argument—one from the side of positive law, the other from that of natural law—do not meet so directly in the essay that Benjamin published in 1921 as they do in some notes he drew up a year earlier, which culminate in a schematization of what Benjamin calls "my moral philosophy" (6: 107). Outside of these notes Benjamin has nothing good to say about "moral philosophy." Even in an early essay entitled "Moral Instruction," which is explicitly "anchored" in the *Critique of Practical Reason* (2: 48), the function of moral philosophy is purely transitional: it leads to a "new instruction in history" (2: 54). In the essay "On the Program of the Coming Philosophy," from around 1917, Benjamin makes the same point in relation to the Kantian conception of morality as a whole: "With a new concept of knowledge, not only the concept of experience but also that of freedom will therefore experience a decisive transformation" (2: 165). The idea of moral philosophy comes under similar suspicions, for even when it begins with a repudiation of the calculation of motivations and consequences as morally unacceptable, it is nevertheless based on the very same categorial moment of ground and consequence. This says nothing against morality per se, only against its theory. Just as Heidegger calls *Lebensphilosophie* (philosophy of life) a pleonasm, equivalent to "the botany of plants," so Benjamin says of "moral philosophy" that it is a "stupid tautology" (6: 93).[23] The context for this remark is a reflection on the tripartite character of the *Critiques* that Cohen adopts for his "system of philosophy": "'Morality' [as] title of the second part of the system" (6: 93). The suggestion is that morality retains its systematic position but precisely not as "philosophy."

And yet, despite his repudiation of the term, Benjamin describes the central task of his "moral philosophy" in some notes he developed in response to an essay he came across in the first volume of the journal *Blätter für religiösen Sozialismus* (Pages for religious socialism). As editor of the journal, Carl Mennicke entered into a conversation with Paul Tillich around the question of the right to use *Gewalt* in the context of the recent Kapp-Lüttwitz putsch, in which decommissioned military regiments sought to overthrow the fledgling Weimar republic and replace it with some form of monarchial governance.[24] A series of general strikes erupted in major German cities, especially Berlin, which thwarted the putsch. For the inauguration of the journal, Mennicke convened a formal discussion of the question raised by his conversation with Tillich and asked a noted jurist of the time, Herbert Vorwerk, to present his views.[25] From Benjamin's perspective, Vorwerk's rather dry paper, "Das Recht zur

Gewaltsanwendung" (The right to apply *Gewalt*), is less notable for what it contains than for what it neglects. In identifying his oversight, Benjamin may have been led to consider whether a comparable lacuna could be found among proponents of legal positivism, and this, in turn, may have contributed to the methodological procedures governing "Toward the Critique of *Gewalt*." Benjamin's reflection on "The Right to Apply *Gewalt*" represents, in any case, an abbreviated version of a program for the coming moral philosophy.

Vorwerk identifies two forms in which the question raised by Mennicke and Tillich can be considered: "The question should be considered first from a purely juridical, then from an ethical perspective." The juridical perspective corresponds to the premise of legal positivism, which presents the legal order as the sole basis for a decision concerning the use of *Gewalt*, understood here as coercive force: "The jurist can speak of a right to something only where the objective legal order grants authorization to an individual. The question of the right to apply *Gewalt* therefore leads to the question of the relation of the legal order to *Gewalt*."[26] From a specifically juridical perspective, two answers to the question under consideration are possible, one of which gives the State a limited right to use force, the other an unlimited right under the condition that its own laws are followed: "(1) Only the State has a right to apply *Gewalt*. (2) Everywhere where the State applies *Gewalt* (that is, naturally under the observation of a legal order), it also has a right to do so."[27] After dismissing the "right to revolution" as a residue of premodern theories of State power, which cannot be consistently formulated, Vorwerk draws on the now-famous definition of the modern State that Max Weber had proposed in the previous year: "The State is the human community that, within a specific domain—this 'domain' is one of its characteristics—claims (successfully) the monopoly of legitimate physical violence [Gewaltsamkeit]."[28] According to Vorwerk, the "ethicist" does not so much dispute this premise as take a certain distance from its consequences when he admits some space, however small, for the use of extralegal force. And just as Vorwerk specifies two answers to the question from the side of legal positivism, he does the same from the side of an ethics, which, for its part, corresponds to the perspective of natural law: "The ethicist can then take up a different position in relation to this juridically consistent establishment of a monopolistic right to the use of force for the organized inner-State and inter-State community. He can (1) deny the right to apply force and its organs (ethical anarchism);

(2) the right to apply force, in contrast to jurists, also to recognize for individuals and classes (in the popular community of the State), if 'injustice' is done by the established legal order."²⁹

Without making any distinction between jurist and ethicist, Benjamin formalizes the four responses that Vorwerk identifies and produces a schema of all possible "standpoints" under the condition that the range of potential rights-bearers be reduced to the State, on the one hand, and the individual, on the other: "(a) Deny that the State and the individual have the right to apply force; (b) unconditionally recognize that the State and the individual have the right to apply force; (c) recognize that the State has the right to apply force; (d) recognize that only the individual has the right to apply force" (6: 105). According to Benjamin, Vorwerk fails to see that the last "standpoint" is independent of the others; even worse, "he does not even clarify its logical possibility as a particular standpoint but, rather, calls it an inconsistently one-sided application of ethical anarchism" (6: 106). The term "one-sided" refers to the following remark, which alludes to the events surrounding the Kapp-Lüttwitz putsch: "Both conceptions are multiply intertwined with contemporary events. It seems as though ethical anarchism is in many cases applied in a one-sided way only toward the State, not toward whatever fights against it. Here, in truth, is the second assumption."³⁰ In other words, the contention that individuals alone have a right to apply coercive force would potentially apply only to the participants in the putsch, not to the participants in the general strike that prevented the right-wing factions from reestablishing a hierarchical-nondemocratic regime. Benjamin, by contrast, unambiguously affirms only the last "standpoint," and he does so in such a way that the action of the strikers could be disentangled from the assumption that the strike was done in service to the legal regime because it acted against its opponents. That Vorwerk misconstrues the last "standpoint," for Benjamin, is no accident: it exceeds the jurist's perspective, even when he adopts the optic of the ethicist. And because the "standpoint" is not based on a higher principle such as the life, liberty, or property, it can be described in terms of "ethical anarchism" only if the epithet "ethical" is either suspended or radically transformed:

> The exposition of this standpoint belongs to the tasks of my moral philosophy, in connection with which the term *anarchism* may very well be used for a theory that does not deny the moral right of *Gewalt* as such but, rather, only

denies the right to every human institution, community, or individuality that claims a monopoly on it or concedes for itself the right to it from whatever perspective, even if only in principle and generally—instead of honoring it in the particular case as a gift of divine power, as **plenary power** [anstatt sie als eine Gabe der göttlichen Macht, als *Machtvollkommenheit*]. (6: 107)

By heavily emphasizing *vollkommenheit* (completion, perfection), Benjamin indicates that the term *Machtvollkommenheit* (plenary power) forms the nucleus of his coming "moral philosophy." There is good reason to suppose that it represents the core of "true politics" as well. For Kant, "true politics" is constitutively progressive. Its goal lies in the extension of the sphere of right to the point where it interpenetrates all affairs of State and stretches across the entire globe. The sole condition of such progress is that pure practical reason be honored under the rubric of morality: "True politics cannot take a step without having already honored morality" (K, 8: 380). For Benjamin, by contrast, *Gewalt* alone is to be honored, which is presumably what the "true politician" does. In contrast to both the "political moralist," who glorifies power in the general form of the State, and the "moral politician," who postpones the institution of the principles of right under the doctrine of "permissive law," the "true politician" permits no delays and honors *Gewalt* only in particular cases, not in general. In other words, *Gewalt* is recognized as a gift, not made into a possession: it is the *datum*, if not the *factum*, from which "the ethics of pure *Gewalt*," to alter the title of Cohen's treatise, takes its point of departure. Under these two conditions—the singularity of the case, the acknowledgment of its gift-character—*Gewalt* emerges as the good to which the "Notes Toward a Work on the Category of Justice" enigmatically refers: a good that cancels its possession-character and thereby makes all other goods possessionless.

Of course, Benjamin does not call *Gewalt* a good. At the same time, he does not condemn it as an evil, much less as a necessary evil, which assumes the form of sovereign power in response to the inherently sinful or radically evil character of human beings. As Benjamin notes at the beginning of "Toward the Critique of *Gewalt*," his inquiry is divided into two sections: an analysis of *Gewalt* according to its meaning, followed by a critique of "the sphere of its application according to its value" (2: 181). Because the dogma shared by the two schools of law presumes that, in order for the use of force to be evaluated as good, it must be mediated by something—a historically sanctioned institution in the case of positive law, a generally valid principle in the case of natural law—the specifically critical dimension of the inquiry undertakes a transvaluation of *Gewalt* that frees its value from both rational principles and historical statutes. *Gewalt* thus emerges as a good but—and this is essential—neither a good-in-itself, akin to the "good will" with which Kant's "groundwork" for his own "moral philosophy" begins (K, 4: 394), nor a good-for-something-else: a "shock," for example, that rejuvenates a sclerotic system, to use the vitalistic imagery of Georges Sorel, from whom Benjamin borrows his description of the "proletarian general strike." *Gewalt* escapes this alternative between good-in-itself or good-for something-else only under the condition that its temporality be that of total transience: its appearance coincides with its disappearance. Or, to draw on the titles of Benjamin's proposed "Politics," the emergence of pure *Gewalt* goes hand in hand with the "dismantling" of fateful *Gewalt*.

"Plenary Power"

If *Gewalt* is the good to which no possession-character accrues, then it can be understood as the post-lapsarian residue of the Tree of Knowledge of Good and Evil: it is eminently usable; but as soon as it is held as a possession, it no longer is the good that it otherwise would be. The expulsion from Eden can thus been seen as the event in which *Gewalt* begins the process of its "monopolization." And the two schools of law provide complementary rationales for this transformation: general principles are the means of storage, for the advocates of natural law; established statutes, for the proponents of positive law. The point, then, is not to reject *Gewalt* as akin to "forbidden fruit," which must be shunned as an evil in itself; rather, the "fruit" should not be eaten, that is, stored up. For Benjamin, the abstention from taking possession of *Gewalt* consists in honoring it in each singular situation "as a gift of divine power, as **plenary** *power*."

"Plenary power" (*Machtvollkommenheit*) can be provisionally understood as the perfection of "power of attorney" (*Vollmacht*). The latter transfers "full power" of a private person to his or her representative under the condition that the representative does not personally possess the powers in question. The recipient of "plenary power" is in a similar situation: the "full power" that is received does not belong to its recipient. The difference is that in "plenary power" the "fullness" of power is not limited either to private right or to public power. Foreign to the tradition of Roman law, the doctrine of *plenitudo potestatis* arose within the field of canon law in response to a question of decisive importance for the founding of the Church on the "rock" of Saint Peter: how does the power of the bishop of Rome differ from that of the other bishops? Insofar as this question is tantamount to the singular problem of papal power, an answer is required for the establishment of the Church as a representative of divine dominion in its temporal dimension: "It was [the twelfth-century mystic-monk] Bernard de Clairvaux who provided the classic formulation of the papal *plenitudo potestatis*. . . . 'According to your canons, some are called to a share of responsibilities, but you are called to the fullness of power.'"[31] The pope does not simply have more power than the other bishops; rather, his power is of a categorically different kind than theirs. *Plenitude* qualifies *potestas* for this reason: in every other case of *potestas*, its division implies its diminishment, whereas in the extreme case of *plenitudo potestatis* power can be infinitely divided without being lost in the least. Just as God gives his power to the pope without any diminishment of his omnipotence, so the pope can distribute powers to bishops without losing any of the power he is given. A similar structure can then be discerned in the field of international law under the premises of the absolutist State: the sovereign has the power to name a diplomat as plenipotentiary without losing any of his own divinely derived power. For a number of reasons—including alterations in the theory of sovereignty and concomitant developments in the technology of telegraphic communication—the diplomatic office of the plenipotentiary fell into disuse among European States in the course of the nineteenth century, and it is presumably this phenomenon to which Benjamin refers in "Toward the Critique of *Gewalt*," when he briefly discusses the decay of the nonviolent mode of concurrence that takes the form of interstate diplomacy (2: 193).

With one stroke, however, Benjamin distinguishes his doctrine of plenary power from that of the pope and the absolutist State: the field of its reception is so exceedingly narrow that it approaches zero. Thus he emphasizes that the "divine gift" of *Gewalt* is limited to "the particular case." For the pope of the "catholic" church and the sovereign of the absolutist State, the gift of plenary power is given for the duration of their earthly existence. The pope and the prince can thus be said to possess the power they receive. The doctrine of *plenitudo potestatis* Benjamin briefly articulates in his notes on Vorwerk's paper moves in the opposite direction— not out of a contrarian spirit but in keeping with the doctrine itself. For, if the condition under which perfect power can be given is that the recipient has no power of his or her own, then the recipient is in no position to give power away. On the contrary, the self-conscious act of giving it would presuppose that it was in his or her possession prior to its donation. What Benjamin says of translation in the essay with which he prefaced his contemporaneous translation of Baudelaire's *Tableaux parisienne* (Parisian scenes) accords with what he suggests in his notes on Vorwerk's paper: just as a translation is itself untranslatable (4: 15), so a plenipotence cannot delegate plenary power. And for similar reasons: just as the translatability of a work is inversely proportional to the "information" it can be said to convey, so the plenipotentiary character of an agent is inversely proportional to the extent of his or her "own" *potestas*. The smaller the domain over which one exercises *potestas*, the greater the possibility of receiving plenary power in any given situation. The proletariat is by definition the sole legitimate recipient of genuinely *plenary* power. To the Kantian program of extending the sphere in which the concept of right is applicable until it finally covers the entire surface of the earth, Benjamin responds with a counterprogram: reduce the sphere of *potestas* to the point where the gift of plenary power can be honored as such. If the "moral politician" is the proponent of an ever-expanding sphere of right, even when the expansion is forever delayed on the basis of "permissive laws," the "true politician" is the exponent of its immediate contraction.

A Technical Term

"Toward the Critique of *Gewalt*" makes no use of the term *plenary power*, which Benjamin probably encountered while researching a *Habilitationsschrift* on scholastic philosophy. The Kantian character of the essay requires that *plenary power* be replaced by its Kantian equivalent, namely "pure *Gewalt*," which is similarly structured. The following claim, drawn from the final paragraph of Benjamin's essay, applies equally well to plenary power in the absence of an ecclesiastical structure, morality in the Kantian sense, and pure *Gewalt*, as Benjamin presents it: "Not equally possible and not equally urgent is, for human beings, the decision when pure *Gewalt* in a specific case was real [wirklich]" (2: 202–3). Just as an instance of purely moral action cannot be ascertained as such, neither can a case of pure *Gewalt*. And just as the structure of pure practical reason is traversed by a separation of power terms, so, too, is that of pure *Gewalt*.

The same is also true of the political counterpart to pure practical reason that Kant wishes to establish under the rubric of "right." In justifying the extension of the axiom of right beyond the sphere of its immediate application, Kant postulates a *Gewalt* that must be distinguished from physical "power" (*Macht*). In elucidating the character of pure *Gewalt* at the end of his own contribution to the critical enterprise, Benjamin likewise postulates a "force" (*Kraft*) that remains distinct from *Gewalt* and is constitutively incapable of appearing as such: "For only mythic *Gewalt*, not divine, allows itself to be known with certainty, be it therefore in incomparable effects [Wirkungen], because the expiating force [entsühnende Kraft] of *Gewalt* does not come to light for human beings" (2: 203).

Because the "expiating force" that distinguishes pure from impure *Gewalt* recedes from appearances, it is impossible to determine in any given instance whether *Gewalt* is pure or impure—which immediately raises the question of whether there is such thing as pure *Gewalt* in the first place. In response to a similar question, Kant can give a well-worn answer: it does not matter whether or not there has even been a case of morally pure action, for such action is nevertheless demanded. If, however, pure *Gewalt* is immediately related to the category of justice, and if, as Benjamin claims in his "Notes," of 1916, justice, unlike the other virtues, cannot be demanded, he is barred from formulating a similar response. This impasse prompts him to conclude "Toward the Critique of *Gewalt*" by introducing a series of technical terms for what he has previously discussed in their absence. Among the recommendations that Benjamin proposes in *One-Way Street* for the production of "thick books" is the following: "Terms for concepts are to be introduced that never appear in the entire book except in those places where they are defined" (4: 104). "Toward the Critique of *Gewalt*" is not a thick book by any means; but Benjamin incorporates in reverse the stylistic principle that he thus satirizes, as the conclusion of the essay introduces terms for concepts that are absent from the body of the essay: "All mythic, law-positing *Gewalt* is reprehensible, which can be called turning-on [schaltende] *Gewalt*. Reprehensible also is law-preserving, administrative [verwaltende] *Gewalt*, which serves it. Divine *Gewalt*, which is an insignia and seal, never a means, of sacred enactment, may be called presiding [waltende] *Gewalt*" (2: 203).

To the extent that *walten* resonates with *Walter*, as Werner Hamacher and Jacques Derrida have pointed out, the final sentence of "Toward the Critique of *Gewalt*" subtly modifies a passage in one of Benjamin's plans for his *Habilitationsschrift*, where, under the heading of "name," only two appear: "Walter Benjamin, *Adonai*" (6: 21).[32] Instead of using the traditional formula for the introduction of a technical term, "I call such-and-such *x*," he thus embeds himself in the term whose meaning he stipulates—or pretends to stipulate, for "pure" and "divine" are probably less ambiguous qualifications of *Gewalt* than is *walten*. Of the three terms Benjamin introduces, only the middle one goes without saying: "administrative power" (*verwaltende Gewalt*) is a commonly used term for the governing forces that uphold the laws in any given legal order, and his final definition functions as a translation of his own terminology into colloquial form. The other two terms frequently appear in such phrases such as *walten und schalten mit* (to dispose over) and *frei schalten und walten können* (to be able to do as one pleases).[33] So frequently do *walten* and *schalten* appear together that the technical meaning of each derives from their separation from the other: there is no "walten *und* schalten," in other words—no faculty of freedom, understood as the ability to do whatever one pleases, such that the concept of right must be indefinitely extended as a means for keeping this faculty in check. More importantly, however, the separation of *walten* from *schalten* indicates that the former means the opposite of the latter. The most appropriate translation for *schalten* is "turning-on," as in the "enabling" (*anschalten*) of an apparatus. In this case, the apparatus that gets started is that of law, which then requires "administrative power" for its maintenance. To the extent that any use of "administrative power" is a drain on the system, a new "turning on" is required every once in a while, with the result that the power cycle repeats itself ad infinitum—unless and until the system of power, grid and all, is "turned off" (*ausgeschaltet*) for good. And this is the resulting sense of *walten*. In "On Painting or Sign and Mark" Benjamin adopts Husserl's term *ausschalten* (switching or turning off) in describing what happens when an "absolute mark" appears: "the resistance of the present to the future and past is switched off [*ausgeschaltet*]" (2: 605). Pure *Gewalt* is similarly absolute: without any relative mark of its "expiating force," it turns off the current of power that appears as violence, as force, and then again as administrative power. Benjamin avoids the doubly technical term *ausschalten*—introduced into philosophy less than a decade earlier, drawn from technological innovation only a few decades older—by replacing it with the ancient word *walten* and thus obviating any suggestion that the current of legal power can be turned off simply by reversing the direction of the apparatus through which it is enabled.

Toward a Higher *Gewalt*

If the technical term that Benjamin introduces into the final sentence of "Toward the Critique of *Gewalt*" is a placeholder for *ausschalten*, then the idea of pure *Gewalt* is traversed by a powerful tension. On the one hand, the concept accords with a "Kantian typic" and indeed with the very same typos that guides Benjamin's exegesis of the Book of Genesis in "On Language as Such and on Human Language." According to Kant, divine intuition, about which one can speak only in problematic terms, is purely spontaneous, whereas human intuition is wholly receptive. Only in combination with the spontaneity of the understanding does receptivity yield knowledge. Similarly, according to Benjamin's "little treatise," the creative word is wholly spontaneous, whereas the language of names is receptively spontaneous and spontaneously receptive. The same is true of pure *Gewalt* in human hands: it, too, is receptively spontaneous and spontaneously receptive. The character of receptivity in the latter case lies in "honor," which is presumably what "true politicians" do: they neither justify nor glorify *Gewalt* but only pay it homage as "a divine gift, as plenary power." On the other hand, insofar as receptive spontaneity is less pure than pure receptivity, the Kantian typic fails to capture the character of a *Gewalt* that would "preside" (*walten*) solely by "switching off" or "disabling" the current of power. In the "little treatise" on language, the slightly impure character of receptive spontaneity expresses itself in Adamic naming, which is then said to be responsible for the sadness that befalls things even in the absence of their "overnaming." And in "Toward the Critique of *Gewalt*," the slightly impure character of pure *Gewalt* expresses itself in destruction, which is as spontaneous as creation.

Nowhere is Benjamin more emphatic about the destructive character of pure *Gewalt* than in a fragment he wrote in the early 1920s under the title of "Welt und Zeit" (World and time): "A genuine divine *Gewalt* can manifest itself as *other than destructive* only in the coming world (of fulfillment). Wherever, by contrast, divine *Gewalt* enters into the earthly world, it breaths destruction" (6: 99). Although Benjamin underlines "other than destructive" in the previous passage, the accent falls more strongly on "manifest itself," which implies that there is another mode of pure *Gewalt* that abstains from manifesting itself and is thus indistinguishable from *Gewaltlosigkeit* (nonviolence). In the "earthly world," as opposed to the "coming world," according to Benjamin, the lack of *Gewalt* goes unmarked and thus unnoticed: "In this world divine *Gewalt* is higher than divine lack of violence [Gewaltlosigkeit]; in the coming world, divine lack of violence higher than divine *Gewalt*" (6: 99). There is little trace of this other mode of *Gewalt* in the sole section of Benjamin's planned "Politics" that he published. Perhaps it would have occupied a place in the subsection entitled "Teleology Without Final Purpose," as a dimension of the "without" to which the title refers. Or perhaps it have become an element of the "continuation" of the critique for which Benjamin collected a substantial list of works on the foundations of legal theory (WBA, MS 1858). A discernable formulation of the idea can be derived in any case from Benjamin's abbreviated program for his "moral philosophy." Insofar as the "honoring" of *Gewalt* as a "divine gift" requires its manifestation as *Gewalt*, the absence of its manifestation involves a corresponding dishonor or disgrace—or, in other words, a certain shame, the coloration of which "gushes from outside onto the one who is ashamed and extinguishes the disgrace and simultaneously withdraws him from all disgracing" (6: 69). Benjamin emphasizes the "violent" character of this situation: the incomparable coloration negates a negation, insofar as it "annihilates [vernichtet] . . . a sublime indeterminacy" (6: 69). As for the agent of this "annihilation," it, too, can be identified only by means of negation: "not by a lower power [Gewalt]" (6: 69).

Benjamin had good reason to abstain from discussing the coloration of shame in conjunction with his critique of mythic *Gewalt*, for there is something mendacious about the premise on which rests the description of the event of shame, namely, that the "tone" of human skin is "almost decolored" (6: 69). But the coloration in question is not altogether overlooked, for "Toward the Critique of *Gewalt*" draws attention to its biological substrate—the flow of blood, which turns bright red as soon as it is exposed to the open air. Mythic *Gewalt*, according to Benjamin, is characterized by the shedding of blood, whereas divine *Gewalt*—as exemplified by the rebellion of Korah, when the earth opens up and swallows the rebels (Num. 16:30)—is equally deadly but nevertheless bloodless: "A deep connection between the nonbloody and the expiatory character of this [divine] *Gewalt* cannot be misrecognized. For blood is the symbol of mere life [bloßes Leben]" (2: 199). Benjamin does not further develop the character of the "deep" connection he thus identifies, nor does he specify the provenance of his disturbing thesis. The following passage from the Book of Leviticus presumably lies somewhere in the background: "For the life of the flesh is in the blood, and I [the Lord] have assigned it to you for making expiation for your lives upon the alter; it is the blood, as life, that effects expiation" (Lev. 17:11).[34] In Benjamin's revision of the passage from Leviticus, blood does not *mean* "mere life," for, as he elsewhere notes, "a symbol does not mean anything [bedeutet nichts] but, rather, is, according to its essence" (6: 21). In this sense, for Benjamin, blood is indeed "the life of the flesh," as the passage from Leviticus states. So far from being an expiatory force, however, blood inclines in the opposite direction: where no blood flows, there is—perhaps—expiation.[35] A question is then implied in the "deep" connection Benjamin briefly uncovers: how can the absence of bloodshed show itself as such—that is, as the absence of violent bloodshed and not merely the absence of blood being shed? A life other than "mere life" would make itself apparent wherever this nonappearance itself appears.

A cryptic remark that Benjamin includes among his color studies connects the word *red* to the process of natural exposure, which leaves an indelible mark and thereby indicates the transient character of the affected thing: "Rust [Rost] is etymologically connected with red [Rot]" (6: 123).[36] From his contemporaneous study of Hebrew, Benjamin would also have known that *dam* (blood), like Adam, derives from *edom* (red). In the coloration of shame—which is altogether distinct from shameful feelings, whatever their a priori object may be—blood momentarily appears without being shed. And it appears as the absence of blood "really" appearing. Instead of being a permanent stain, it verges on the "absolute mark," which "turns off the resistance of the present to the future and past," thus allowing past, present, and future to converge in the "now." Thus arises a situation, which, for its part, is the stance or state of justice: nonderivable and thus "original," universal yet nonuniversalizable. "Truth consists in the 'now of knowability'" (6: 46), Benjamin writes in a contemporaneous note on the theory of knowledge. The "absolute mark," which "gushes" from the outside, allowing blood to appear without shedding, would be a witness for truth under two conditions: the mark is indeed that of shame and not of rage, which is also red; and the mark escapes every conceivable color concept, including that of redness. The only way to guarantee the latter condition is to add a coda that erases the "reality" of the mark: the coloration in question, comparable to a "colorless" heaven (2: 178), comes from pure fantasy.

Conclusion

The Shape of Time

Psychology

The critiques of psychologism undertaken by Hermann Cohen and Edmund Husserl were not intended to eliminate psychology as a legitimate science; rather, each of them sought to demonstrate that the object of psychological research is different from that of philosophical inquiry and that the confusion of the two was detrimental to both. Under the general title of psychology, Cohen originally intended to add a fourth volume to his "system of philosophy" in which he would show in detail that the "systematic unity of consciousness" lies in universal culture: "culture alone can give unity to consciousness."[1] With little time left in his life, however, Cohen abandoned his plans to write a "Psychology" and wrote, instead, *Religion der Vernunft aus den Quellen des Judentums* (Religion of reason from the source of Judaism), the final pages of which identify a different source for the unity of consciousness, namely "peace" (*Friede*), which can henceforth be called the principle of the coming psychology. Not only is peace, according to Cohen, the "epitome of all morality," it is also the "emblem of the messianic age [Wahrzeichen des messianischen Zeitalters]": "The Messiah is therefore called the 'Prince of Peace' (*sahr shalom*). The unity of human consciousness is thus expressed by the peace of the soul."[2] And as for the critique of psychologism that Husserl develops in the prolegomenon to *Logical Investigations*, it doubtless makes a categorical separation between logic and psychology; but so far from doing away with psychology, Husserl famously adopts a term from Franz Brentano and describes phenomenology as a form of "descriptive psychology" (Hu, 18: 3).³ In a fragment from around 1919 that bears the title of "Psychology," Benjamin borrows Husserl's terminology and presents the science in terms that would be familiar to Cohen, even if they are antithetical to the purely ideal character of the messianism that finds consummate expression in his *Religion of Reason*:

> Psychology is, so to speak (if this is a final epistemological category), a descriptive science, not an explanatory one: the perception therein described is a pure and indeed the pure (apocalyptic) perception of the human being. Whatever remains of the human being after the moral catastrophe, after return and purification. This is not "inner" . . . but rather outer: the perception of the human being, which he gives to fellow human beings. This, however, is only pure, only outward, only entirely perceptible, and therefore only entirely perception after the moral restitution of the human being. (6: 65)

The critique of psychologism in Benjamin's hands thus gives rise to the idea of a pure psychology, which begins at the precise moment when human beings appear exactly as they are: without any shadows or adumbration, therefore "apocalyptic" in the original sense of the word. The principal task of the coming psychology is to discover the situation that allows everyone to be as purely perceptible as the rainbow, which at one point Benjamin, alluding to the Book of Genesis, calls "the symbol of peace" (6: 123).

As if it were an addendum to the exegesis of the Book of Genesis Benjamin undertook in "On Language as such and on Human Language," he then presents the object of pure psychology in Adamic terms: "The relation of the human form [Menschengestalt] to language, i.e., as God, who formed him linguistically, acts [wirkt] in him, is the object of psychology. To this also belongs the living body in that God immediately—and perhaps incomprehensibly—acts [wirkt] linguistically therein" (6: 66). The fragment says nothing further about the possibility of divine incomprehension, which guarantees that the "living body" (*Leib*) retains a trace of opacity even in light of apocalyptic restoration. One thing is clear, though: the opacity has nothing to do with the difference between the epistemic perspective of the "I" and that of the other: "The psychic life [Seelenleben] of the other is not perceived, in principle, differently from one's own; it is not inferred but, rather, seen in the living embodiment that belongs to it as a psychic life" (6: 65). To the extent that there is an irreducible degree of opacity, it does not lie in the living body but, rather, in the substantial character of the "corpus" (*Körper*). And in a six-part schematization of "the psycho-physical problem" that Benjamin wrote in the early 1920s, he draws a distinction between living body and corpus on the basis of the distinction that Cassirer develops in his *Concept of Substance and Concept of Function*: "In general, this then can be said: so long as we know by way of perception, we know our own corpus, which, in contrast to our living body, extends itself without corpus, formed boundaries. This corpus is not the final substrate of our being but is nevertheless a substance in contrast to the living body, which is only a function" (6: 79). As a function of "our being," the living body is identical to psychic life and is "formed" for this reason; as a substance of "our being," by contrast, the corpus lies below or beyond and cannot for this reason be "purely perceived."

As Benjamin continues the schematization of the psycho-physical problem, it soon becomes evident that the origin of the problem does not so much lie in the supposed synthesis of *psychē* and *physis* as in certain differences that characterize *physis* alone. The living body and the corpus are both "natural," and yet they are by no means identical to each other: the former is identical to the psychic life, whereas the latter consists entirely in differences, beginning with the difference between pain and pleasure, which it "sensibly presents" (6: 79), and culminating in sexuality, where it expresses the "differential instrument of vital reactions" (6: 81). But—and here is the problem incarnate—the differential function of the corpus depends on the functional character of the living body, and this dependence opens up a logical space for another category, which distinguishes the living body from psychic life and at the same time makes the corpus into the locus of spiritual identity. The supplemental category is called "the person," who is at once agent of, and subject to, the order of law: "Herein lies the difficult problem, that the 'nature' whose affiliation with the corpus has been asserted, nevertheless refers in the strongest manner to the limitation and particularity of the living being. That limited reality, which is constituted through the founding of a spiritual nature in a body, is called the person. The person is in fact limited but not formed" (6: 80). The absence of bodily form means that the limitation of the person tends to express itself in formless expansion: "It [the person] therefore derives its particularity, which one may attribute to it in a certain sense, not from itself, as it were, but rather from the radius of its maximal extension," which Benjamin ambiguously identifies with "the people" (6: 80).

The task of politics can then be determined in relation to the "difficult problem" that Benjamin identifies in his reflections on the science of psychology: reduce the person to a structural possibility of the psychophysical relation rather than mendaciously inflate it into a spiritual reality in its own right. Insofar as the extension of the person is symbolized in the extension of the corpus, the portent of the relevant reduction appears in its shrinkage, which similarly reduces the intensity of "the people." The definition that Benjamin proposes for the term *politics* in "World and Time" indicates by way of negation the character of such shrinkage: the condition under which the task of politics can be carried out is that there be no expansion of human affairs. Or, as Benjamin writes: "My definition of politics: the fulfillment of nonincreased humanness [die Erfüllung des ungesteigerten Menschhaftigkeit]" (6: 99). The fulfillment does not correspond to a demand or a law, and it does not concern humanity in its moral dimension, much less an even higher dimension of humanity, whose "seal" lies in the total isolation of the corpus, which, in Benjamin's view, was "created for the fulfillment of commandments" (6: 82). Rather, the fulfillment in question concerns only the living body, minus the "zone of immediate divine efficacy" (6: 99), to which Mosaic law presumably applies. And the fulfillment without expansion presumably consists in the state of happiness: "The order of the profane is to be erected in relation to the idea of happiness" (2: 203), Benjamin asserts in the so-called "Theological-political Fragment," which was probably written in conjunction with the critique of Ernst Bloch's *Spirit of Utopia* that was itself an integral part of his plans for his "Politics."[4] And it is for similar reasons that Benjamin intended to preface or conclude his "Politics" with a critical commentary on another work he associated with the idea of "utopia" (6: 119), namely Paul Scheerbart's "asteroid-novel" *Lesabéndio*. Whatever else can be said about the world inhabited by Lesabéndio and his fellow asteroidians, this much is certain: their proper names derive from the first sounds they emit, not from a procedure of designation akin to the establishment of technical terms, and the laws governing the waxing and waning of their bodies are unlike those under which the bodies on the surface of the earth fall. This makes the novel utopian: the proper names of the asteroidean are disentangled from designation, and the shape of their bodies is immune to the juridical process through which their earthly counterparts expand. And for Benjamin, the clarity of Scheerbart's style—which has as little to do with the staging of moral conflicts among his characters as with the production of psychological tension in his readers—becomes the appropriate vehicle for "descriptive psychology" in the sense sketched above: as "pure (apocalyptic) perception," it is a prolegomenon to pure phenomenology.

Philology

Apocalyptic psychology prepares a place for messianic philology. Soon after Benjamin completed "Toward the Critique of *Gewalt*," he came across a literary work that is even more thoroughly permeated by the theme of corporeal expansion and contraction than Scheerbart's little novel: an artistically rendered fairy tale entitled "Die neue Melusine" (The new Melusine), which Goethe wrote around the time he completed *Elective Affinities*, but which he published only in the last of his novels, *Wilhelm Meisters Wanderjahre* (Wilhelm Meister's journeyman years). Whereas Melusine is traditionally represented as a nixie or water sprite, the new Melusine belongs to a society of terrestrial dwarfs, who were created so that the wonders of the subterranean world would be duly admired. Melusine is charged with the task of saving her people from ever-greater diminution by mating with a terrestrial "knight." With the assistance of a magic ring, she is able to increase her size and again shrink to her original form, at which point she can enter into the casket that the decidedly nonknightly narrator of the tale carries with him, as he aimlessly wanders about Europe.[5] For at least a decade Benjamin expressed an interest in writing a commentary on Goethe's tale, occasionally in association with his "Politics" (*GB*, 2: 15), occasionally in conjunction with his "habilitation" projects (6: 23); but the most extensive description of his plans can be found in a letter to Scholem from February 1921, where he describes the circumstances in which his intention first formed:

> I have had (as I did in Switzerland) some thoughts about philology. Clear to me was always the seductive side. It seems to me—and I do not know whether I understand it in the same sense as you—that philology, like all historical modes of research, but here to the highest degree, promises the enjoyments that the Neoplatonists sought in the ascesis of contemplation. Perfection instead of completion, guaranteed extinguishing of morality (without smothering its fire). It presents a side of history [Geschichte], or better yet, a layer of the historical [eine Schicht des Historischen] for which the human being is perhaps able to acquire regulative, methodical concepts as well as

constitutive, elementary-logical ones; but the connection [Zusammenhang] among them must remain closed to the human being. I define philology not as a science or history of language but, rather, in its deepest layer as a *history of terminology*, in which one is then certainly concerned with a highly enigmatic concept of time and very enigmatic phenomena. If I am not mistaken, and without being able to develop it, I also get a sense of what you mean when you say that philology stands close to history from the side of the chronicle. The chronicle is history interpolated in principle. In its form philological interpretation in chronicles simply brings the intention of the content to light, for its concept interpolates history. What kind of project this could be has now become vivid to me in a work that has seized me in the deepest way and spurred me to interpolation. It is "The New Melusine" of Goethe. Are you familiar with it? If not, it is unconditionally recommended that you read this story, which can be found in [Goethe's] *Journeyman Years*, for itself, that is, *without* the frame in which it can be found, just as I came across it by accident. Should you be familiar with it, I can perhaps indicate a few things.—I don't know whether you can do anything with the oracular statements about philology. Rest assured that it is altogether clear to me that one must attain an approach to this matter that is other than "romantic." (I see once again in your letter. Chronicle—interpolation—commentary—philology. That is *one* connection [*einen* Zusammenhang]. . . . When the sage is not dealing with the Bible, he certainly won't direct his philology toward the end of the aforementioned series but, rather, toward its beginning.) (*GB*, 2: 137–38)

The letter Scholem wrote to Benjamin in which he discusses the series "chronicle-interpolation-commentary-philology" has unfortunately been lost; but some of the "thoughts on philology" that occupied Benjamin during his time in Switzerland are preserved in a document entitled "Methodischen Arten der Geschichte" (Methodical modes of history), which succinctly captures the intersection of two lines of inquiry he pursued as a student: the first line takes its point of departure from his conversation with Scholem in August 1916, during which they discussed the meaning of "course [Ablauf], series of years, and as the final starting point, direction" (S, 1: 390); the second line derives from Benjamin's effort to "gain entrance" into Husserl's "school" (*GB*, 1: 301–2). With respect to the conversation in the summer of 1916, Scholem places the following note in his diary: "This is a thought complex that I very much want to think about again" (S, 1: 390). And Benjamin, too, returns to the "thought complex" in the aforementioned document, which begins by recalling, with only slight modifications, two of the major terms that came under discussion: "Generally speaking, a history is a unidirectional process [ein einsinniger Verlauf]" (6: 93). Three modes of historiography are then distinguished from one another: pragmatic history, "phenomenon-history," and philology. About the first, Benjamin only notes that it "proceeds temporally, in battles." This is contrasted with "phenomenon-history," which "concerns the series of phenomenological (not temporal) presuppositions of phenomena; it, too, is unidirectional (areas of application, for example, the history of art and nature)" (6: 93). The objects of "phenomenon-history" are thus phenomena in the phenomenological sense: essences or essentialities, which are presupposed in any discussion of history as the structures that make it possible to say that the history of a natural kind or an artistic genre is concerned with the *same* natural kind or the *same* artistic genre. The processional character of "phenomenon-history" has nothing to do with the flow of time but consists, rather, in the serial arrangement of the discrete essences or essentialities that are therein identified and described. As for the object of the even higher mode of historiography, namely philology, it is characterized by a degree of continuity in its object that is categorically different than the continuity attributed to the flow of time, as it is experienced by consciousness, or the ordering of phenomena qua essences. Around the time in which Benjamin sketched the three modes of historiography he created a term to describe "truth" that captures, to a certain extent, the higher mode of continuity that characterizes the objects of philological inquiry: "thi(n)ck [denkicht]" (*GB*, 1: 409). The philological object is so thick with thinking that its *terminus*, limit, horizon, or "definition" can lie only in the completion of the entire "terminological process":

> Philology is concerned with the process that is neither essentially temporal nor discloses essentially separated phenomena: the terminological process. Philology is the history of transformation [Verwandlungsgeschichte]; its unidirectionality rests on the fact that terminal[ogy] does not become a presupposition but, rather, the material of a new [terminology], and so forth. In philology the object has the highest continuity. Unidirectionality is particularly modified in philology, since it inclines in the end toward the cyclical. This history has an end but no goal. (Example: intellectual history [Geistesgeschichte], history of the Enlightenment). (6: 93–94)

To the extent that every new terminology is a transformation of a previous one, there can be no room for neologisms that derive from "judgments of designation" (6: 9). Any attempt to create new terms damages the continuity that characterizes the object of philology. For this reason, Benjamin calls the introduction of new terminologies "worrisome" (1: 217), and for related reasons he singles out Kant, whose terminology immediately acquires the kind of continuity that characterizes the object of philology (*GB*, 1: 389). The fact of philology—to recall a formula from Hermann Cohen—is therefore of compelling interest. If every new terminology is a transformation of a previous one, then the original terminology could not have developed in a rectilinear fashion, as new discoveries give rise to new words; rather, the original terminology must have continually transformed itself. In this context, philology is not so much evidence of an original language of names as a guarantee that there is, after all, such a language "in" every human language. Beyond a few remarks about the manner in which the three modes of historiography can and cannot be combined with one another, Benjamin goes no further in the notes he drafted in Switzerland. As he then reflects on an expansion of "Toward the Critique of *Gewalt*," and simultaneously searches for an appropriate topic for a *Habilitationsschrift*, he comes across two items that propel him further: Scholem's lost letter, which introduces the terminology of "interpolation," and Goethe's "New Melusine," which is concerned with nothing so much as cycles of enlargement punctuating a unidirectional history of shrinkage.

Benjamin places a remarkable condition on his otherwise unconditional recommendation that Scholem familiarize himself with "The New Melusine": the tale must be isolated from the framework in which it originally appeared. Much of what Benjamin writes about philology in his letter to Scholem is enigmatic—and described as such—but the reason for this recommendation is straightforward enough: the content of the tale should not be subsumed under that of the novel, whose subtitle, "the renunciants," betrays its dominant theme. The narrator of the tale is primarily characterized by his failure to fulfill the demand for renunciation that the new Melusine imposes on him: "With incredible skill she withdrew from my arms, and I could not even place a kiss on her cheek. 'Restrain yourself from such outbreaks of passion, if you do not wish to forfeit a happiness which lies very close to you but which can be seized only after a few tests.'"[6] If "The New Melusine" is encountered as part of the *Journeyman Years*, there is a powerful temptation to interpret it as a counternarrative of failed renunciation, in which the withdrawal and reduction of the erotic object is correlated with the advancement and enlargement of the erotic drive. Loss would thus emerge as the constitutive characteristic of the object, and renunciation its dialectical affirmation. A purely negative conclusion can thus be drawn from Benjamin's recommendation that "The New Melusine" be abstracted from the novel in which it appears: its content cannot be understood in terms of loss. It follows that the aim of philology, as the "love of words," does not consist in the restoration of whatever has been lost—whether it be the erotic object or the words of the author. Benjamin then borrows a term from Scholem as a replacement for the philological program of restoration: "interpolation."

Interpolation

In the winter semester of 1919–20, while look for an appropriate dissertation topic, Scholem attended Moritz Geiger's lectures on mathematics and philosophy. At the end of a letter from January 1920 that outlines the chapters for a volume tentatively entitled "Politics," Benjamin asks Scholem about the content of Geiger's lectures, and immediately adds that he is considering whether or not he should ask Käthe Holländer to tutor him in mathematics (*GB*, 2: 109). Scholem's response to Benjamin's query has been lost; but there is little reason to suppose that he was drawn to Geiger, whom he generally ignores, and there are at least two interrelated reasons to assume that he viewed Geiger with a certain suspicion: Geiger's grandfather, Abraham, developed the theoretical framework for the form of liberal German Judaism Scholem had long before rejected, and by 1920, as an expression of this rejection, he had decided to pursue a doctorate in Jewish philology rather than continue his study of mathematics. This decision, it appears, was the principal motivation for his choice of Munich over Göttingen: the latter had an outstanding faculty of mathematics, whereas the former had an enticing collection of kabbalastic material.[7] Geiger, as it happens, went in the opposite direction, for soon after delivering the lectures that Scholem attended, he moved to Göttingen, having been chosen over Heidegger for a vacant position in the faculty of philosophy.[8] The fact that Scholem attended Geiger's lectures in Munich indicates, however, that he retained a residual interest in mathematics, even as he began preparing himself as a philologist. And in the concept of interpolation his two interests momentarily intersect, for *interpolation* is as much a mathematical as a philological term.

In philology, *interpolation* means, roughly speaking, either the inclusion of something for which there is no direct textual evidence but which can be justified by extraneous considerations or the recognition of an element as a later insertion; in mathematics, it means the process of finding a value of a function between two known values by a procedure other than the law that is given by the function itself. In both cases, the *-polation* part of *interpolation* refers to an act of "polishing": texts that have been corrupted in the course of their transmission are "polished" by the philologist, so that they accurately recover the original intention of the author; and trigonometric or logarithmic tables, which can obviously list only a few values, must be "polished" by anyone who wants to determine a missing value.[9] Every act of interpolation presupposes and thus ultimately refers to a certain discontinuity, which the philologist or mathematician is called upon to rectify. The simplest forms of interpolation within the field of philology are the addition of elements that are left out and the identification of inauthentic accretions. Since kabbalistic texts include multiple forms of abbreviation and have undergone uncertain processes of transmission, they solicit interpolation on both accounts.

The simplest form of interpolation in the field of mathematics is the drawing of a straight line through two points. The generalized version of this process is called "linear interpolation," which connects a set of points by means of straight lines. The obvious disadvantage of linear interpolation lies in the fact that the resulting curve generally takes a sharp turn at every such point, which not only means that the curve has no direction at these points but also that it is not therefore everywhere differentiable. Higher-order methods such as polynomial interpolation, by contrast, produce more polished curves. The philological counterpart to polynomial interpolation is the fully edited and emendated text, which flows without fail from first to last letter under the sign of authorial intention. The presupposition of philology, so conceived, is that the intention may have been lost or distorted, but that a carefully controlled use and recognition of interpolation can nevertheless recover or restore it. In encountering the collection of kabbalistic writings in the library at the University of Munich, while attending lectures on mathematics by Geiger, Scholem may have been drawn to the conjecture that philological and mathematical modes of interpolation are equivalent in at least one respect: both tend toward greater degrees of smoothness by means of procedures other than the law that is given by the function (mathematics) or the tradition (philology) itself.

Whatever "spurs" Benjamin to interpolation in the case of Goethe's tale, however, it cannot be either the poor quality of the text or an uncertain process of transmission. No writer in the German literary tradition stands less in need of philological interpolation, as it is usually understood, than Goethe. This is especially true of "The New Melusine," which presents itself as exemplary polishing of the "popular" genre of fairy tale. There is, to be sure, something unpolished about *Wilhelm Meister's Journeyman Years*, which continues to baffle its readers; but Benjamin extracts the tale from its setting, and the novel in any case provides a succinct theory of its unevenness in one of the aphorism that interrupts the narrative: "The master presents his work as finished after only a few strokes; worked out or not, it is nevertheless complete."[10] Benjamin may even be alluding to this aphorism at the beginning of his letter to Scholem, when he describes the ecstasis accompanying philology as an inversion of the master's—which is to say, Wilhelm Meister's—characteristic gesture: "perfection instead of completion." Philology seeks neither to complete a text nor make its rough patches smooth; rather, "philological interpretation," as Benjamin emphasizes, consists in recognizing interpolation after interpolation and thus reducing the text to a point where, like Melusine in the absence of a rejuvenating "knight," it begins to vanish. So close is the relation between interpretation and interpolation that Scholem appears to have mistaken one for the other. Thus, when he edited the letter for inclusion in his volume of Benjamin's correspondence, he reads "philological interpolation in chronicles" as "philological interpretation in chronicles."[11] Regardless of which edition successfully recovers his intention, the letter unambiguously identifies the first term of the "series" in which interpolation occupies the second place: *chronicle*. The chronicle is reduced history, in which every addition is recognizable as an interpolation, for none is based on an authenticating intention. The essence of philology—its "concept," as Benjamin notes—lies in identifying chronic accretions. And with the fact of philology, there emerge some "very enigmatic phenomena" and "a highly enigmatic concept of time."

The chronicle is distinguished from other modes of historiography in that it "chronicles" occurrences without any intention of determining how they follow from one another, much less how one epoch is related to another. The spatial equivalent of the daily chronicle is the nightly manifold of stars, which, by soliciting interpolation, give rise to the constellations that can be read as a text in the absence of an author. A stellarlike discontinuity characterizes the units of the chronicle, even as it continues to present the course of time in calendrical terms. The chronicle can therefore be described as the generic equivalent of what Benjamin calls "chronology" (*Zeitrechnung*) in the following remark, which Scholem twice records: "The problem of historical time is already posed through the peculiar form of historical chronology" (S, 1: 402; 2: 601). A chronicle without diurnal interpolation is chronology as such. Both of the documents where the previous remark appears connect it to a "difficult remark" with which Benjamin began a conversation with Scholem during an afternoon in August 1916: "The years are countable but, in contrast to most countables, not numerable" (S, 1: 390). The conversation turned away from the structure of chronology to the problem of determining the precise shape of time. The "natural" way to represent the shape of time is clear: as a rectilinear flow from past to future. Kant claims something to this effect at the very heart of the *Critique of Pure Reason*: "we cannot represent time without, in *drawing* a straight line (which is to be the external figurative representation of time), attending merely to the action of the synthesis of the manifold" (K, B 154). Perhaps with the help of Hausdorff's "epistemo-critical essay," Scholem and Benjamin respond to Kant's claim by describing it as uncritical in the extreme: "It is a thoroughly metaphysical assertion that time is like a straight line."[12] Two further suggestions regarding the shape of time emerge in the course of the conversation, the last of which is left in a highly uncertain state: "Perhaps it is a cycloid or something else, which has no direction at many points. (Where there are *no* tangents.)" (S, 1: 390).

Origin of a Kabbalah Scholar

The conversation that Benjamin and Scholem conducted in a villa on the outskirts of Munich in the summer of 1916 exercised so powerful an influence over Scholem that it gave direction to all of his future scholarship. Not only did it prompt him to work out a "mathematical theory of truth" (S, 1: 417); it also established the character of the research he would undertake after he decided to abandon his mathematical studies and devote himself to Jewish philology instead. Nowhere is the effect of the conversation more starkly evident than in an open letter Scholem wrote to the publisher Zalman Schocken in October 1937, shortly before he delivered the series of lectures that would form the nucleus of his groundbreaking study, *Major Trends in Jewish Mysticism*, which is dedicated to the memory of Walter Benjamin. In recounting the unlikely set of events that led him into the study of the Kabbalah, Scholem begins with the year in which the conversation was held and refers to mathematics under the general rubric of rationalism: "Three years, which were determinative for my entire life, 1916–1918, lay behind me: a very excited mode of thinking led me as much to the most rationalistic skepticism about my objects of study as to an intuitive affirmation of mystical theses, which lay precisely on the border between religion and nihilism."[13]

Scholem draws the letter to a close by slightly altering the series of terms he had identified in the early 1920s: *chronicle-interpolation-commentary-philology*. Projective geometry takes the place of interpolation, and the Kabbalah occupies the position that was originally reserved for chronicle: "In the strange concave mirror of philology, for contemporary humanity, the mystical totality of the system, whose existence disappears precisely when it is projected into historical time, can become visible at first and in the purest way only in the legitimate orderings of commentary." By way of geometric projection (rather than "interpolation"), Kabbalah (rather than "chronicle") gives way to the "orderings of commentary," which interrupt the apparently natural flow of "historical time." And the entire series of terms can be described by the last of its members, namely "philology." At the very end of the letter to Schocken, Scholem resumes his reflection on the conversation with Benjamin in the summer of 1916 and presents the structure of history in accordance with the last of the shapes that came under discussion: "Today, as on the first day"—which means precisely, August 20, 1916—"my work lives in this paradox, in such a hope for a correct responsiveness [Angesprochensein] from the mountain, for that most invisible, smallest dislocation of history, which allows truth to break from the illusion of 'development.'"[14] To the extent that every development has a discernable direction, it can be represented by a curve that is everywhere differentiable. When, however, the "displacement" or "deferral" (*Verschiebung*) of history is set against the idea of its "development," it follows as a matter of course that the shape of time can be captured only by "something . . . which has no direction at many points." At those nondifferentiable points in the course of time, an integral and thus "mountainous" communicability arises in an otherwise leveled landscape of "pure chatter" (S, 2: 197).

"The Turn of Time"

The "highly enigmatic concept of time" that emerges from Benjamin's reflections on the same series goes even further. Its point of departure lies in the idea of "temporal plastics" (2: 120), which Benjamin introduces at the end of his dense inquiry into the spatio-temporal order that can be generated from Hölderlin's late poetry: "The phrase 'turn of time' plainly captures the instant [Augenblick] of persistence, the moment [Moment] of inner plastics in time" (2: 119). The "plastic" character of time does not appear to mortals, who experience time as a flux, and it appears to the gods only insofar as they succumb to their own Apollonian plasticity and thus become mere objects which the poets handle. Here is the structural place for the "bringing"—not the birth—of another god, about whom, however, Benjamin has almost nothing to say. If, as he would later tell Scholem, he found himself unable to build on the "magnificent foundation" (*GB*, 3: 521) that he laid out in his early years as a student, especially his "aesthetic commentary" (2: 105) on Hölderlin's poetry, at least one reason for this can be discerned from the letter in which he first discusses the series *chronicle-interpolation-commentary-philology*: he rashly began with the third element of the series rather than with the first: "When," however, "the sage is not dealing with the Bible, he certainly won't direct his philology toward the end of the aforementioned series [commentary] but, rather, toward its beginning" (*GB*, 2: 137)—namely "chronicle." And in the absence of the second term of the series, "interpolation," philology can concern itself only with a "temporal process" in which its "unidirectionality . . . inclines in the end toward the cyclical" (6: 94). To quote once again from the crucial passage from Scholem's account of the conversation in the summer of 1916: "Perhaps [the shape of time] is a cycloid or something else, which has no direction at many points. (Where there are *no* tangents.)" The cycloid is, of course, generated by a cyclical movement. With the inclusion of interpolation in the series, the concept of time with which the philologist is concerned alters: it becomes considerably more enigmatic, as the "plastic" character of time, which finds partial expression in the Bergsonian idea of duration, inclines toward another kind of shape altogether.

The elements of the alteration are the following: whereas higher-order methods of interpolation in mathematics tend to produce smoother curves, philological interpolation moves in the opposite direction and generates ever greater degrees of roughness. The highest degree of roughness can be found in a curve that takes a sharp turn at *every* point—not simply, as in Scholem's letter to Schocken, at certain luminous points, which would represent the moments in which a distorted and inverted image of the mystical whole becomes visible. So construed, the course of time that accords with the philological mode of historiography acquires the "highest degree" of continuity precisely because its connectivity cannot be intuited, imagined, or phenomenalized in any sense. The "turn of time" can thus appear only under the condition that the natural-mythological attitude be "turned off." The absence of lines tangent to the curve under consideration means that there is no law whereby it can be "touched" (4: 19). In other words, the course of time, so construed, is inviolate. In still other words, it is innocent. For this reason, it acquires the function that Benjamin describes in a fragment entitled "Die Bedeutung der Zeit in der moralischen Welt" (The meaning of time in the moral world): for the criminal, time appears as a "storm of forgiveness" (6: 98). And for the same reason, time is altogether different from the process of history. In a few dense notes Benjamin probably wrote in conjunction with his plan to write a dissertation on Kant's theory of history, he distinguishes the categories of history from those of nature. Unless the two sets of categories are distinguished from each other, history cannot gain independence from what Kant calls "natural history." If, so the argument goes, the broadest category of nature is causality, then the highest category of history can be defined as "guilt" (*Schuld*), which doubtless produces a nexus of effects but which is not itself an effect that can be attributed to a prior cause. Benjamin's interest in the theme of "inherited" or "original" sin, on the one hand, and his reflections on the religious character of capitalism, on the other, derive from a basic thesis that finds succinct expression in these early notes: the category of guilt gives history its directionality—toward ever-greater guiltiness. Unlike causal interaction, the historical process is therefore irreversible: "In order to guarantee the unidirectionality of every occurrence, the highest category of world history is guilt [Schuld]. Every world-historical moment is indebted and indebting [Jedes weltgeschichtliche Moment verschuldet und verschuldend]" (6: 92).

The "unidirectionality" or "one-senseness" (*Einsinnigkeit*) of world history becomes obscured when a second stratum of sense is added, which would hold out the illusion of reversibility; but the addition is in this case a subtraction, for the "historical process" then succumbs to the "ambiguity" (*Zweideutigkeit*) that, for Benjamin, characterizes myth. In myth, there is always an illusory way back; in history, only a one-way street. The process of history is therefore constitutively at odds with the course of time: the first is unidirectional, the second without any direction whatsoever.[15] Thus does

Benjamin correct himself in the letter to Scholem in which he first discusses "The New Melusine." Philology is not so much concerned with "history" (*Geschichte*) as with "a layer of the historical" (*eine Schicht des Historischen*). He approaches this enigmatic layer by way of the "Kantian typic" (2: 160): one side consists in "regular, methodical" concepts, the other in "constitutive, elementary-logical" categories. Each of the two sides is "perhaps" accessible to human beings—but certainly not the "connection" (*Zusammenhang*) that binds them together and thus makes the layer into *one*. The connection withdraws from both the *Wesensschau* of phenomenology and the generative synthesis of epistemo-critique. The highest regulative concept is identified in the previously discussed notes: it is the concept of guilt, which governs history; one of the "constitutive, elementary-logical" concepts—and there is perhaps only one—comes under discussion in the notebook that Scholem borrowed from Benjamin in October 1916, namely the category of justice, which characterizes "the existent" (S, 1: 401). The connection between the two sides of the layer, as Benjamin emphasizes, "must remain closed to the human being." Its name, however, is "time," which philology does not disclose but for which it provides surety.

To discover the contradirectional in history—which is likewise "contrasensical," where history rubs against the grain—becomes the philological problem par excellence. Thus, at the very beginning of the methodological reflections Benjamin included among the folders for his *Arcades Project*, he—as if he were amplifying Scholem's open letter to Schocken—identifies the "deviations" from the path that leads to the so-called "true north," as opposed to the magnetic north pole, with the "data that determines my course," and then adds: "I construct my calculation on the differentials of time, which, for others, disturb the 'great line' of the investigation" (5: 570; N, 1, 2). A differential of time in this case is precisely *not* its direction but, on the contrary, its deviation from the very direction that nonphilological modes of historiography are trying not so much to discover as to enforce, in conflict with compasses that always only point elsewhere.

During their conversation in August 1916 Scholem doubtless discussed with Benjamin the so-called Weierstraß function, which he could have also discovered in a variety of other sources, written and oral. The professor of mathematics under whom Scholem studied in 1916, Konrad Knopp, was in the process of writing papers not only on the Weierstraß function but also on the Koch curve, which is generated by the process of endlessly dividing every segment of the line into three equal parts and replacing the middle part by an equilateral triangle.[16] "What is, above all, striking in the Koch curve," Ernesto Cesàro writes in a paper from 1905, "is that *it is in all of its parts similar to itself*. This endless interlocking of a figure in itself gives us the image of what Tennyson somewhere called the *inwardly infinite*, which is, after all, the only infinite that we can conceive of in nature."[17] Knopp was promoted to professorial rank on the basis of a contribution to the theory of convergent series in which he showed that the limit procedures of Cesàro and Otto Hölder were equivalent. It goes without saying that he would have been familiar with Cesàro's paper, and it is far from unlikely that he would have discussed its insight with some of his students, including Scholem. And what an insight it is: there are certain nondifferentiable curves in which each segment is similar to every other segment.[18] If the course of time can be captured by a curve of this kind, its concept can be aptly described as "highly enigmatic," for every time, down to the smallest unit, would be similar to every other time and to time as a whole. The question of philological interpretation as interpolation then becomes: how to uncover these "very enigmatic phenomena," which recede from intuition? Answer: by means of a thoroughgoing *epochē*. Since every time is like every other time, every stretch of time can be identified with a technical term introduced into philosophy by Leibniz, who also invented the notation generally used for the differentiation and integration of continuous curves: *monad*. Just as every monad mirrors the universe as a whole, so does every time recapitulate all of time.[19] History, interpolated in the form of a "constellation," acquires the monadic character of time by virtue of an *epochē* whose unity is of a higher "power" than that of any activity of thinking that directs itself toward immanent objects of thought: "Where thinking suddenly halts [einhält] in a constellation saturated with tensions, it imparts to this constellation a shock through which it crystallizes as a monad" (1: 701–2).

Every time recapitulates, without ever exactly repeating, all of time. The circular or cycloid character of the "eternal return of the same" is thus broken open—without time taking on a telos in the process. At the end of certain versions of the document from which the remark above is drawn, "Über den Begriff der Geschichte" (On the concept of history), Benjamin cites a passage from an unnamed scientist, who describes the short stretch of time in which human beings have existed on earth, and then makes the following comment: "The now-time [Jetztzeit], which, as a model of messianic time, embraces the history of all humanity in an enormous abbreviation [ungeheueren Abbreviatur], coincides precisely with *the* figure that the history of humanity makes in the universe" (1: 703). The "enormous abbreviation" that characterizes "now-time" derives from the shape of time: every segment of time, no matter how small, is like all of time; the smaller the segment, the more "enormously" does its recapitulative character appear. The appearance makes the "now" of time—which is to say, its shape—recognizable. In order for historiography to correspond with time without succumbing to myth, the philologist cum historian must abstain from attributing to the "historical process" the essential characteristic of time: continuity. Since the directionality of this process owes its origin to guilt, which always deepens, the "times" in which the historical continuum collapses cannot fail to be redemptive.

If traditional philosophical terms can be used to capture the character of time, without disclosing its shape, they would be the ones through which Benjamin determines the highest task of the coming philosophy: that of identifying "a sphere of total neutrality in relation to the concepts of object and subject" (2: 153). So conceived, messianic time is not another time; it is just time—time and nothing but "plastic" time. The paradisal character of space, toward which the painterly plane tends, accords with the messianic character of time, which is thus charged with tension and can be called "full" because every stretch of time contains all of time. In "World and Time" Benjamin says of the first category: "there is nothing continuous" (6: 99); the opposite condition characterizes the other concept under consideration, namely time, which, as Benjamin briefly suggests, should actually be called "the coming world." By virtue of its nondirectional continuity, time—"turned" by "now"—is the "coming world."

Reference Matter

Appendix: Translations

The Rainbow: Dialogue on Fantasy
Dedicated to Grete Radt

[7: 19] MARGARETHE: It is early in the morning, I was afraid to disturb you. And yet I could not wait. I want to tell you a dream before it fades away.

GEORG: How delighted I am when you come to me in the morning—because I'm then entirely alone with my images and do not expect you at all. You've gone through the rain, which has refreshed you. Now tell.

MARGARETHE: Georg—I see that I cannot. A dream doesn't allow itself to be said.

GEORG: But what have you dreamed? Was it beautiful or terrible? Was it an experience, one with me?

MARGARETHE: No, nothing like this. It was entirely simple. It was a landscape. But it glowed in colors; I have never seen such colors. Even painters know nothing of them.

GEORG: They were the colors of fantasy, Margarethe.

MARGARETHE: The colors of fantasy, so it was. The landscape shimmered in them. Every mountain, every tree, leaves: they had infinitely many colors in them. Indeed, infinitely many landscapes. As if nature vivified itself in a thousandfold innateness [Eingeboren-Sein].

GEORG: I know these images of fantasy. I believe that they stand within me when I paint. I mix the colors, and I then see nothing but color. I'd almost say: I am color.

MARGARETHE: So it was in my dream. I was nothing but seeing. All other senses were forgotten, vanished. Even I myself [7: 20] did not exist, nor my understanding, which discloses things from the images of the senses. I was not a viewer, I was only viewing. And what I saw were not things, Georg, only colors. And I myself was colored in this landscape.

GEORG: What you describe is like ecstasy [Rausch]. Remember what I told you about that strange and exquisite feeling of drunkenness which I know from earlier times. I felt myself altogether weightless in these hours. Above all, I perceived only what let me be in things—their properties, through which I permeated them. I was myself a property of the world and hovered over it. The world was filled by me, as it was filled by color.

MARGARETHE: Why have I never found in those images made by painters such glowing, pure colors, the colors of the dream? For the place from which they come, their source—fantasy, and which you compare to ecstasy, the pure reception in self-forgetting—that is the soul of the artist. And fantasy is the most inward essence of art. Never have I seen this more clearly.

GEORG: Even if it were the soul of the artist, it is not yet for this reason the essence of art. Art creates. And it creates objectively, that is, in relation to the pure forms of nature. Think about this—and often you have done this with me--: to the forms. Nature creates according to an infinite canon, which grounds infinite forms of beauty. They are forms, they all rest in the form, in the relation to nature.

MARGARETHE: Do you mean that art copies nature?

GEORG: You know that I do not think so. It is true that at bottom the artist always only wants to grasp nature; he wants to receive it purely, to know it formally. But the inner, productive forms of reception rest in the canon. Consider painting. It does not proceed from fantasy, from color; rather, it proceeds from the spiritual, the creative, from form. Its form is to grasp living space. To construct it according to a principle, for the living is never received except via generation. The principle is its canon. And so often I have thought this over, I've found that, for painting, this principle is spatial infinitude, just as, for the plastic arts, it is spatial dimension. Color is not the essence of painting; rather, the essence is the plane [die Fläche]. In it, in depth, space lives according to its infinitude. In the [7: 21] plane, the existence of things unfolds itself *toward* space, not actually *in* space [*zum* Raum, nicht eigentlich *in* ihm]. And color is only the concentration of the plane, the forming of an image of infinitude into it. Pure color is itself infinite; but in painting only its reflection appears.

MARGARETHE: How are the colors of the painter distinguished from those of fantasy? And isn't fantasy the ultimate source of color?

GEORG: So it is, although this is amazing. But the painter's colors are relative as opposed to the absolute color of fantasy. Pure color is only in intuition; the absolute exists only in intuition. Painterly color is only a reflection of fantasy. In painterly color fantasy actually turns into creation; it makes transitions with light and shadow; it is impoverished. The spiritual basis in the image is the plane, and if you truly have learned to see, this is what you see: the plane illuminates the color—not the other way around. Spatial infinitude is the form of the plane; it is the canon, and color emanates from it.

MARGARETHE: You will not be so paradoxical as to say that fantasy has nothing to do with art. And its canon may be spiritual; it may mean a creative forming of vividness—which is related, of course, to nature only in infinite possibilities—nevertheless, the artist also receives. The simple beauty, the vision, the gladdening of pure seeing appears to him not less but, on the contrary, far more and far more deeply than it appears to us others.

GEORG: How do you understand the appearing of fantasy? Do you mean it as model [Vorbild] and the creating as a copy [Abbild]?

MARGARETHE: The creator knows no model and therefore none in fantasy. I do not mean it as a model but rather as an archetype [Urbild]. As the appearing into which he merges, in which he remains, which he never leaves and which originates from fantasy.

GEORG: The muse gives the artist the archetype of creation. You've spoken the truth.—And what is this archetype other than a surety bond [Bürgschaft] of the truth of his creation, the guarantee [Gewähr] that he is one with the unity of spirit, from which mathematics originates no less than plastic arts, history no less than language. What else does the muse pledge for the poet by means of the archetype, what other than the canon itself, the eternal truth, which lies at the basis of art. And that ecstasy which rushes through our nerves during moments of the greatest spiritual clarity, the consuming ecstasy of [7: 22] creating—this is the consciousness of creating in the canon, in accordance with the truth we fulfill. In the writing hand of the poet, in the painting hand of the artist, in the fingers of the musician, in the movement of the one who makes forms, in the singular impulse, the complete immersion in gesture, which he, divinely animated, intuits in himself—himself, the one who forms an image, as a vision, his hand led by the hand of the muse—it is here that fantasy, as an intuition of the canon in the viewer and in the things, holds sway. As unity of both in the intuition of the canon. Only this holding sway [Walten] of fantasy leads from the ecstasy of the delighted spectator, about which I spoke, to the ecstasy of the artist. And only where the artist strives to make the archetype into a model; where he wants to take possession of the spiritual in an unformed manner, by contrast, formlessly intuiting something, the work becomes fantastic.

MARGARETHE: If, however, we say that fantasy is the gift of pure receptivity in general, aren't we stretching its essence into the immeasurable? For fantasy is then in every movement that is done in an entirely pure manner, entirely self-forgetting, in intuition, so to speak—in dance and song and walking and language just as much as in the pure seeing of color. And yet why did we want to see fantasy primarily in the essence of color?

GEORG: Certainly there is also in us a pure intuition of our movement, especially of our generating things in any number of ways, and it is on the basis of this intuition, I believe, that the fantasy of the artist rests. But color nevertheless remains the purest expression of the essence of fantasy. For, among human beings, no creative capacity corresponds to color. Line is not purely received, because we can transform it by a movement in the mind, and tone is not absolute, because we have the gift of voice. They are not of the pure, the untouchable, the appearing beauty of color.—Of course, I see that there arises in the face a particular region of human senses to which no creative capacity corresponds: color perception, smell, and taste. See how clearly and sharply language designates this. It says the same thing about these objects as about the activities of the senses themselves: they smell and taste. But from their colors: they look [sie sehen aus]. For one never says anything of the kind about objects when one wants to designate their pure form. Do you get an intimation of the secret, deep region of spirit that begins here?

MARGARETHE: Hasn't this intimation come earlier to me than to you, Georg? [7: 23] Still, I want to lift colors purely from the secret realm of the senses. For, the more deeply we climb up that second realm of receiving senses, to which no creative capacity corresponds, the more vexed its objects become in terms of their substantiality, and the less the senses are permitted to sense pure properties. One cannot receive them for themselves alone, with a pure, detached sense, but, rather, only as a property of a substance. But color originates, for this reason, in the most inward region of fantasy, because it is only a property; in nothing is it substance or does it even refer to substance. Therefore only this can be said of it—that it is a property, not that it has a property. For this reason colors have become the symbols for those who are bereft of fantasy. In color the eye is purely turned toward the spiritual; color spares the path of whomever creates in nature by means of form. In pure reception it allows sense immediately to encounter the spiritual, upon harmony. A viewer is entirely in the color; to see it means to immerse the gaze in a foreign eye, where the viewer is swallowed—in the eye of fantasy. Colors see themselves; pure viewing is in them, and they are its object and organ at the same time. Our eye is colorful. Color is generated from viewing and colors from pure viewing.

GEORG: You have beautifully said how the genuinely spiritual essence of the senses, how reception, appears in color—how color, as this spiritual, this immediate, is the pure expression of fantasy. Only now do I understand what language says when it speaks of what things look like [Aussehen]. It points toward the vision of color. Color is the pure expression of world-intuiting [Weltanschauen], the overcoming of the viewer. By means of fantasy it touches itself with smell and taste, and the noblest human beings will freely develop fantasy in the entire region of their senses. I, for one, believe that chosen spirits receive purely from themselves certain fantasies of smell, even of taste, as others do fantasy of color. Don't you remember Baudelaire?[1] These uttermost fantasies will even be a surety of innocence, since only pure fantasy, from which they flow, is not desecrated by mood and by symbols.

MARGARETHE: You call innocent that region of fantasy in which sensations still live in themselves, purely as properties in themselves, [7: 24] still unclouded [*ungetrübt*] in the receiving spirit. Isn't this sphere of innocence the sphere of children and artists? I now see clearly that both children and artists live in the world of color. That fantasy is the medium in which they receive and create. A poet once said: "If I were made of material, I would color myself."[2]

GEORG: To create by receiving is the perfection of the artist. This reception out of fantasy is no reception of the model but, rather, of the laws themselves. It would unite the poet with his figures [Gestalten] in the medium of color. To create entirely out of fantasy would mean to be divine. It would mean to create entirely from the laws, immediately and free from all the relation to them by means of forms [Formen]. God creates out of an emanation of his essence, as the Neoplatonists say, for this essence would be nothing other than fantasy, from whose essence the canon arises. Perhaps the poet recognized this in color.

MARGARETHE: Thus only do children dwell entirely in innocence, and in blushing they themselves return to the existence of color. In them fantasy is so pure that they are able to do so.—But look, it's stopped raining. A rainbow.

GEORG: The rainbow. Look at it; it is only color, nothing in it is form. And it is the emblem [Sinnbild] of the canon, as it divinely arises from fantasy, for in it the consequence of beauty is that of nature. Its beauty is the law itself, no longer beautiful in nature, no longer transformed in space, no longer equal by means of sameness, symmetry, and rules. No longer beautiful by means of forms derived from the canon—no, beautiful in itself. In harmony, for it is canon and work at the same time.

MARGARETHE: And does not all beauty return to this bow as an emblem in which the order of beauty appears as nature?

GEORG: So it does. The canon stands in pure intuition and appears only in color. For in color nature is spiritual, and considered from its spiritual side it is purely colorful. It is really an archetype of art in accordance with its existence in fantasy. Nature lives most inwardly in color, as the community of all things that are not creating, not created. Nature received in pure intuition. All objectivity of art goes back to it.

MARGARETHE: If I could tell you how familiar color is to me! [7: 25] A world of remembrance surrounds me. I am thinking of children's colors. How, in childhood, it is everywhere received in a pure manner as the expression of fantasy. Dwelling within harmony, on nature, in innocence. The motley and monochromatic, the beautiful strange technique of my oldest picture books. Do you know how, in these books, everywhere contours were blurred in a rainbow-like play of colors, how heaven and earth were stroked in India ink with invisible colors! How colors, as if with wings, always hovered over things, they really colored and swallowed. Think of the many games of children that point toward the pure intuition in fantasy! Soap bubbles, tea games, the damp colorfulness of the magic lantern, India ink, decal pictures. Color was always as blurred as possible, dissolving, nuanced in an entirely monotonous manner, without transitions of light and shadow. Wooly sometimes, as in the motley wool used for sewing. There were no sets as in the colors of painting. And doesn't it appear to you that this, the proper world of color, color as a medium, as spaceless, was best presented by motliness? A dispersed, spaceless infinitude of pure reception—thus was the art world of the child formed. Its only extension was height.—The perception of children is itself dispersed [zerstreut] in color. They do not deduce. Their fantasy is untouched.

GEORG: And all these things about which you speak are only various sides of one and the same color of fantasy. It is without transitions and yet plays in countless nuances; it is moist, blurring things in the coloring of their contours. It is a medium, pure property of no substance, motley and yet monochromatic, a colorful fulfillment of *one* infinite by fantasy. It is the color of nature, of mountains, trees, rivers, and valleys—but above all of flowers and butterflies, of the sea and the clouds. The clouds of fantasy are so near because of color. And the rainbow is to me the purest appearance of this color that thoroughly spiritualizes and animates nature, leads its origin back [zurückführt] to fantasy, and makes it into a silent, intuited archetype of art. Finally, religion places its holy kingdom in the clouds and its blessed region in paradise. And for his altarpiece Matthias Grünewald painted the halos of angels in the colors of the rainbow, so that the soul shines through the holy figures as fantasy. [7: 25]

MARGARETHE: Fantasy is also the soul of the dream world. The dream is pure reception of appearance in the pure sense. I began by speaking of dreams;

now I could tell you about my dream even less than before; but on your own you have seen its essence.

GEORG: The ground of all the beauty that alone appears to us in pure reception lies in fantasy. It is beautiful, indeed it is the essence of beauty that we can do nothing else but receive the beautiful, while the artist can live only in fantasy and immerse himself in the archetype. The deeper the beauty enters into a work, the deeper it is received. All creation is imperfect; all creation is unbeautiful. Let us be silent.

Probably written in 1915 or 1916, unpublished in Benjamin's lifetime.

The Rainbow; or, The Art of Paradise
From an old manuscript

[7: 562] It is a difficult question: whence comes the beauty of nature? For it must [be] entirely different than the beauty of painting and the plastic arts, because these are not the imitation of natural beauty. And yet, nature also has its beauty not by mere chance and not at disparate places [zerstreuten Stellen]; rather, its beauty is from its own spirit. This is shown by the fact that certain good people, who are not deformed into artificial simplicity, can nevertheless dwell in nature and find it beautiful—children, above all. Nature, however, is not beautiful via its space, neither via nearness nor distance; neither via immensity nor tininess; neither via abundance nor poverty. Space is not beautiful by nature, and art rests on space (since its beauty is transformed [umgestaltet]), but nature does not. In all such respects nature is not beautiful from mere intuition, but only from a sentimental and edifying perspective in which one represents something like the Alps and the immensity of the sea.

Beauty rests on concentration and every beauty of art on concentration of form. Only form is given to human beings for the expression of their spirit, and on its completion rests every one of their creations in spirit. Art proceeds from generation [Erzeugung], and truth is always encountered in art, since it is concerned with generated spirit. Space is actually the medium of generation in art, and it is creative only to the degree that it productively presents what is spiritual about space. Space has no spiritual appearance other than in art. In the plastic arts space is generated in a certain manner, and no one will doubt that this happens. Space is made an object in relation to its dimensionality. Natural space, in accordance with vision, is undimensional, dull, a nothing, if it is not enlivened in an empirical and lower manner. Given this restriction, three-dimensional space is inconceivable. Plastic art [Plastik] has to do with a space that emerges via generation and is therefore as conceivable as it is unlimited.

But painting also spiritually generates space; its form-generation originally bases itself on space as well, but it generates it on the basis of a different form [*sic*]. Plastic art is directed toward the existence of space, painting toward its depth. However, is depth not to be considered [7: 563] as a dimension? Spatial depth, which is generated in painting, concerns the relation of space to objects. This is mediated by the plane. The spatial nature of things develops itself in the plane, and it does so nonempirically, concentrated. Not the dimension but, rather, the infinitude of space is construed in painting. This happens via the plane, in such a way that things do not absolutely develop their dimensionality, nor their extension *in* space but, rather, their existence *toward* space. Depth yields [ergibt] infinite space. The form of concentration is thereby given; but this now requires its fulfillment, the satisfaction of its tension, a presentation of an infinite that is in itself, no longer dimensional and extended. Objects demand a form of appearance that is grounded purely on their relation to space, a form of appearance that does not express their dimensionality, but rather their contural tension (not their structive [struktive] but, rather, their painterly form), their existence in depth. For without this, the plane does not come into concentration; it remains two-dimensional and gains only a graphic, perspectival, illusionistic depth—not, however, a depth as nondimensional relational form of spatial infinitude and object. The requisite form of appearance, which constitutes this, is color in its artistic significance.

The criterion of the colorfulness of an image lies in the extent to which the color develops the content of infinity from the spatial form of the object, the extent to which it places an object in the plane, giving it depth from out of itself.

Clothing and decoration not mentioned in the dialogue.

Probably written in 1915 or 1916, unpublished in Benjamin's lifetime.

Notes on an Afternoon Conversation

[S, 1: 390] We spent an entire afternoon discussing a very difficult remark: the series of years is indeed countable [zählbar] but not numerable [numierbar]. Which led us to [the meaning of] course [Ablauf], numerical series, and above all, as a last point of departure, direction. Is there a direction without a course? "Direction is the variant measure of two straight lines." This is a thought complex that I will very much reflect on again. Time is indeed a course, but is time directed? For it is a thoroughly metaphysical assertion that time is like a straight line; perhaps it is a cycloid or something else, which has no direction at many points. (Where there are *no* tangents.) The same problem occurs among the series of numbers. Here, though, the numbers [Zahlen] somehow bear in themselves numerals [Nummern]; but in the proper sense of the word, they are not numerable, since numerability presupposes exchangeability, and this applies neither to numbers nor to years: they are in no way exchangeable. Beyond this, it became clear that direction is a determination in relation to the following dimension. There is a direction of two straight lines only in relation to a surface, a direction of two surfaces only in relation to space. Direction somehow is connected in the most intimate way with dimensionality. A direction of two straight lines is their variant measure in relation to the surface. Very difficult things also emerge with respect to the direction of crooked lines through the insertion [Einsetzung] of the tangent direction as curve direction, which is doubtless arbitrary yet satisfying to the human spirit and is something that appears self-evident.

[S, 1: 391] All this, of course, is closely bound up with the problem of history.

Recorded in Gershom Scholem's diary on August 24, 1916.

From a Notebook Walter Benjamin Lent to Me [Gershom Scholem]: "Notes Toward a Work on the Category of Justice"

[S, 1: 401] To every good, limited as it is by the spatio-temporal order, there accrues a possession-character. But the possession, as something caught in the same finitude, is always unjust. No order of possession, however articulated, can therefore lead to justice.

Rather, this lies in the condition of a good that cannot be a possession. This alone is the good through which goods become possessionless.

In the concept of society one tries to assign a possessor to the good that cancels its possession-character.

Every socialist or communist theory misses its goal precisely because the claim of the individual ranges over every good. If, for individual A, there is a need z that can be satisfied with good x, and if one therefore believes that good y, which is equal to x may and should be given for the sake of justice to individual B in order to calm the same need, one errs. For there is an entirely abstract claim of the subject in principle to every good, a claim that in no way leads back to needs but, rather, to justice, and whose ultimate direction does not point toward the possession-right of the person but possibly toward a good-right of the good.

Justice is the striving to make the world into the highest good.

The aforementioned thoughts lead to this conjecture: justice is not a virtue among other virtues (humility, love of one's neighbor, loyalty, bravery); rather, it grounds a new ethical category that one must not perhaps even call a category of virtue but, rather, a category of the same order as virtue. Justice does not appear to refer to the good will of the subject but, instead, constitutes a state [Zustand] of the world. Justice designates the ethical category of the existent, virtue the ethical category of the demanded. Virtue can be demanded; justice in the final analysis can only be as a state of the world or as a state of God. In God all virtues have the form of justice; the epithet *omni* in omnigracious, omniscient, and so forth points in this direction. Only a fulfillment of what is demanded [S, 1: 402] can be virtuous; only a guarantee [Gewährleistung] of the existent (*no longer perhaps* determining through demands, nevertheless in any case not any arbitrary one) can be just.

Justice is the ethical side of the battle; justice is the power [Macht] of virtue and the virtue of power. Responsibility for the world we have is protected from the instance of justice.

The Lord's Prayer: Lead us not into temptation but, rather, redeem us from evil (*several words unreadable*), is the prayer for justice, for the just state of the world.[1] The empirical act is related to the moral law somehow as a (non-deducible) fulfillment of the formal schema. In reverse, law [Recht] is related to justice as the schema to its fulfillment. Other languages have designated the enormous gulf that essentially separates law from justice.

ius themis mishpat
fas dikē tsedek

The problem of historical time is already posed through the peculiar form of historical chronology. The years are countable but, in contrast to most countables, not numerable.

Recorded in Gershom Scholem's diary on October 8 and 9, 1916.

Notes

Introduction

1. Gershom Scholem, *Walter Benjamin: Die Geschichte einer Freundschaft* (translated as *Walter Benjamin: The Story of a Friendship*), 65.

2. See Theodor Adorno, *Zur Metakritik der Erkenntnistheorie* (translated as *Against Epistemology*), which first appeared in 1956, a year after the edition of Benjamin's writings that he co-edited with Scholem under the title of *Schriften*.

3. See Edmund Husserl, "Vorlesungen zur Phänomenologie des inneren Zeitbewusstseins" (translated as *The Philosophy of Internal Time-Consciousness*); Heidegger, *Sein und Zeit* (translated as *Being and Time*), 1. For a thorough discussion of Benjamin's relation to Husserl, see Uwe Steiner, "Phänomenologie der Moderne."

4. The subtitle of two of the major works of the Marburg school identifies the project as "epistemo-critique." See Cohen, *Das Princip der Infinitesmal-Methode und seine Geschichte; ein Kapitel zur Grundlegung der Erkenntniskritik*; and Cassirer, *Substanzbegriff und Funktionsbegriff: Untersuchungen über die Grundfragen der Erkenntniskritik* (translated as *Substance and Function*). For a discussion of the history of term, which Cohen invented in order to avoid the disadvantages of *Erkenntnistheorie* (theory of knowledge, epistemology), see Kurt Eedel's introduction to the reissue of Cohen's *Kants Theorie der Erfahrung*. An informative discussion of the term in Benjamin's work can be found in Gary Smith, "Thinking Through Benjamin."

5. See Benjamin's brief letter to Fritz Radt from May 1915, in which he mentions Cassirer's seminar (*GB*, 1: 266); see also the brief discussion in a curriculum vitae, 6: 215.

6. See in particular, Heinrich Rickert, *Der Gegenstand der Erkenntnis*.

7. Heidegger and Rickert, *Briefe 1912 bis 1933*, 11.

8. Benjamin is more amused by Rickert's lecture series ostensibly on "logic": "In his company is now the entire literary Freiburg; he reads momentarily a

sketch of his system as an introduction into his logic, which justifies a new and complete philosophical discipline: philosophy as complete life. (The woman as its representative.) As interesting as it is problematic" (*GB*, 1: 112).

9. For a description of Geiger's teaching persona, see Wolfgang Trillhaas's *Aufgehobene Vergangenheit*: "He was a genius of improvisation" (72). Further remarks in this regard can be found in Hermann Zeltner's essay "Moritz Geiger im Gedächtnis," which is also an excellent introduction to his various contributions to aesthetics and psychology. In the first paragraph of his preface to a major collection of Geiger's aesthetic writings, Klaus Berger makes a brief comparison between Benjamin and Geiger, emphasizing their common concern with the object (in Geiger, *Die Bedeutung der Kunst*, 8).

10. See Geiger's *Habilitationsschrift*, which he completed under Theodor Lipps, "Methodologische und experimentelle Beiträge zur Quantitätslehre." Geiger's major contribution to the philosophy of mathematics is *Systematische Axiomatik der euklidischen Geometrie*, which David Hilbert, among others, valued very highly. As Geiger's letters to Husserl indicate, his seminars included such topics as the foundations of arithmetic, especially in the form developed by Gottlob Frege (see Husserl, *Briefwechsel*, 2: 94). Geiger's inquiry into the theory of the unconscious can be found in his last contribution to the phenomenological yearbook, "Fragment über den Begriff des Unbewussten und die psychische Realität." Geiger also threw himself into the philosophical debates that were ignited by the confirmation of Einstein's general theory of relativity, and in a published lecture from 1921 he explained that "from the perspective of the theory of relativity, space and time are by nature just as subjective as colors" (Geiger, *Die philosophische Bedeutung der Relativitätstheorie*, 38). Perhaps the most revealing document for Geiger's intellectual disposition in 1916 is an article in memory of Ernst Meumann, where he discusses the cultural conditions of the moment, "Zur Erinnerung an Ernst Meumann."

11. Herbert Spiegelberg, *The Phenomenological Movement*, 172. Spiegelberg was studying at this time in Munich under Alexander Pfänder, and his account of the early years of phenomenology is an indispensable resource in reconstructing the complexity of the philosophical context during the time in which Husserl's "school" began to split; see also the more recent work of Dermot Moran, *Introduction to Phenomenology*, 60–90. With respect to Geiger's place in the phenomenological movement, see especially Wolfhart Henckmann's afterword to Geiger, *Bedeutung der Kunst*, 549–79. For a study of Geiger's aesthetics, see Alexandre Métraux, "Zur phänomenologischen Ästhetik Moritz Geigers."

12. See Moritz Geiger, *Beiträge zur Phänomenologie des ästhetischen Genusses*; Max Scheler, *Der Formalismus in der Ethik und die materiale Wertethik* (translated as *Formalism in Ethics and Non-Formal Ethics of Value*); Alexander Pfänder, *Zur Psychologie der Gesinnungen*; and Adolph Reinach, *Die apriorischen Grund-*

lagen des bürgerlichen Rechtes (translated as "The A Priori Foundations of Civil Law"). Reinach's important treatise, which can be considered the foundation of "speech act theory," would later draw Benjamin's attention (WBA, MS 1850).

13. Geiger, *Beiträge zur Phänomenologie des ästhetischen Genusses*, 655–58. For Geiger's brief but incisive exposition of the eidetic character of phenomenological research, which distinguishes it from both inductive and deductive science, see *Beiträge*, 567–73.

14. See Geiger, *Beiträge zur Phänomenologie des ästhetischen Genusses*, 645.

15. Geiger served in the military from 1915 to 1918. In 1916, he also taught criminal psychology to members of the civil police force. Because of his military service, he was briefly allowed to retain his position in Göttingen after the Nazi seizure of power. One of his students at the time, Saunders Mac Lane, describes his condition during the early months of the Hitler regime in "Mathematics at Göttingen," 1136.

16. See Paul Linke, "Das Recht der Phänomenologie," which is discussed extensively in Chapter 2.

17. At the opening of *La Voix et le phénomène* (translated as *Speech and Phenomenon*) Derrida famously analyzes the "reductive" character of the "soliloquy that guards the silence." A discussion with Derrida concerning the relation between Husserl's idea of monologue and Benjamin's "panologic" response was the starting point for this study.

18. Quoted in Werner Kraft, "Friedrich C. Heinle," 21.

Chapter 1: Substance Poem Versus Function Poem

1. For a sustained reflection on this practice, see Samuel Weber's *Benjamin's -abilities*.

2. See, for example, Marin Heidegger, *Erläuterungen zu Hölderlins Dichtung* (translated as *Elucidations of Hölderlin's Poetry*), 15–16; Heidegger also uses the term *das Gedichtete* in his lectures on one of Hölderlin's late poems, which his editors called "Der Ister."

3. Friedrich Nietzsche, *Die Geburt der Tragödie* (translated as *The Birth of Tragedy*), § 5, reprinted in *Sämtliche Werke*, 1: 43.

4. For a wide-ranging discussion of psychologism and its discontents around 1900, see Martin Kusch, *Psychologism*.

5. For insightful examinations of Benjamin's complicated relation to Heinle, see Marion Picker, *Der konservative Charakter*, esp. 91–96; Astrid Deuber-Mankowski, *Der frühe Walter Benjamin und Hermann Cohen*, 221–22; and Uwe Steiner, *Die Geburt der Kritik aus dem Geiste der Kunst*, 91–167.

6. For Cohen's critique of the concept of function, especially as it is formulated in Cassirer's treatise from 1910, see especially *Ästhetik des reinen Gefühls*, 1: 68–69. Although Cassirer does not use the term *cognitive commentary* in his

impressive studies of the development of the modern sciences, the term is nevertheless an accurate description of its intent; see especially the first volume of his groundbreaking study in the history of science, *Das Erkenntnisproblem in der Philosophie und Wissenschaft der neueren Zeit*.

7. The title of "Blödigkeit" is very difficult to capture in English. The Grimm brothers' dictionary identifies three Latin terms as equivalents: *infirmitas* (frailty), *hebetudo* (bluntness or dullness), and *timiditas* (timidity); see *Deutsches Wörterbuch*, 2: 141. Of these three suggestions, the middle one is perhaps the most intriguing, especially since it brings out the *blöd* (stupid) with which the word begins. For an exposition of "Blödigkeit" with *hebetudo* in mind, see Avital Ronell, *Stupidity*, 4–9. As with some other words in this volume, the unsatisfactory character of the translations forces me to retain the original. A translation of Hölderlin's poems can be found in *Poems and Fragments*.

8. See, for example, the fragment of Hölderlin that describes the differences among poetic modes, *Sämtliche Werke*, 4: 270; translations of Hölderlin's poetological writings can be found in *Essays and Letters on Theory*. No one in 1914 was in a position to recognize the extent of Hölderlin's contribution to the debates about the fate of Kantian critique. A few years after Benjamin completed "Two Poems," Franz Rosenzweig discovered and then published "the oldest system Fragment of German idealism" (see Rosenzweig, *Das älteste Systemprogramm des deutschen Idealismus*). And the now famous fragment on intellectual intuition entitled "Urteil und Sein" (Judgment and being) or "Seyn/Urtheil" (Being/judgment) was discovered only in 1960 (*Sämtliche Werke*, 4: 216).

9. Henri Bergson, *Matière et mémoire* (translated as *Matter and Memory*), 232. On the "almost unthinkable" character of duration, which means that language tends to "crush" it, see Suzanne Guerlac, *Thinking in Time*, 91–93.

10. See Georg Lukács, *Die Theorie des Romans* (translated as The *Theory of the Novel*), 127–28. Lukács's treatise was written in the winter 1914–15 and first published in the *Zeitschrift für Ästhetik und allgemeine Kunstwissenschaft* in 1916.

11. In support of his analysis Benjamin cites the opening lines of "Chiron" as an intensification of what is already said in "the thinking day": "Where are you, Contemplative, which always must / Go to the side at times, where are you light? [Wo bist du, Nachdenkliches! das immer muß / Zur Seite gehen zu Zeiten, wo bist du, Licht?]" (quoted at 2: 119). Since the context of Benjamin's analysis is precisely the shape of time, and since, moreover, he emphasizes the alliterative character of *den denkenden Tag* (the thinking day), it is remarkable that he does not discuss the intensification of alliteration in *zur Seite* (to the side) and *zu Zeiten* (at times), where *Seite* (side) thus emerges as a shaping of time. Apropos the sideway-character of time, especially in light of these lines from "Chiron," see Werner Hamacher, "Parusie, Mauer," and my discussion of Hamacher's analysis in the context of Benjamin's essay, "Toward Another Teichology."

12. Gundolf, *Dichter und Helden*, 13

13. In a letter to his publisher Friedrich Wilmans from September 1803 Hölderlin uses the term *Orientalischen* (oriental). The letter would have been available to Benjamin in Hellingrath's edition of *Sämtliche Werke: Historisch-kritische Ausgabe*, 5: 329–30. (Volume 5 of the six-volume edition appeared in 1913.) The "oriental," for Hölderlin, stands in contrast with "Greek art, which is foreign to us," and he expresses the hope that he will "bring out the oriental, which it has denied and its art-mistake, which it occurs, improve."

14. See, for instance, Martin Buber, *Drei Reden* (translated in *On Judaism*), esp. 79: "The decisive connection between God and the human being is, for the oriental, the act, for the occidental, faith." Among the many discussions of Judaism as "oriental," Jakob Wassermann's "Der Jude als Orientale" is among the most revealing.

15. For a complementary analysis of Benjamin's essay, which concludes with a reading of this crucial yet indecisive sentence, see Samuel Weber, *Targets of Opportunity*, 123–33.

16. See Scholem, *Walter Benjamin*, 73–74.

17. Benjamin may have said something to this effect in a conversation with Scholem in June 1919. The discussion revolved around the interpretive principle governing the book with which Cohen first launched the Marburg school, *Kants Theorie der Erfahrung*, and then reformulates in *Das Princip der Infinitesimal-Methode*. "The rationalists," Scholem writes in diary, referring to Cohen, "consider everything that is an object to be *absolute*, which means that not only the Bible (also Hölderlin, says Walter) *may* and should be legitimately commented on with violence, but also Aristotle, Descartes, Kant, and everyone else" (S, 2: 227; cf. Scholem, *Walter Benjamin*, 78). By placing Hölderlin in the company of the Bible, Benjamin indicates the direction of the "aesthetic commentary" he wrote some five years earlier: toward doctrine, which momentarily teaches what the limit term—call it "life"—means.

Chapter 2: Entering the Phenomenological School

1. At the end of his dissertation *The Concept of Art Criticism in German Romanticism* Benjamin repeats this reversal (1: 104): an analysis of art as a pure medium of reflection turns into a conclusion in which the Hölderlinian word *sobriety* invokes, however indirectly, what was meant from the beginning.

2. Friedrich Hölderlin, "Hälfte des Lebens," in *Sämtliche Werke*, 2: 117.

3. Also quoted in Paul Linke's essay, "Das Recht der Phänomenologie," 196–97.

4. Husserl, *Briefwechsel*, 2: 108.

5. There can be little doubt that MS 506 should be dated around 1916: among the works listed—which includes Cassirer's *Concept of Substance and Concept of*

Function—there is the following note: "Linke-Elsenhans polemic about concepts in *Kant-Studien*," which refers to the essay that prompts Benjamin's "Eidos and Concept" and is discussed extensively in the following section.

6. A now canonical discussion of the new theory of intuition that Husserl begins to develop in *Logical Investigations* can be found in Emmanuel Lévinas's dissertation, which was first published in 1930; see Lévinas, *Théorie de l'intuition dans la phénoménologie de Husserl* (translated as *The Theory of Intuition in Husserl's Phenomenology*).

7. This study does not consider the general sense of the phenomenological or transcendental reductions within the context of Husserl's work in general; rather, it is limited to the only published exposition of the reduction that would have been available to Benjamin during his years as a student, namely *Ideas I*. It would therefore exceed the limits of the argument developed here to compare Benjamin's work from about 1915 to 1926 with the following paragraph from Husserl's posthumously published *Krisis der europäischen Wissenschaften* (translated as *The Crisis of the European Sciences*), in which a certain messianicity is inseparable from the completion of the phenomenological-transcendental reduction: "Perhaps it will come to light that the total phenomenological attitude and the corresponding *epoché* is called upon [berufen] to effect, by virtue of its essence, a complete personal transformation [Wandlung], which might be compared to a religious conversion, but which even beyond this, bears within itself the significance of the greatest existential conversion that is given to humanity as its task" (Hu, 6: 140).

8. Jean Héring, *Phénoménologie et philosophie religieuse*; for a brief account of Héring's contributions to phenomenology, see Jacek Surzyn, "Jean Héring." Héring wrote his first dissertation in Göttingen under Husserl and Adolf Reinach, and he later became a pastor. In Héring's class at the Faculty of Protestant Theology at the university in Strasbourg, Emmanuel Lévinas experienced his first account of phenomenology, and it was Héring who first gave him a copy of Heidegger's *Sein und Zeit*; see Ethan Kleinberg, *Generation Existential*, 27–33.

9. The quotation in this passage stems from Jean Héring, "Bemerkungen über das Wesen, die Wesenheit und die Idee," 522.

10. Husserl's brief comments on the distinction between essence and concept appears in *Ideas*, under the title "The Objection of Platonic Realism: Essence and Concept" (Hu, 3: 48–50); the canonical discussion of abstraction and concept formation in Husserl's published work of the time can be found in the second of the *Logical Investigations*. A more recent consideration of the topic under discussion in Benjamin's paper appears in Jitendranath Mohanty, "Individual Fact and Essence in Husserl's Philosophy."

11. An account of Linke's work can be found in the informative essay of Reinhold Smid, "'Münchener Phänomenologie'—Zur Frühgeschichte des Begriffs," esp. 133–35. For Scholem's appraisal of Linke, see *Von Berlin nach Jerusalem* (trans-

lated as *From Berlin to Jerusalem*), 114. It is worth noting that, unlike Scholem, Smid links Linke with "Munich phenomenology" and especially with Geiger's version of phenomenology. Geiger may have brought Linke's essay to Benjamin's attention.

12. Linke, "Das Recht der Phänomenologie," 215.
13. Ibid.
14. Finke's highly influential essay, "Die phänomenologische Philosophie Edmund Husserls und die gegenwärtigen Kritik," appeared in *Kant-Studien* in 1933 and can be considered, to a certain extent, as a continuation of Linke's "Law of Phenomenology."
15. Linke, *Grundfragen der Wahrnehmungslehre*, 368.
16. Gershom Scholem, *Walter Benjamin*, 65.
17. Héring, "Bemerkungen über das Wesen, die Wesenheit und die Idee," esp. 523–34.
18. Benjamin's "Theses on the Problem of Identity" are discussed more extensively in Chapter 6.
19. See Martin Grabmann, *Mittelalterliches Geistesleben*, 1: 116–25; Grabmann first noted the erroneous attribution in 1922 in a brief article for the *Archivum Franciscanum Historicum*.
20. See Martin Heidegger, "Neuere Forschungen zur Logik," *Literarische Rundschau für das katholische Deutschland* (He, 17–43); a translation of the essay can be found in Martin Heidegger, *Becoming Heidegger*, 30–43. That Benjamin would wish to consult this out-of-the-way item can be seen as evidence that he remembered Heidegger from his time in Freiberg.
21. By contrast, Heidegger's sycophantic relation to Rickert is perceptible in his analysis of the *modus essendi* first in phenomenological terms and then in terms of Rickert's epistemological constructivism; that it is the ultimate result of an "'infinite process' . . . in and through which the x of the object is supposed to be radically dissolved into form and form-system" (He, 318).
22. For a discussion of the intellectual situation to which Thomas of Erfurt contributed, see the introduction to the recent edition of his work translated into English, *Grammatica speculativa*.
23. Samuel Weber discusses this passage in connection with "The Task of the Translator," in his *Benjamin's -abilities*, 72–73.
24. For a succinct account of Rickert's theory of "sense," see the essay with which he launched the journal *Logos*, "Der Begriff der Philosophie." Heidegger's dissertation from 1913, *Die Lehre vom Urteil im Psychologismus: Ein kritisch-positiver Beitrag zur Logik*, poses the question, "What is the sense of sense [Was ist der Sinn des Sinnes?]" (He, 171).
25. Beyond an essay he published in *Die literarische Welt* in 1926, under the title "Outlook into Children's Books" (4: 609–15), there is almost nothing in his

subsequent work that directly draws from his studies of color; similarly, there is almost no trace of his interest in Scotian speculative grammar.

26. The studies, fragments, and remarks that Benjamin developed in relation to the themes of color and fantasy have been extensively discussed by Heinz Brüggemann, especially in his *Walter Benjamin über Spiel, Farbe und Phantasie*; see also his essays "Fragmente zur Ästhetik / Phantasie und Farbe," and "Walter Benjamins Projekt 'Phantasie und Farbe' in romantischen Kontexten." Brüggemann's studies do not place Benjamin's color studies in the context of his contemporaneous philosophical inquiry; thus, neither neo-Kantianism nor phenomenology is considered, even though the latter in particular is thoroughly saturated by color studies. Drawing on the work of Max Imdahl, whose *Habilitationsschrift* entitled *Farbe* is largely concerned with French aesthetic theories and takes its point of departure from the identification of differential "epistemes" that Foucault proposed in *The Order of Things*, Brüggemann discusses Benjamin's idea of pure receptivity in relation to the debate over what Ruskin famously called the "innocence of the eye," in his *Elements of Drawing* from 1856 (for a vivid exposition of conception in art-historical terms, see Jonathan Fineberg's *Innocent Eye*). Brüggemann draws on the authority of Ernst Gombrich, who cites a variety of psychological inquiries into the early childhood perception that dispute the validity of the "innocent-eye" thesis (*Walter Benjamin über Spiel, Farbe und Phantasie*, 138–39). To which one may respond by citing the authority of Merleau-Ponty, who described the relation between experience, including that of children, and "objective thought," in the following manner:

> To see is to enter a universe of beings which display themselves, and they would not do this if they could not be hidden behind each other or behind me. In other words: to look at an object is to inhabit it, and from this habitation to grasp all things in terms of the aspects which they present to it. . . . Thus every object is the mirror of all others. When I look at the lamp on my table, I attribute to it not only the qualities visible from where I am, but also those which the chimney, the walls, the table can "see"; but the back of my lamp is nothing but the fact which it "shows" to the chimney. . . . The complete object is translucent, being shot through from all sides by an infinite number of present scrutinies which intersect in its depths leaving nothing hidden. (Merleau-Ponty, *The Phenomenology of Perception*, 79)

27. In dating Benjamin's fragment "Über Scham" (On shame) (6: 69–71), which includes a long quotation from Goethe's *Farbenlehre*, Tiedemann and Schweppenhäuser refer to a letter Benjamin sends to Scholem in 1918, in which he reports on his study of Goethe and quotes the following passage: "Before I hope to read the *Farbenlehre*, I want to return to the meteorology with which I once concerned myself" (*GB*, 1: 488). Tiedemann and Schweppenhäuser then identify the precise moment when Benjamin read Goethe's meteorological writings (nr. 646 on the list printed in volume 7), note that Benjamin's essay on Goethe's

Wahlverwandtschaften (Elective affinities) includes three passages from the *Farbenlehre* (1: 132, 148, 160), and conclude that he must have read the *Farbenlehre* in 1919 "at the earliest" (6: 675). There is a major philological problem with this dating process, however: the fact that the editors have to make a guess. They must do so because the *Farbenlehre*, unlike the meteorological writings, does not appear on Benjamin's carefully prepared list of books he read, and yet he must have read it. A plausible conclusion: he read it much earlier, and indeed read it so intensively that, unlike the meteorological writings, he did not require an extensive rereading in preparation for his essay on Goethe's novel. And there are at least two good reasons to suppose that Benjamin read the *Farbenlehre* earlier than Tiedemann and Schweppenhäuser propose: his longstanding interest in producing a major project on color, and the fact that his list of books and essays concerning color (WBA, MS, 530) include ones that specifically address Goethe's color theory.

28. Johann Wolfgang von Goethe, *Zur Farbenlehre* (orig. publ., 1812, translated as *Theory of Color*), reprinted in *Werke*, part 2, 1: 1.

29. The book under review is Goethe, *Farbenlehre*, ed. and intro. Hans Wohlbold, who was a disciple of Steiner's.

30. Goethe, *Werke*, part 2, 1: 56; § 135. In the same vein, see § 775: "It is no wonder," Goethe writes, "that energetic, healthy, crude people take particular delight in this color [yellow-red]. It has been noted among savage peoples everywhere. And if children, left to themselves, begin to illustrate, they will spare vermilion and minium" (*Werke*, part 2, 1: 314); see also § 835, which is discussed below.

31. In response to certain statements made by Richard Müller-Freienfels, in "Gefühlstöne der Farbenempfindungen," Benjamin claims that "mere color" has sufficient power to "dissolve" powerful feeling: "color sensations are [in Müller-Freienfels' view] too weak for the dissolution of more significant feeling; for that, color representations are required. But the latter . . . appear to him in an entirely false manner, to be formed only by way of objects, not to be awoken by mere colors" (6: 122). Benjamin finds support his claim by citing Goethe—not, however, Goethe as scientist. Coming across a striking example of synesthesia in Goethe's late collection of poems *Westöstlichen Divan*—"A resounding play of colors [Ein erklingend Farbenspiel]" (*Werke*, part 1, 6: 189)—Benjamin presents his own studies as an alternative to the opening section of the *Farbenlehre*: "The unifying element, discussed here with reference to the light and darkness of dawn, the harmonically resounding play of colors, is a thoroughly suitable [term] for the colors of fantasy, which play between rise and fall" (6: 122–23).

32. Goethe, *Werke*, part 2, 1: 332; § 835.

33. Ibid., 1: 72; § 175.

34. Ibid., 1: xxxi; introduction.

35. Particularly important for Benjamin's conception of gender and sexual difference in his early writings is "Metaphysik der Jugend" (Metaphysics of

youth), esp. 2: 91–96. In this regard, see in particular Eva Geulen, "Toward a Genealogy of Gender in Walter Benjamin's Writing," and Sigrid Weigel, *Body- and Image-Space*.

36. Goethe, *Werke*, part 2, 2: 69.

37. The remark from Goethe can be found in his *Werke*, part 2, 1: 263. Just as the passages previously considered distinguish the "cultured" human being from its "natural" counterpart, including the child, so this one distinguishes the human being *tout court* from the purely natural ape.

38. See Franz Kafka, "Ein Berich für eine Akademie" (Report for an academy), which was originally published in Martin Buber's journal, *Der Jude*, in 1917 and is reprinted in *Gesammelte Werke*, 1: 234–45.

39. Thus Benjamin writes about nature "retreating" from the surface of human beings, such that an "almost decolored [entfärbten] tone" (6: 69) remains. That Benjamin could not find a way to present the color of the human surface in any other way may have led him away from developing the thoughts that he sketches in "On Shame." This is discussed further in the final section of the last chapter.

40. As an indication of what Benjamin learned when he entered into Husserl's school, it would be worthwhile to compare "On Shame" with some remarks on Nietzsche's concept of shame that Benjamin makes in a letter to Ludwig Strauss from November 1912: "[*Thus Spoke Zarathustra*] is dangerous without limits, wherever Nietzsche hides himself in a heightened philistinism. Everywhere in the biological sphere, and most obliquely and worst of all in the concept of shame. He wants by all means to see in it something valuable, even holy. And yet shame is thoroughly natural without remainder; it designates the place where the spiritual recedes from the natural" (*GB*, 1: 78).

41. It would be of some interest to compare Benjamin's reflections on shame with those of Jean-Paul Sartre, especially in the following passage from *L'Être et le néant* (translated as *Being and Nothingness*): "If there is an Other . . . then I have an outside. I have a *nature*. My original fall is the existence of the Other. Shame—like pride—is the apprehension of myself as a nature although that very nature escapes me and is unknowable as such" (*Being and Nothingness*, 352). For Sartre, however, the fundamental character of this existential "falling," comparable to original sin, lies in guilt (see *Being and Nothingness*, 531). In 1913, Max Scheler published a phenomenological inquiry under the title "Über Scham und Schamgefühl," reprinted in *Gesammelte Werke*, 10: 67–154.

42. For Heidegger's famous analysis of the workplace, see *Sein und Zeit*, 66–88.

43. Ibid., 268.

44. In his *Metamorphoses of the Phenomenological Reduction*, Jacques Taminiaux demonstrates the structural affinity between "categorial intuition" in the sixth of the *Logical Investigations* and "the call of conscience" in *Being and Time*. Heidegger himself seems to have come to the conclusion that there was some-

thing suspect about the entire discussion. One of his marginal remarks reads: "But is everything up until now only an assertion?" (*Sein und Zeit*, 445).

45. Heidegger, *Sein und Zeit*, 285.
46. Ibid., 299.
47. See ibid., 285.
48. About these notes, Scholem writes without proof: "the absolute rejection of the doctrine of original sin is recognizable in Benjamin's notes on this doctrine" (S, 2: 28, 199). For a discussion of Benjamin's idea of original sin in relation to the notes on "Capitalism as Religion" (6: 100–103), see Werner Hamacher, "Schuldgeschichte."
49. For a reading of "Destiny and Character" in terms of their respective sign systems, see especially Timothy Bahti, "Theories of Knowledge."
50. Benjamin's phrase "On the Occasion of Lyser" obviously refers to the occasion of obtaining copies of Johann Peter Lyser's illustrated children's books. Scholem describes Benjamin's joy in this acquisition around 1918 or 1919 (*Walter Benjamin*, 85–86), which is when the editors of the collected writings date the text under discussion (6: 702). The essay Benjamin eventually wrote in relation to Lyser is, as mentioned above, "Outlook into Children's Books" (4: 609–15), which begins with a quotation from Friedrich Heinle.

Chapter 3: "Existence Toward Space"

1. "The Rainbow" is not Benjamin's only dialogue; in fact, it appears to be his last, as if its impasse were decisive with respect to the form. In presenting his thoughts in dialogue form, Benjamin was following a contemporaneous trend. Among the major exemplars, two are particularly important in this context: Martin Buber's *Daniel: Gespräche von der Verwirklichung* (translated as *Daniel: Dialogues on Realization*), which Benjamin discussed with its author without having actually read the text (see GB, 1: 218–19), and Georg Lukács, *Die Seele und die Formen* (translated as *Soul and Form*), esp. the dialogue on "Lawrence (sic) Sterne," 265–324.
2. For an informative discussion of Benjamin's life-long interest in the phenomenon of dreaming, see Burkhardt Lindner's afterword to his edition of Benjamin's *Träume*, 135–68.
3. See Friedrich Nietzsche, *Die Geburt der Tragödie*, § 1; reprinted in *Sämtliche Werke*, 1: 25–27.
4. Hermann Cohen, *Ästhetik des reinen Gefühls*, 1: 106. For an examination of Benjamin's relation to Cohen's aesthetics from a different perspective, which pays particular attention to his essay on Goethe's *Elective Affinities*, see Astrid Deuber-Mankowsky, "Der schöne Schein und das Menschenopfer."
5. Cohen, *Ästhetik des reinen Gefühls*, 1: 85.

6. Ibid., 1: 78.
7. Ibid.
8. For a brief discussion, see Jerome Pollitt, "The Canon of Polykleitos and Other Canons."
9. Cohen, *Ästhetik des reinen Gefühls*, 1: 107–9; see K, 5: 234.
10. Cohen, *Ästhetik des reinen Gefühls*, 1: 70.
11. The poet in question is Friedrich Heinle, a line of whose Margarethe had just cited: "Wäre ich aus Stoff, ich würde mich färben [If I would made out of material, I would color myself]" (quoted at 7: 24). Benjamin obviously admires this line; he quotes it often, and it serves as an indication of a paradisal condition: the materialized self that exists in the medium of color.
12. See Georg Wilhelm Friedrich Hegel, introduction to the *Vorlesungen über die Ästhetik* (translated as *Lectures on Fine Art*), which is reprinted in *Werke*, 13: 13.
13. For a brief comparison of Benjamin to Baumgarten, see my essay "Is There an Answer to the Aestheticizing of the Political?"
14. This remark scarcely does justice to the idea of clouds in Benjamin's work; for a detailed reflection, see Werner Hamacher, "The Word *Wolke*."
15. The decision as to whether *Urbild* should be translated as "archetype" or "primordial image" depends largely on a choice of the interpretative context in which "The Rainbow" should be placed. If the context is defined by Plato and Neoplatonism, *archetype* is appropriate; if it is defined by vitalism, particularly in the version propounded by Ludwig Klages, then *primordial image* is more appropriate. Of particular importance for the latter context is Klages's "Vom Traumbewußtsein." In a letter to Klages in 1920 Benjamin indicates that he "recently, by chance" (*GB*, 2: 114) came across a copy of "Vom Traumbewußtsein" and inquires into its sequel. In this connection, see John McCole, *Walter Benjamin and the Antinomies of Tradition*, esp. 229–40; see also Irving Wohlfarth, "Walter Benjamin and the Idea of Technological Eros," as well as Richard Block, "Selective Affinities." Since Benjamin specifically mentions a Neoplatonic doctrine in "The Rainbow," it seems as though "archetype" is the more apt translation, although "primordial image" is by no means ruled out.
16. The title of the relevant section of *Ideas* thus bears the title: "Being as Consciousness, Being as Reality: Principal Difference in the Modes of Intuition" (Hu, 3: 95). By replacing the term *conscious* with *spiritual* (*geistig*) and arrives at a similar formulation: "something spiritual" has no adumbrations.
17. See Moritz Geiger, *Beiträge zur Phänomenologie des ästhetischen Genusses*, 626–29. In this regard, Geiger continues the work of his Doktorvater, Theodor Lipps, whose *Ästhetik* includes an examination of the psychology of depth under the following slogan: "it is *my* depth" (*Ästhetik*, 1: 523). Depth, for Lipps, is an essential dimension of individuation.
18. Benjamin's use of the term "form of appearance" (*Erscheinungsform*) may

owe something to the opening section of *Capital*, in which Marx analyzes the difference between the content of particular kind of object and the form in which this object appears: "Exchange value in general can be only the mode of expression, the 'form of appearance' [die Ausdrucksweise, die 'Erscheinungsform'] of a content that is different from it" (*Kapital*, 1: 51; translated as *Capital*). Kant, for his part, uses *form of appearance* only in an obscure passage in the *Nachlass* (K, 18: 56), and it is altogether absent from Hegel. It might also be of some interest to compare Benjamin's color studies with Wittengenstein's, especially with regard to the problem of depth; see, for instance, Wittgenstein, *Remarks on Color*, 52.

19. It is worth comparing this situation with a favorite topos of Benjamin's: the so-called *Vexierbild*, which could be translated as the "spot the object game." On this theme in Benjamin, see especially Rainer Nägele, *Literarische Vexierbilder*.

20. See Jacques Derrida, *La Vérité en peinture* (translated as *Truth in Painting*), esp. 21–168 ("The Parergon").

21. While editing the Italian edition of Benjamin's writing, Giorgio Agamben unexpectedly discovered that Herbert Blumenthal (later, Belmore) lived in his neighborhood. Around 1917 Benjamin had suddenly broken off their friendship and withdrawn from all contact for no discernable reason. (A distant view of this event can be found in Scholem, *Walter Benjamin*, 57–58.) Blumenthal-Belmore kept all the writings and letters Benjamin had sent him, including the "Rainbow" texts, and meticulously maintained a portfolio of Benjamin's subsequent writings—all, so it seems, as a monument to massive ambivalence. The "Rainbow" dialogue to which he refers in the note quoted in the text may be different from the copy Blumenthal-Belmore acquired.

22. For a detailed description of the exhibition, see Georg Brühl, *Herwarth Walden und "Der Sturm."*

Chapter 4: "The Problem of Historical Time"

1. See especially Martin Buber, *Drei Reden*, which appeared in 1911 and to which Scholem repeatedly refers in his brief lecture (S, 1: 111–12). An extensive discussion of Scholem's relation to Buber can be found in David Biale, *Gershom Scholem*, esp. 43–47, 56–61, and 81–86. A wide-ranging discussion of Benjamin and Scholem in this period can be found in Erich Jacobsen, *Metaphysics of the Profane*. A brief study of the young Scholem, under the title "A Jewish Zarathustra, 1914–1918," appears in *A Life in Letters, 1914–1982*, 7–21; see also the reflections scattered throughout Skinner's edition of the diaries, which is published under the title *Lamentations of Youth*. With regard to the latter, though, it is important to note that it has numerous errors in translation, including some that are relevant to this chapter, such as the events of August 1916. An incisive reflection on Scholem's early life undertaken in connection with the publication

of *Lamentations of Youth* can be found in Paul Reitter's aptly titled essay "Irrational Man."

2. See, for example, Geiger's contribution to the second congress for "aesthetics and art scholarship," in which he presented a paper on "Phänomenologische Ästhetik," esp. 32–33. What Geiger says about the tragic in 1924 will find a reverberation in the *Habilitationsschrift* that Benjamin submits to the philosophical faculty of the university in Frankfurt soon thereafter: "Here it must be established whether behind the unitary name of the tragic there does not lie hidden entirely different things, whether during antiquity and today the tragic really means the same, or whether it concerns different phenomena, for which obviously the analysis of essence [Wesensanalyse] would have to come out differently" (35–36).

3. For a discussion of Benjamin's conception of Judaism in light of his letters to Schoen, see Gary Smith, "'Das Judentum versteht sich von selbst.'"

4. The audacity of this letter can be seen by comparing it with that of a well-known writer of the period, namely Franz Kafka, who had recently received the Fontane prize. Upon receiving a request from Buber to submit his work, Kafka was initially hesitant—but only because he felt "too burdened and insecure to speak up in such company, even in the smallest way" (Kafka, *Briefe, April 1914–17*, 146). Soon afterward, he sent Buber a number of stories, two of which eventually appeared in *Der Jude*, including "Report for an Academy." Benjamin, for his part, was not in general adverse to working with Buber. A decade after his letter of rejection, he published his account of Moscow in another journal that Buber edited, *Die Kreatur*.

5. In the preface to the first issue of *Der Jude* Buber describes the clarifying power of war: "From all the letters from the field, from all the conversations with those who have returned, I receive the same impression—that of a strengthening of relationships to Judaism through the clarifying the gaze and the solidification of the will" (Martin Buber, "Die Losung," 2). For two astute analyses of Benjamin's letter to Buber, see Marion Picker, *Der konservative Charakter*, 68–80; and Samuel Weber, "Der Brief an Buber vom 17.7.1916."

6. In Scholem's reconstruction of his visit, Dora's contribution to the conversation, particularly with regard to Hegel and the deducibility of the world, is elided; see Scholem, *Walter Benjamin*, 44.

7. For an account of the conversation between Scholem and Benjamin, see Gershom Scholem, *Walter Benjamin*, 13. By an odd coincidence, another pair of friends was conducting an (epistolary) conversation that could be brought into a productive dialogue with the one under consideration in this chapter. While posted on the front, Franz Rosenzweig and Eugen Rosenstock-Huessy exchanged a series of letters about the significance of the calendar, including the following reflection, which Rosenzweig recorded in his letter to Rosenstock-

Huessy on July 17, 1916: "The 'calendrical' is the form of this philosophizing. Therefore the pure calendar . . . stands in the system where the idealists (*Critique of Pure Reason*, [Fichte's] *Wissenschaftslehre*, transcendental idealism) place logic" (Rosenzweig, *Briefe und Tagebücher*, 2: 221; translated in *Judaism Despite Christianity*, edited by Rosenstock-Huessy). For an indispensable discussion of the relation among Benjamin, Scholem, and Rosenzweig, see Stéphane Mosès, *Angel of History*.

8. For a discussion of the disposition of the manuscript entitled "Notizen zu einer Arbeit über die Kategorie der Gerechtigkeit," see Hermann Schwepenhäuser, "Walter Benjamin über Gerechtigkeit."

9. Scholem, *Walter Benjamin*, 45.

10. For reasons that will be discussed later, it appears doubtful that Benjamin said "*series* of years." In the written record, there is no remark of this kind.

11. In the authoritative, multilinguistic *Dictionary of Mathematics* edited by Günther Eisenreich and Ralf Sube, the word *Ablauf* is nowhere to be found. Nor does the word appear in the equally reliable *Mathematisches Wörterbuch*, edited by Josef Naas and Hermann Ludwig Schmid.

12. The difficulty of capturing the term *Ablauf* can be indicated by the fact that in his fine translation of the essay Stanley Corngold is forced to use three English words in the course of only four sentences: *course*, *unfolding*, and *exposition* (*Selected Writings*, 1: 33). Sometime during Scholem's visit to Munich in August 1916 Benjamin describes "Two Poems" as the only "essential" work he has so far completed (S, 1: 391).

13. Felix Hausdorff (under the name Paul Mongré), *Das Chaos in kosmischer Auslege*, 11; for an insightful discussion of the relation between Hausdorff's philosophical and mathematical work, see Moritz Epple, "Felix Hausdorff's Considered Empiricism." Hausdorff plays only a minor role in Scholem's diaries, as they have been edited and published. Scholem is far more enthusiastic about the mathematico-philosophical reflections of Edgar Zilsel, who had recently completed a dissertation entitled *Anwendungsproblem: Ein philosophischer Versuch über das Gesetz der grossen Zahlen und die Induktion*; see, in particular, the concluding remarks of "Potpourri Regarding a Mechanistic World-Image" (S, 1: 354). Nevertheless, Scholem would have known that, in comparison with a mathematician of Hausdorff's stature, Zilsel must be considered a neophyte at best. Scholem may have tempered his interest in Hausdorff's philosophical theories under the presumption that he had to first absorb his mathematical work, which requires no small effort.

14. Hausdorff, *Das Chaos in kosmischer Auslege*, 154.

15. According to Julia Ng, the inside notebook cover of Scholem's notebook recording Frobenius's lectures on analytic geometry is inscribed as follows: "The definition of different *directions*."

16. Benjamin presumably refers to these lectures when he asks at the end of the letter in which he also announces the near completion of his essay on language, "What has become of your mathematics lecture?" (*GB*, 1: 345). Knopp's lectures had already been published in German under the title *Funktionstheorie* (translated as *Theory of Function*). Scholem brought a copy of Knopp's original version with him to Jerusalem. For an overview of Knopp's work, which only briefly touches on his work during his last years in Berlin, see Erich Kamke and Karl Zeller, "Konrad Knopp."

17. Konrad Knopp, "Bemerkungen zur Struktur einer linearen perfekten nirgends dichten Punktmenge."

18. Konrad Knopp, "Einheitliche Erzeugung und Darstellung der Kurven von Peano, Osgood und v. Koch."

19. Konrad Knopp, "Ein einfaches Verfahren zur Bildung stetiger nirgends differenzierbarer Funktionen."

20. Karl Weierstraß's original paper has been translated as the first "classic" of fractal mathematics, under the title "On Continuous Functions of a Real Argument That Do Not Have a Well-Defined Differential Quotient." According to Julia Ng, to whom the Scholem archive at the Hebrew University in Jerusalem kindly granted access to Scholem's mathematical notebooks: "He wrote down the Weierstraß function on an insert in pencil to some loose leaves at the back of Knopp's winter semester 1915–16 course on ordinary differential equation." It is perhaps worth noting here that the mathematical concept of continuity has changed significantly since the early years of the twentieth century. It is not so much a feature of the function as of the underlying space and should thus be considered a topological property. A relatively accessible discussion of continuity in the context of contemporary mathematics can be found in Martin Crossley, *Essential Topology*, 1–14.

21. See Benoit Mandelbrot, *Les Objets fractals*, which was heavily revised as *The Fractal Geometry of Nature*. Edgar's previously mentioned volume, *Classics on Fractals*, begins with the following three papers: Weierstraß's discussion of a continuous yet nowhere differentiable function (the subject of Knopp's 1918 paper); Cantor's description of a linear perfect yet nowhere dense sets of points (the subject of Knopp's 1916 paper); and von Koch's geometric version of a nowhere differentiable yet continuous function (the subject of Knopp's 1917 paper). In his last years in Berlin, in short, Knopp was well on his way to developing a theory of fractals. The following thesis, drawn from his book on the theory of functions, prevented him from seeing the "fractal geometry of nature," to cite the title of Mandelbrot's groundbreaking work: "Natural phenomena themselves posses an intrinsic regularity" (*Theory of Functions*, 1: 86). Benjamin, for his part, was attracted to certain natural phenomena that do not "possess an intrinsic regularity." Among the irregular shapes that drew Benjamin attention are clouds, vortices, and seashores.

22. Scholem claims that it was Dora who persuaded Benjamin to change the location of their meeting and invite him to her husband's villa; see *Walter Benjamin*, 38–39. It would exceed the limits of a note—and this volume as a whole—to discuss Scholem's troubled relation with Dora, whose father, Leon Kellner, was an important early supporter of Theodor Herzl and thus, in a certain sense, a "father" to Scholem's Zionism. Scholem's diaries include chilling accounts of the tension between the two, with little indication as to what provoked them. One of the most intense of these interactions is signaled in Scholem's diaries with the unqualified statement: "One must unfortunately admit: Walter is not a just man" (S, 2: 230).

23. For Noeggerath's attempt to present the Kantian category of relation in mathematical terms, see "Synthesis und Systembegriff in der Philosophie," 52–57. It is presumably to this section that Benjamin describes as "*highly* significant" (*GB*, 1: 364). I am in debt to Anna Glazova for procuring for me a copy of Noeggerath's 1916 dissertation.

24. For Benjamin's references to Jentzsch, see *GB*, 1: 241–42; Scholem's mathematical notebooks include references to Jentzsch's dissertation. In addition to his close association with Georg Heym, it appears as though Jentzsch was acquainted with Fritz Heinle as well. For a preliminary study of Jentzsch's life and mathematical work, see Peter Duren, Anne-Katrin Herbig, and Dmitry Khavinson, "Robert Jentzsch."

25. Scholem learned of Benjamin's relation to Schoenflies early in their friendship; see Scholem, *Walter Benjamin*, 28. The best source of information about Schoenflies is the book about him commissioned by his family and produced by Thomas Kaemmel and Philipp Sonntag, *Arthur Schoenflies*.

26. Arthur Schoenflies, *Entwicklung der Mengenlehre und ihrer Anwendungen*, 2–3. Discussions of Benjamin's use of the set-theoretical term *power* (*Mächtigkeit*) can be found in Chapter 6.

27. Henoch Berliner taught number theory and higher algebra at the university in Bern, having produced a dissertation entitled *Involutionssyteme in der Ebene des Dreiecks*.

28. The Scholem archive includes some twenty-eight mathematical notebooks, which Scholem presumably brought with him to Jerusalem in the event that he had to fall back on his original plan of teaching mathematics and physics in Palestine.

29. In a diary entry from 1918, Scholem indicates that he understands the historical limits of Newtonian science. At issue is the methodology of Hermann Cohen, whose epistemology, according to Scholem, is too closely connected with Newtonian science: "How would Cohen have spoken two thousand years ago, when there was not yet any Newton, and how about today, when Newton perhaps is surpassed?" (S, 2: 170).

30. See Scholem, *Walter Benjamin*, 45. With regard to this information, I am indebted to Paul Reitter, who asked representatives of the Scholem archive whether the document exists, and to Julia Ng, who confirmed that it is indeed missing.

31. Scholem, for his part, located two lacunae, one of which is specifically mathematical: "Of course, the treatise still lacks fundamental parts, such as, for example, one concerned with the symbolic in language (which will certainly be treated in a second treatise) and the theory of the sign and of writing, which, in my view, lead to the ultimate depths, since they throw open decisive questions in mathematics and the philosophy of religion" (Gershom Scholem, *Briefe an Werner Kraft*, 17).

32. A translation and discussion of this essay can be found in Heidegger, *Becoming Heidegger*, 59–72.

33. For a discussion of Benjamin's response to Heidegger's *Habilitationsschrift*, see Chapter 2.

34. That Heidegger did not completely divorce himself from the line of thought developed in this early essay can be seen by the fact that he cites it in *Being and Time* immediately after making a similar claim about the philosophical import of the theory of relativity: "As a first attempt at interpreting chronological time and the 'historical number,' see the Freiberg 'habilitation'-lecture of the author" (*Sein und Zeit*, 418). Heidegger then singles out Simmel's *Problem of Historical Time* in the same context. A useful discussion of the context of Heidegger's essay can be found in John van Buren, *The Young Heidegger*, 97–99.

35. For an insightful and informative description of Rickert's work on the historical sciences that is framed by Heidegger's attempt to find a position from which to separate his work from his teacher's, see Charles Bambach, *Heidegger, Dilthey, and the Crisis of Historicism*, esp. 83–126.

36. Scholem does return to Heidegger's essay later in his diaries, but only to the section that concerns the concept of time in historical scholarship (S, 2: 65).

37. The draft of the "Epistemo-Critical Preface" published in the *Gesammelte Schriften* is even more stamped by Benjamin's reflections on the theme of mathematics and language than the published version, which, despite its "boundless chutzpah" (*GB*, 3: 14), represents a certain diminution of the intensity that characterizes the original version. Thus, as part of the description of "philosophical style," Benjamin writes in the draft, but not in the published text, the following: "Entirely renouncing the unbroken course of the intention, which characterizes mathematical thinking, is its first characteristic" (1: 926). Then again: "Contemplation knows no end. This is valid not in the sense in which it would also be attributed to mathematical thinking, which is never exhausted by its objects" (1: 926). And still again: "[The method of philosophical knowledge] once more shows itself as different from that of mathematics" (1: 928).

Chapter 5: Meaning in the Proper Sense of the Word

1. Gottlob Frege, "Über Schoenflies: 'Die logische Paradoxen der Mengenlehre,'" in *Nachgelassene Schriften*, 191; the article under examination is Arthur Schoenflies, "Über die logischen Paradoxieen der Mengenlehre," from 1906.
2. Bertrand Russell, *The Principles of Mathematics*, 101–7.
3. Ibid., 105; for the "doctrine of types," see 523–28.
4. Schoenflies, "Über die logischen Paradoxieen," 22.
5. Ibid., 21. Schoenflies's proposed solution is only one of dozens. For a catalog, see Alexander Rüstow, *Der Lügner*, 130–32. Benjamin cites Rüstow's study in his reflections on the correlated paradox of the Cretan liar (6: 57).
6. Frege, *Nachgelassene Schriften*, 191. On the importance of this distinction in Frege's work, see the lucid discussion of Cora Diamond, *The Realistic Spirit*, 113–44. In *Zur Wissenschaftstheorie Walter Benjamins* Lieselotte Wiesenthal associates Benjamin's early theory of language with Wittgenstein's roughly contemporaneous *Logische-philosophische Abhandlung* (translated as *Tractatus Logico-Philosophicus*). Since language, for Benjamin, unlike Wittgenstein, does not have a "picture" character, the attempt is unsuccessful; but to the extent that the *Tractatus* rigorously respects Frege's distinction between object and function, there is some basis for the association of these otherwise divergent "little treatises."
7. Frege, *Nachgelassene Schriften*, 194.
8. Gerhard Hessenberg, *Grundbegriffe der Mengenlehre*, 144. Benjamin may have been familiar with Hessenberg's textbook, which includes many chapters intended for nonexperts, including its extensive discussion of Russell's paradox.
9. Hessenberg, *Grundbegriffe der Mengenlehre*, 143–44; for Russell's version, see *Principles of Mathematics*, 101.
10. See especially Frege's now famous essay "Über Sinn und Bedeutung" (1892), which is usually translated as "On Sense and Reference" and can be found, among many other places, in Frege, *Funktion, Begriff, Bedeutung*, 40–65; translated in *Philosophical Writings*. For an extensive discussion of the context of Frege's "context principle," see Michael Dummett, *The Interpretation of Frege's Philosophy*, 360–427.
11. Schweppenhäuser and Tiedemann publish the brief "Attempt at a Solution of Russell's Paradox" after the notes entitled "Judgment of Designation," but there is no apparent reason for this decision, and purely content-related reasons argue against it.
12. For further discussion of this passage, see Chapter 2.
13. The title of the fragment, evidently from Benjamin's hand, runs "Notes Toward the Continuation of the Work on Language" (7: 785); but as the editors of Benjamin's papers correctly note, it is more likely that many of the manuscripts predate the "little treatise" (*GB*, 1: 343) he ultimately drafted in lieu of his inability to work through the theme of mathematics and language.

14. For an argument that Benjamin's term *Mitteilung* should be translated as "imparting" rather than "communication" (as is done here), see Samuel Weber, *Benjamin's -abilities*, esp. 31–52.

15. According to Tiedemann and Schweppenhäuser, the two fragments concerning the "word-skeleton" were written in 1920 or 1921 (6: 641–420); but they recognize that they could have been written around the time of the treatise on language, which internal evidence suggests is the case: it would be odd if Benjamin were to take up the example of the triangle four or five years later, much less that he would use the very same phrases. For this reason, I consider the fragments under consideration as part of the reflections around the solution to Russell's paradox.

16. See the eighth paragraph of the first of Husserl's six *Logical Investigations*, "Expressions in the Solitary Life of the Soul." In this context, it is again necessary to refer to Derrida's famous analysis of the first investigation at the beginning of his monograph from 1967, *La Voix et le phénomène*, esp. 34–52.

17. The parallel argument with respect to time runs: "Time is a necessary representation that grounds all intuitions. In regard to appearances in general one cannot remove time, although one can very well take the appearances away from time. Time is therefore given a priori. In it alone is all actuality of appearances possible. The latter could all disappear, but time itself (as the universal condition of their possibility) cannot be removed" (K, A 31; B 46).

18. See Husserl's discussion in *Ideas* of fantasy as the "paradoxical" source and medium of eidetic science (Hu, 3: 129).

19. Ernst Barthel, "Die geometrischen Grundbegriffe (Parallelproblem)."

20. See Schoenflies, *Entwicklung der Mengenlehre*, 18–19. Cantor's original formulation of the problem is in geometric terms:

> If we clothe this problem in geometrical dress and understand by a *linear* manifold of real numbers any thinkable collection of infinitely many distinct real numbers, then one can ask under *how many* and what classes the linear manifolds fall, if manifolds of the same power are put into one and the same class and manifolds of different power into different classes. By an inductive process . . . we are led to the proposition that the number of classes which arise according to this principle of classification of linear manifolds is finite and indeed that it is equal to *two*. (Georg Cantor, *Gesammelte Abhandlungen*, 132)

21. The so-called "continuum hypothesis" vexed Cantor ever since he discovered the two classes of infinite sets. An account of the origin of the "continuum hypothesis" can be found in Moore, *Zermelo's Axiom of Choice*, 39–51. The reason Cantor could not prove his conjecture was discovered in the course of the twentieth century: Kurt Gödel showed that given the nine axioms of Zermelo-Fraenkel set theory, it is impossible to show that the conjecture was false, after which Paul J. Cohen proved that it would impossible (with the same axioms)

to show that it is true. It is therefore in the exact sense of the term undecidable within the standard axioms of set theory.

22. In a series of philosophical-mathematical reflections, some of which are addressed to major figures in the Catholic Church, including a future pope, Cantor seeks to defend the "actual infinite." Transfinite numbers, he argues, "are delimited forms [abgegrenzte Gestaltungen] or modifications (*aphorismena*) of the actual infinite" (*GA*, 395–96). The term *aphorismenon* is drawn from Aristotle, who, as is well known, recognizes only potential infinity (*apeiron dunamei*), which is opposed to *apeiron hos aphorismenon*, that is, a "separated infinity" that would be in some sense actual (*Physics*, 208a6; *Gesammelte Abhandlungen*, 396). For Cantor's exposition of the three "positions" from which the actual infinite can be considered, see especially his paper, "Über die verschiedenen Standpunkte in bezug auf das aktuelle Unendliche," reprinted in *Gesammelte Abhandlungen*, 370–76.

23. Schoenflies, *Entwicklung der Mengenlehre*, 208–10.

24. For a discussion of Benjamin's familiarity with Cantorian set theory, see the previous chapter.

25. For a discussion of this passage, see an essay of mine that errs on numerous other points, "The Genesis of Judgment," 91–93.

26. For an analysis of the emphatic "but" (*aber*) in Benjamin's "bold" suggestion, along with an analysis of the gender politics at work in the treatise as a whole, see Carol Jacobs, *In the Language of Walter Benjamin*, 107–12.

27. Werner Hamacher formulates the law of irony according to which the paradisal language must always already be inhabited by its "parody" (2: 153); see Hamacher, "Afformative, Strike," 136–38.

28. See especially, Kierkegaard, *Concept of Anxiety*, 47.

29. See Husserl, *Logical Investigations*, first investigation, § 1 (Hu, 19, 1; 24). On the use of the term *entanglement* in this context, see Jacques Derrida, *La Voix et le phénomène*, 22.

30. Kierkegaard, *The Concept of Anxiety*, 45.

Chapter 6: Pure Knowledge and the Continuity of Experience

1. Ludwig Feuerbach, *Grundsätze der Philosophie der Zukunft* (translated as *Principles for the Philosophy of the Future*), § 41; *Werke*, 1: 270–71.

2. Friedrich Nietzsche, *Jenseits des Gut und Böse* (translated as *Beyond Good and Evil*), §§ 1–23; *Sämtliche Werke*, 5: 15–39.

3. See the discussion among theology students in chapter 14 of Mann's *Doktor Faustus*, which reproduces in part his own views at the time, as they were expressed in "Gedanken im Krieg" (Thoughts in war), from November 1914. Benjamin describes Mann's essay as "despicable" (*GB*, 1: 348), an assessment that Mann later came to share.

4. See, for instance, one of the early chapters of Dilthey's ground-laying *Einleitung in die Geisteswissenschaften: Verusch einer Grundlegung für das Studium der Gesellschaft und der Geschichte* from the 1880s entitled "The sciences of spirit [are] an independent whole, alongside the natural ones" (Wilhelm Dilthey, *Einleitung*, 1: 5–16).

5. See, for instance, the second of Buber's "three discourses," which poses the following question: "It there a true Jewish religiosity?" (*Drei Reden*, 12).

6. For Hegel's concept of experience, see the introduction to the *Phänomenologie des Geistes* (translated as *Phenomenology of Spirit*); reprinted in Hegel, *Werke*, 3: 79–81.

7. On the perilous character of experience, see especially Jean-Luc Nancy, *The Experience of Freedom*, 18–20.

8. See Lev Shestov, *Potestas clavium*, esp. 293–97; Jean Héring, "Sub specie aeterni: Réponse à une critique de la philosophie de Husserl."

9. On the relation between Rickert and Husserl, see Kern Iso, *Husserl und Kant*, 32–33.

10. Heinrich Rickert, "Vom Begriff der Philosophie," 12. Rickert proposes that the opposition between "realities and values" (*Wirklichkeiten und Werten*) replace the traditional opposition between subject and object. Only in this way can the "problem of the world" be solved, and we can, in turn, "hope to gain a *Weltanschauung* that is not merely an explanation of reality" (13). According to Rickert, Kant misrecognized the concept of "sense" when he founded epistemology on the transcendental unity of apperception (31).

11. For a broad discussion of Benjamin's relation to historicism, see John McCole, *Walter Benjamin and the Antinomies of Tradition*.

12. See Klaus Reich, *Die Vollständigkeit der kantischen Urteilstafel* (translated as *The Completeness of Kant's Table of Judgment*).

13. Gottlob Frege, *Funktion, Begriff, Bedeutung*, 40; *Grundgesetze der Arithmetik*, 2: 254–55.

14. At the same time as Benjamin was seeking to "fix" the concept of identity, Franz Rosenzweig was pursuing a similar line of thought. Written during his deployment as a soldier in the War, the "Urzelle" (primordial cell) for the *Stern der Erlösung* (translated as *The Star of Redemption*) seeks to transcend the principle of identity that begins with the Ionian philosophy and culminates in the absolute idealism of Hegel; a translation of *Urzelle* can be found in Rosenzweig, *Philosophical and Theological Writings*, 48–72. Rosenzweig's guide in the endeavor is the late Schelling, who, in his *Philosophische Untersuchungen über das Wesen der menschlichen Freiheit* (translated as *Philosophical Investigations into the Essence of Human Freedom*), identified an *Ungrund* (un-grounding) at the basis of divine existence. If self-identity cannot be attributed to God—except perhaps

when the theogonal process is complete and He is, after all, "all in all"—it certainly cannot be predicated of any finite being.

15. Benjamin's line of argument could be productively compared with the one Hölderlin pursues in a text that he would not have known, since it came to light only after his death, namely the fragment the editor of the Stuttgart edition of his works called "Judgment and Being," in which Hölderlin asks how the "I" can say "I" (see *Sämtliche Werke*, 4: 216–17). That Hölderlin may have been on Benjamin's mind when he composed his "Theses on the Problem of Identity" is indicated by a curious term he introduces, "the a-identical [das Aidentische]" (6: 27), which perhaps reflects the Hölderlinian term *aorgic*, understood as the opposite of anything "organic," that is, formed or otherwise limited.

16. A celebrated essay of Rickert's, "Das Eine, die Einheit und die Eins" would also be useful in "fixing" the concept of unity. Benjamin read the essay in 1913, without being particularly impressed by it (*GB*, 1: 154).

17. In a series of notes contemporaneous with the completion of his dissertation, Benjamin explicates perception in terms of *Deuten* (interpretation) as opposed to *Bedeuten* (meaning); similarly, Benjamin briefly identifies perception with reading (6: 32–33). Sometimes, however, perception is equivalent to "pure seeing," as it is discussed in the studies on color (6: 65). The difference between the two concepts of perception seems to lie in the context in which they emerge: perception per se is comparable to "pure seeing," whereas perception is understood as "interpretation" when it is distinguished from experience, which is essentially continuous. For a brief discussion of perception as reading, see Howard Caygill, *Walter Benjamin*, esp. 3–4.

18. Cohen's *Religion der Vernunft* (translated as *Religion of Reason*) appeared posthumously in 1919, shortly after Benjamin wrote "On the Program of the Coming Philosophy."

19. G. W. F. Hegel, introduction to the *Phenomenology of Spirit*; reprinted in *Werke*, 3: 22.

20. Benjamin had recently acquired a copy of Kant's correspondence (*GB*, 1: 358), and, as Scholem reports, enjoyed reading it out loud (S, 2: 221–22). On the problem of the "gap in the critical system," see especially Eckart Förster, *The Final Synthesis*, 48–74; and my own *Late Kant*, 154–58.

21. Gottfried Leibniz, "Discours de la métaphysique" (Discourse on metaphysics), reprinted in *Sämtliche Schriften*, series 6; 4: 1540; a translation can be found in Leibniz, *Philosophical Papers and Letters*. In the section of the "Epistemo-Critical Preface" (1: 228) entitled "Monadology," Benjamin refers to this text, which is generally considered Leibniz's first attempt to give a public version of his monadological system. It is in the "Supplement" that Benjamin begins to draw on Leibnizian terminology for the formulation of his "doctrinal" elements,

including the "monadic" character of any unity that is of a higher power than the unity that is generated by the "spontaneity of the understanding" (2: 210).

22. Isaac Newton, *Principia mathematica*, general scholium, 974.

23. The quotations stems from the lectures Max Planck held at Columbia University; see *Acht Vorlesungen*, 117. Benjamin may also have gained some knowledge of the complexity of Einstein's program of "generalizing" the special theory of relativity by reading Ernst Barthel's "Der astronomische Relativismus und sein Gegenstück." Among Benjamin's early acquaintances was Hans Reichenbach, who in his youth was among the first to reflect on the philosophical implications of Einstein's work; see especially *Relativitätstheorie und Erkenntnis apriori* from 1920. Moritz Geiger was also keenly interested in the same problem; see his lecture from 1921, published as *Die philosophische Bedeutung der Relativitätstheorie*. Ernst Cassirer was similarly interested in the theory of relativity, appending a discussion of its philosophical importance to the second edition of *Substanzbegriff und Funktionsbegriff*, which has been translated as *Substance and Function and Einstein's Theory of Relativity*.

24. In later writings Benjamin occasionally discusses the theory of relativity and even associates his own form of research with Einstein's, both of which are akin to Husserlian phenomenology:

> This integration of domains that tears down the limits of established bodies of knowledge and established modes of thought [des Fachwissens und des Fachdenkens] by driving toward a unity and continuity of intuition stands in strict opposition to the traditional form of such unity: the system. If, more specifically, the system claims to find that unity, that continuity also in the object, it is striking, in reflecting on the line of thought developed here [a book by a contemporary 'neo-Darwinist'], how closely it approaches the work of other major contemporaneous researchers precisely by breaching traditional dreams of a system. Husserl posits discontinuous phenomenology in the place of the idealistic system; Einstein posits a finite discontinuous world-space in the place of the infinite continuous one. (4: 536)

"Finite . . . world-space" refers to the general theory of relativity, while "discontinuous world-space" alludes to quantum theory. In another essay of the same period, "Experience and Poverty," Benjamin again implies a certain affinity between the theory of relativity and his own theoretical efforts: Einstein should be seen as a "constructer" (2: 215), along the lines of Klee and Scheerbart.

25. With assistance from Emile Meyerson's *De l'explication dans les sciences* (translated as *Explanation in the Sciences*) Benjamin briefly reiterates his thesis concerning the priority of explanation over observation near the beginning of the "Epistemo-Critical Preface" (1: 213). It would be an understatement to say that both physics and the philosophy of the physical sciences underwent a major convulsion from 1900 to the early 1930s. In a conversation with Werner Heisenberg in the mid-1920s Einstein, as it happens, proposed a thesis remarkably simi-

lar to the one that Benjamin advances in "Attempt at a Proof that the Scientific Description of a Process Presupposes its Explanation": "It is quite wrong to try founding a theory on observable magnitudes alone. . . . In reality the very opposite happens. It is the theory which decides what we can observe" (quoted in Werner Heisenberg, *Physics and Beyond*, 63). According to Heisenberg, the conversation in which Einstein made this assertion prompted him to develop the outlines of what would later be called the "uncertainty principle." Benjamin, for his part, derived at least one major concept from the "discontinuous" science of the quantum, namely "superposition," which describes the character of particles that exist only when they are measured, prompting their corresponding wave functions to collapse: "The sensation of the newest and most modern is as much a dream-form of events as the eternal return of the same. The perception of space that corresponds to this perception of time is superposition" (5, 1023). Benjamin discusses Arthur Stanley Eddington's description of superposition in a famous letter to Scholem about the nature of Kafka's world (*GB*, 6: 110–11).

26. See Hermann Cohen, *Das Princip der Infinitesimal-Methode*; see also Cohen, *Logik der reinen Erkenntnis*, 121–43. Scholem is particularly taken with Cohen's understanding of reality in terms of the differential, which runs counter to his general contempt for Cohen's appropriation of mathematical terms for philosophical purposes (S, 2: 169–70). On Benjamin's relation to Cohen's "infinitesimal method," see especially Hamacher, "Intensive Sprache."

27. Johann Gottlieb Fichte, *Grundlegung zur gesamten Wissenschaftslehre* (translated in Fichte, *The Science of Knowledge*), reprinted in *Werke*, 1: 212. For Benjamin's exposition of the basic idea of the "Wissenschaftslehre," see his dissertation on *The Concept of Art Critique in German Romanticism*, 1: 18–25 ("Reflection and Positing in Fichte").

28. The continuity of Benjamin's connection between experience and continuity can be measured by a remark in his *Passagenwerk* (translated as the *Arcades Project*): "What distinguishes experience from lived experience is that the former cannot be isolated from the representation of a continuity, a sequence" (5: 964; m2a, 4).

29. Schoenflies, *Entwicklung der Mengenlehre*, 208–10. The concept of hypercontinuity emerges in relation to the debates about Cantor's continuum hypothesis, which were not resolved, as explained above (note 21 to Chapter 5), until the 1960s.

30. In his dissertation Benjamin draws this locution from the edition of Hölderlin's Pindar commentaries that appeared under the title "Untreue der Weisheit: Ungedruckte Handschrift aus den Sammlungen auf Stift Neuberg," in the first volume of *Das Reich*, 309.

31. Paul Linke, "Das Recht der Phänomenologie," 203; see the discussion of Linke's article in Chapter 2.

Chapter 7: The Political Counterpart to Pure Practical Reason

1. Hermann Cohen, *Ethik des reinen Willens*, 471–583.
2. As co-editor of Benjamin's collected writings, Hermann Schweppenhäuser discusses the "Notes Toward a Work on the Category of Justice" and provides an editorial apparatus, in "Walter Benjamin über Gerechtigkeit."
3. See Gershom Scholem, *Walter Benjamin*, 45; see the discussion in the fourth chapter above.
4. Scholem's revealing document of the differences between his and Benjamin's idea of justice was not published in the two-volume edition of his *Diaries* but appeared, rather, in Eric Jacobson's *Metaphysics of the Profane*, 174–80. Jacobson also notes that the circumstances surrounding the publication of Benjamin's "Notes" are odd: "It also seems highly unlikely that Scholem would simply forget this text" (298).
5. For a discussion of this condition in relation to Benjamin's reflections on marriage in his essay on Goethe's *Elective Affinities*, see my essay, "Marital, Martial, and Maritime Law."
6. In the *Digest*, the following juridical pronouncements are definitive. According to Gauis, "whatever is subject to divine right [diuini iuris] is no one's thing. . . . Public goods are considered to be no one's, for they belong corporately to the whole community" (Justinian, *Digest*, 1, 8, 1). According to Marcion: "Things sacred or religious or sanctified are no one's things. Things sacred are then those that have been consecrated by an act of the whole people. . . . Being religious is a quality that every person can impose on a site of his own free will by burying a corpse in a place one owns" (*Digest*, 1, 8, 6–8); and according to Ulpian, "we use the term 'sanctified' of things that are neither sacred nor profane but are still confirmed by some kind of sanction. Thus, laws are sanctified, for they are supported by a kind of sanction. Anything supported by some kind of sanction is sanctified, even though it is not consecrated to a god" (*Digest*, 1, 8, 9).
7. See Jakob Sigismund Beck, *Commentar über Kants Metaphysik der Sitten*, 150.
8. See John Locke, *Two Treatises of Government*, 303–18.
9. See Martin Heidegger, *Kant und das Problem der Metaphysik* (translated as *Kant and the Problem of Metaphysics*), 208: "In unveiling the subjectivity of the subject Kant shrank back [züruckweicht] from the very ground that he himself had established."
10. A critique of this kind lies dormant in Kant's *Metaphysische Anfängsgründe zur Rechtslehre* (translated as *Metaphysical Foundations of the Doctrine of Right*, in Kant, *Practical Philosophy*), perhaps less in its published version than in its numerous drafts, which Benjamin, for his part, could have encountered in the fine edition of the Königsberg bibliographer, Rudolf Reicke. There is no doubt

that Benjamin was attracted to Kant's last writings, especially the *Doctrine of Right* with its notorious theory of marriage, about which he writes the following at the beginning of his essay on Goethe's *Elective Affinity*: "[It is] the most sublime product of a *ratio* that, incorruptibly true to itself, penetrates infinitely deeper into the real state of affairs than sentimental rationalism" (1: 127). A trace of Benjamin's earlier enthusiasm for Kant's late writings can be found in two of the letters he collected and discussed in *Deutsche Menschen* (Germans). Around the same time, he wrote lengthy notes for a work on the last years of Immanuel Kant, whose eminently cosmopolitan yet exceedingly narrow life would be presented as exemplary of the German bourgeoisie before its total capitulation to the accumulation of capital, the expansion of the state, and the ideology of race. The collapse that Kant himself experienced late in his life as a consequence of his rule-bound existence could thus appear as doubly exemplary: "For this is precisely peculiar play of this form of existence that in its complete decline it outlives itself [auslebt] in the strict ambiguity of the term" (6: 155).

11. Benjamin may have been familiar with the attempt on the part of natural-rights theorists to present Nimrod as the first proprietor, whose division of the earth's surface mirrors in reverse his plans for political union; for a discussion of Nimrod's role in the theory of natural right, see Rudolf Weigand, *Naturrechtslehre*, 336–39.

12. Franz Kafka, "In der Strafkolonie," reprinted in *Gesammelte Werke*, 1: 187; for Kafka's brief account of the reading as a "great failure," see *Briefe, April 1914–1917*, 277.

13. A discussion of the relation between self-complaisance and radical evil can be found in chapter 4 of my *Late Kant*.

14. My discussion of Scholem's "Theses on the Concept of Justice" is indebted to Eric Jacobson, who published them in German with his own English translation; see *Metaphysics of the Profane*, 177–80; the thesis that "Justice is the idea of the historical annihilation of divine judgment" begins the third of the "Theses."

15. Quoted in Jacobson, *Metaphysics of the Profane*, 178.

16. Ibid., 179.

17. Ibid., 178.

18. Benjamin had intended to publish "Toward the Critique of *Gewalt*" in *Die weißen Blätter*; because of its length and difficulty, however, the editor of *Die weißen Blätter* recommended that it be published in the *Archiv für Sozialwissenschaft und Sozialpolitik*, where it appeared in 1921 (see *GB*, 2: 130). As in much of this volume, my inquiry into the philosophical status of "Toward the Critique of *Gewalt*" takes its point of departure from Werner Hamacher's "Afformative, Strike." Three studies of Benjamin's essay are helpful in situating its purpose and provenance: Uwe Steiner, "True Politician," which, oddly enough, fails to note the Kantian character of the title he adopts as his own; Chryssoula

Kambas, "Walter Benjamin liest Georges Sorel"; and Burkhardt Lindner, "Derrida, Benjamin, Holocaust." Zur politischen Problematik des 'Kritik der Gewalt.'" See also Giorgio Agamben's *State of Exception*, which supposes that Carl Schmitt responded to Benjamin's essay with his *Politische Theologie* from 1922. For a profound inquiry into what Kant calls "true politics," see Susan Meld Shell, *Kant and the Limits of Autonomy*, 212–47.

19. See Hermann Cohen, *Ethik des reinen Willens*, 67–68:

Ethics can be considered the logic of the human sciences [Geisteswissenschaften]. It has the concepts of the individual, the totality, as well as the will and action as its problems. All philosophy is connected to a fact of science. The connection to the fact of science is in our view the eternal in Kant's system. Legal science [Rechtswissenschaft] forms the analogue to mathematics. It can be called the mathematics of the human sciences and, above all, of ethics.

20. For Cohen's critique of the *Critique of Practical Reason*, see *Ethik des reinen Willens*, 338: Kant's putative "fact of reason" is only an "analogue of a fact," and it is for this reason that he postpones his exposition of legal science.

21. For Cohen's reflections on the "abstention from judgment" in the context of a remarkable discussion of Shakespeare's *Hamlet*, see *Ethik des reinen Willens*, 589–90.

22. An argument to this effect can be found in Carl Schmitt's "Macht und Recht," which was published in Franz Blei's journal *Summa*.

23. See Martin Heidegger, *Sein und Zeit*, 46.

24. For a detailed description of the Kapp-Lüttwitz putsch, see the study of Johannes Erger of this name.

25. See Mennike's editorial footnote to Herbert Vorwerk's "Das Recht zur Gewaltanwendung," 14.

26. Ibid.

27. Ibid., 15.

28. Max Weber, *Politik als Beruf* (translated as *Politics as Vocation*), 2. Like Vorwerk, Weber emphasizes that the "monopolization" of violence is a distinctive feature of the modern bureaucratic state (10–11).

29. Vorwerk, "Das Recht zur Gewaltanwendung," 15.

30. Ibid.

31. James Burns, ed., *The Cambridge History of Medieval Political Thought*, 282. The classic doctrine of the "fullness of power" is often attributed to Pope Innocent III: "Peter is the only one who was called to enjoy the fullness of power. From him I received the miter of my priesthood and the crown of my royalty; he has established me vicar of Him upon whose habit it is written: 'King of kings and lord of lords, priest for eternity according to the order of Melchisedech'" (quoted in Pauline Watts, "The Donation of Constantine, Cartography, and Papal 'Plenitudo Potestatis' in the Sixteenth Century," 104). Benjamin

was probably familiar with the discussion of *plenitude potestatis* in Nicholas of Cusa's *De concordantia Catholica*: "One lord of the world rules over the others in the fullness of power [potestas plenitude], and in his own sphere he is the equal of the Roman pontiff in the temporal hierarchy on the model of the sacerdotal hierarchy" (*Catholic Concordance*, 216). For an incisive account of the doctrine of *potestas plenitude* in relation to the idea of power per se, see Giorgio Agamben, *La Régne et la gloire*, 158–67. See also Gerhart Ladner, "The Concepts of 'Ecclesia' and 'Christianitas,' and Their Relation to the Idea of 'Plenitudo Potestatis' from Gregory VII to Boniface VIII,'" in his *Images and Ideas in the Middle Ages*, 487–515; and Alfonso Otero, "Die Eigenständigkeit der plenitudo potestatis in den spanischen Königreichen des Mittelalters." It should be noted that Benjamin was reading large amounts of ecclesiastical history during this period, with an eye toward a dissertation on scholastic theories of language.

32. See Werner Hamacher, "The Word *Wolke*," 151; Jacques Derrida, "Force de loi," 1036.

33. A striking use of the phrase *schalten und walten* can be found in the following passage from Jacob Grimm's famous account of early German property law: "It becomes clear how essential the concept of border [Grenze] is connected with that of property. If our own is that over which we do as we please [schalten und walten], such 'doing as we please' presupposes a division of objects" (Grimm, *Kleinere Schriften*, 2: 30).

34. The translation used here is taken from *Tanakh*, 182. The Hebrew word that is translated as expiation is *kippur* (as in Yom Kippur). For an insightful and wide-ranging inquiry into the passage from the Book of Leviticus to which Benjamin alludes (Lev. 17: 11), see David Biale, *Blood and Belief*, 9–43; an exhaustive treatment can be found in Jacob Milgrom, *Leviticus 17–22*, esp. 1472–79.

35. An indispensable discussion of Benjamin's idea of "mere" or "bare life" (*bloßes Leben*) can be found in Giorgio Agamben, *Homo Sacer*, esp. 63–67. The decisive passage from Benjamin's "Toward the Critique of *Gewalt*" in the context of Agamben's analysis is closely related to the theory of blood Benjamin adopts from the Book of Leviticus: "For with mere life the dominance of right [Herrschaft des Rechts] over the living stops" (2: 200). It should be emphasized that this claim precisely does not say: "mere life ends the dominance of right or law over the living," or even "with the emergence of mere life, the dominance of right or law over the living comes to an end." The "with" of Benjamin's formulation is as enigmatic as the passage from Leviticus to which it presumably refers. The passage from Benjamin's text even suggests that the *use* of "mere life"—*with* is an instrumental preposition—brings the dominance of law over life to an end.

36. As Benjamin indicates in the same fragment, the source of his knowledge about the etymological connection between *red* and *rust* is Wilhelm Wackernagel's *Glossar*; the etymological note can be found on page 444.

Conclusion

1. Cohen, *Ästhetik des reinen Gefühls*, 2: 426 and 2: 428.
2. Cohen, *Religion der Vernunft* (translated as *Religion of Reason*), 527.
3. For a wide-ranging discussion of the terms *descriptive* or *describing psychology* in Brentano and Husserl, see Søren Olesen, *Wissen und Phänomenon*, 69–85.
4. The so-called "Politico-Theological Fragment" has been the source of a certain controversy, ever since Benjamin gave it to Adorno in 1938; see editorial remarks of Tiedemann and Schweppenhäuser, 2: 946–49; and see Irving Wohlfarth, "Nihilistischer Messianismus," and especially Werner Hamacher, "Theologisch-Politische Fragment."
5. See Johann Wolfgang von Goethe, *Wilhelm Meisters Wanderjahre*, reprinted in *Werke*, part 1, 25: 131–65; a translation can be found in Goethe, *Conversations of German Refugeees/Wilhelm Meister's Journeyman Years*, 343–58. For a far-reaching discussion of this letter in a broader context than the one developed here, especially with regard to Goethe's story, see Thomas Schestag, "Interpolationen."
6. Goethe, *Werke*, part 1, 25: 134.
7. See the discussion of David Biale, *Gershom Scholem*, 72–73.
8. See Geiger's letter to Husserl from August 1922; Husserl, *Briefe*, 2: 112.
9. See Steven Schwartzman, *The Words of Mathematics*, 117–18.
10. Goethe, *Werke*, part 1, 48: 187.
11. See Benjamin, *Briefe*, 1: 257.
12. Hausdorff, *Chaos in kosmischer Auslese*, 151–58; see the discussion of Hausdorff-Mongré in Chapter 4.
13. Scholem's letter to Schocken appears in the appendix to David Biale, *Gershom Scholem*, 215–16.
14. Ibid., 216.
15. As I have argued elsewhere, Derrida's early writings are driven by a cognate conception of a preestablished disharmony among the elements of historicity ("Derrida and History"). Whereas the disharmonious elements, for Derrida, are the singularity of every datum and the universality of its communicability, for Benjamin, they are time, which has no direction, and history, the unidirectionality of which is governed by guilt.
16. See Konrad Knopp, "Einheitliche Erzeugung und Darstellung der Kurven von Peano, Osgood und v. Koch."
17. Ernesto Cesàro, *Opere scelte*, 467; for a discussion of Cesàro's 1905 paper, see Benoit Mandelbrot, *Fractal Geometry of Nature*, 43–44.
18. This discussion of the shape of time should be compared with two other expositions of time that derive, in part, from Benjamin's work—those of Werner Hamacher and Giorgio Agamben. In "Parusie, Mauer" Hamacher elucidates the proximity of "side" (*Seite*) to "times" (*Zeiten*) in Hölderlin's late poetry: "time can remain only on the side of side [an der Seite der Seite]; it can remain only a

setting-aside, untimely, spatial, unapproachably lateral" (136). And in his reflections on Paul of Tarsus Agamben schematizes messianic time in terms of a cut line and then, after reflecting on the inadequacy of the straight line as an image of time, poses some questions to which this study indirectly responds: "If you represent time as a straight line and its end as a punctual instant, you end up with something perfectly *representable* but absolutely *unthinkable*. Vice-versa, if you reflect on a real experience of time, you end up with something *thinkable*, but absolutely *unrepresentable*. In the same manner, even though the image of messianic time as a segment between two eons is clear, it tells us nothing of the experience of time that remains, a time that begins to end. Where does the gap between representation and thought, image and experience come from? Is another representation of time possible, one that would avoid this misunderstanding" (*The Time That Remains*, 64).

19. See, for instance, Leibniz, "Monadologie," reprinted in *Die philosophischen Schriften*, 6: 612; a translation can be found in Leibniz, *Philosophical Papers and Letters*.

Appendix: Translations

The Rainbow: Dialogue on Fantasy

Walter Benjamin, "Der Regenbogen: Gespräch über die Phantasie," in *Gesammelte Schriften*, vol. 7, *Nachträge*, S. 19–26. © Suhrkamp Verlag Frankfurt am Main: Suhrkamp 1989. Reprinted with permission. Translated by Peter Fenves.

1. The primary reference is doubtless to the second stanza of the programmatic poem from *Les Fleurs du mal*, "Correspondence":

> Comme de longs échos qui de loin se confondent
> Dans une ténébreuse et profonde unité,
> Vaste comme la nuit et comme la clarté,
> Les parfums, les couleurs et les sons se respondent.

> [Like prolonged echoes that mingle in the distance
> In a deep and tenebrous unity,
> Vast as the dark of night and as the light of day,
> Perfumes, sounds, and colors respond to one another.]

Baudelaire, *Les Fleurs des mal*, 40. Benjamin may also be alluding to the chapter of Baudelaire's *Salon de 1846* on color, which includes a passage from E. T. A. Hoffmann's *Kreisleriana* that "perfectly expresses [his] idea": "It is not only in dreaming and in the light delirium that precedes sleeping but also when I am awake, as I hear music, that I discover an analogy and intimate reunion between colors, sounds, and smells. It seems to me that all things have been created by the same ray of light and that they will reunite in a wonderful concert" (Baudelaire, *Salon de 1846*, 129).

2. The poet is Friedrich Heinle, and Benjamin quotes the lines elsewhere in his studies of color and fantasy (6: 121).

The Rainbow; or, The Art of Paradise

Walter Benjamin, "Der Regenbogen; oder, Die Kunst des Paradieses," in *Gesammelte Schriften*, vol. 7, *Nachträge*, S. 562–63. (c) Suhrkamp Verlag Frankfurt am Main: Suhrkamp 1989. Reprinted with permission. Translated by Peter Fenves.

Notes on an Afternoon Conversation

Translated by Peter Fenves from Gershom Scholem, *Tagebücher, nebst Aufsätzen und Entwürfen bis 1923*, ed. Karlfried Gründer, Herbert Kopp-Oberstebrink, and Friedrich Niewöhner in association Karl Grözinger (Jüdischer Verlag: Frankurt am Main, 1995–2000), S, 1: 390–91.

From a Notebook Walter Benjamin Lent to Me [Gershom Scholem]

Translated by Peter Fenves from Gershom Scholem, *Tagebücher, nebst Aufsätzen und Entwürfen bis 1923*, ed. Karlfried Gründer, Herbert Kopp-Oberstebrink, and Friedrich Niewöhner in association Karl Grözinger (Jüdischer Verlag: Frankurt am Main, 1995–2000), S, 1: 401–2.

1. See Matthew 6:9–13; Luke 11:2–4.

Bibliography

Adorno, Theodor W. *Against Epistemology, a Metacritique: Studies in Husserl and the Phenomenological Antinomies.* Trans. Willis Domingo. Cambridge, Mass.: MIT Press, 1983..
———. *Zur Metakritik der Erkenntnistheorie: Studien über Husserl und die phänomenologischen Antinomien.* Frankfurt am Main: Suhrkamp, 1970.
Agamben, Giorgio. *Homo Sacer: Sovereign Power and Bare Life.* Trans. Daniel Heller-Roazen. Stanford, Calif.: Stanford University Press, 1998.
———. *La Régne et la gloire.* Trans. Joel Gayraud and Martin Rueff. Paris: Seuil, 2007.
———. *State of Exception.* Trans. Kevin Attell. Chicago: University of Chicago Press, 2005.
———. *The Time That Remains: A Commentary on the Letter to the Romans.* Trans. Patricia Dailey. Stanford, Calif.: Stanford University Press, 1998.
Allen, Charles Grant. *The Color-Sense: Its Origin and Development.* London: Trübner, 1879.
Bahti, Timothy, "Theories of Knowledge: Fate and Forgetting in the Early Work of Walter Benjamin." In *Benjamin's Ground: New Readings of Walter Benjamin,* ed. Rainer Nägele, 61–82. Detroit: Wayne State University Press, 1988.
Bambach, Charles. *Heidegger, Dilthey, and the Crisis of Historicism.* Ithaca, N.Y.: Cornell University Press, 1995.
Barthel, Ernst. "Der astronomische Relativismus und sein Gegenstück: Allgemeinverständlich Einführung in die Lehre vom Raum." *Archiv für systematische Philosophie* 22 (1916): 54–78.
———. "Die geometrischen Grundbegriffe (Parallelproblem)." *Archiv für systematische Philosophie* 22 (1916): 368–93.
Baudelaire, Charles. *Les Fleurs du mal.* Ed. Henri Lemaître. Paris, Gallimard, 1964.

———. *Salon de 1846*. Ed. David Kelley. Oxford: Clarendon Press, 1975.
Beck, Jakob Sigismund. *Commentar über Kants Metaphysik der Sitte*. Halle: Rengersche Buchhandlung, 1798.
Benjamin, Walter. *Arcades Project*. Trans. Howard Eiland and Kevin McLaughlin. Cambridge, Mass.: Harvard University Press, 1999.
———. Archive. Akademie der Künste, Berlin.
———. *Briefe*. Ed. Gershom Scholem and Theodor Adorno. 2 vols. Frankfurt am Main: Suhrkamp, 1978.
———. *Correspondence of Walter Benjamin*. Trans. Manfred Jacobsen and Evelyn Jacobsen. Chicago: University of Chicago Press, 1994.
———. *Gesammelte Briefe*. Ed. Christoph Gödde and Henri Lonitz. 6 vols. Frankfurt am Main: Suhrkamp, 1995– .
———. *Gesammelte Schriften*. Ed. Rolf Tiedemann and Hermann Schweppenhäuser. 7 vols. Frankfurt am Main: Suhrkamp, 1972–91.
———. *Schriften*. Ed. Theodor Adorno and Gershom Scholem. 2 vols. Frankfurt am Main: Suhrkamp, 1955.
———. *Selected Writings*. Ed. Michael Jennings. 4 vols. Cambridge, Mass.: Harvard University Press, 1996–2000.
———. *Träume*. Ed. Burkhardt Lindner. Frankfurt am Main: Suhrkamp, 2008.
Bergson, Henri. *Essai sur les données immédiates de la conscience*. Paris: Alcan, 1889.
———. *Matière et mémoire: Essai sur la relation du corps a l'ésprit*. Paris: Alcan, 1896.
———. *Matter and Memory*. Trans. Nancy Paul and Scott Palmer. New York: Zone Books, 1988.
Berliner, Henoch. *Involutionssyteme in der Ebene des Dreiecks*. Braunschweig: Vieweg, 1914.
Biale, David. *Blood and Belief: The Circulation of a Symbol between Jews and Christians*. Berkeley, Calif.: University of California Press, 2007.
———. *Gershom Scholem: Kabbalah and Counter-History*. Cambridge, Mass.: Harvard University Press, 1979.
Bloch, Ernst. *Geist der Utopie*. Munich: Duncker and Humblot, 1918.
Block, Richard. "Selective Affinities: Walter Benjamin and Ludwig Klages." *Arcadia* 35 (Fall 2000): 117–36.
Brüggemann, Heinz. "Fragmente zur Ästhetik / Phantasie und Farbe." In *Benjamin Handbuch: Leben, Werk, Wirking*, ed. Burkhardt Lindner, 124–33. Stuttgart: Metzler, 2006.
———. *Walter Benjamin über Spiel, Farbe und Phantasie*. Würzburg: Königshausen and Neumann, 2007.
———. "Walter Benjamins Projekt 'Phantasie und Farbe' in romantischen Kontexten." In *Walter Benjamin und die romantische Moderne*, ed. Heinz Brüggemann and Günter Oesterle, 395–445. Königshausen and Neumann, 2009.

Brühl, Georg. *Herwarth Walden und "Der Sturm."* Cologne: DuMont, 1983.
Buber, Martin. *Daniel: Dialogues on Realization.* Trans. Maurice Friedman. New York: Holt, Rinehart, and Winston, 1964.
———. *Daniel: Gespräche von der Verwirklichung.* Leipzig: Insel, 1913.
———. *Drei Reden über das Judentum.* Frankfurt am Main: Rütten and Loening, 1911.
———. *On Judaism.* Trans. Nahum Glatzer. New York: Schocken, 1967.
———. "Die Losung," *Der Jude* 1 (April 1916): 1–3.
Burns, James, ed. *The Cambridge History of Medieval Political Thought, c. 350–c. 1450.* Cambridge: Cambridge University Press, 1988.
Cantor, Georg. *Gesammelte Abhandlungen.* Ed. Ernst Zermelo. Berlin: Springer, 1932.
Cassirer, Ernst. *Das Erkenntnisproblem in der Philosophie und Wissenschaft der neueren Zeit.* 2 vols. Berlin: Verlag Cassirer, 1911.
———. *Substance and Function and Einstein's Theory of Relativity.* Trans. William Curtis Swabey and Marie Collins Swabey. New York: Dover Publications, 1923.
———. *Substanzbegriff und Funktionsbegriff: Untersuchungen über die Grundfragen der Erkenntniskritik.* Berlin: Verlag Cassirer, 1910.
Cesàro, Ernesto. *Opere scelte.* 2 vols. Rome: Cremonese, 1964–68.
Caygill, Howard. *Walter Benjamin: The Colour of Experience.* London: Routledge, 1998.
Cohen, Hermann. *Ästhetik des reinen Gefühls.* 2 vols. Berlin: Verlag Cassirer, 1912.
———. *Ethik des reinen Willens.* Berlin: Verlag Cassirer, 1904.
———. *Kants Theorie der Erfahurng.* Intro. Gert Eedel. 5th ed. New York: Olms, 1987.
———. *Logik der reinen Erkenntnis.* 2nd ed. Berlin: Verlag Cassirer, 1914.
———. *Das Princip der Infinitesimal-Methode und seine Geschichte: Ein Kapitel zur Grundlegung der Erkenntniskritik.* Berlin: Dümmler, 1883.
———. *Religion der Vernunft aus den Quellen des Judentums.* Leipzig: Fock, 1919.
———. *Religion of Reason out of the Sources of Judaism.* Trans. Simon Kaplan. New York: Unger, 1972.
Crossley, Martin. *Essential Topology.* London: Springer, 2005.
Derrida, Jacques. *L'Écriture et la différence.* Paris: Seuil, 1967.
———. "Force de loi: Le 'Fondement mystique de l'autorité' / Force of Law: the 'Mystical Foundation of Authority.'" *Cardozo Law Review* 11 (July–August, 1990): 919–1045.
———. *Speech and Phenomenon: Introduction to the Problem of the Sign in Husserl's Phenomenology.* Trans. David Allison. Evanston, Ill.: Northwestern University Press, 1973.
———. *Truth in Painting.* Trans. Geoff Bennington and Ian Mcleod. Chicago: University of Chicago Press, 1987.

———. *La Vérité en peinture*. Paris: Flammarion, 1978.

———. *La Voix et le phénomène: Introduction au probléme du signe dans la phénoménologie de Husserl*. Paris: Presses universitaires de France, 1967.

Deuber-Mankowsky, Astrid. *Der frühe Walter Benjamin und Hermann Cohen: Jüdische Werte, Kritische Philosophie, vergängliche Erfahrung*. Berlin: Verlag Vorwerk, 2000.

———. "Der schöne Schein und das Menschenopfer: Zu Benjamins Kritik an Hermann Cohens Ästhetik des reinen Gefühls." In *Walter Benjamin und die romantische Moderne*, ed. Heinz Brüggemann and Günter Oesterle, 501–19. Würzburg: Königshausen and Neumann 2009.

Diamond, Cora. *The Realist Spirit: Wittgenstein, Philosophy, and the Mind*. Cambridge, Mass.: MIT Press, 1991.

Dilthey, Wilhelm. *Einleitung in die Geisteswissenschaften: Verusch einer Grundlegung für das Studium der Gesellschaft und der Geschichte*. Leipzig: Duncker and Humblot, 1883.

Dummett, Michael. *The Interpretation of Frege's Philosophy*. Cambridge, Mass.: Harvard University Press, 1981.

Duns Scotus, John. *Opera omnia*. Ed. Luke Wadding. 26 vols. Paris: Vivès, 1891.

Duren, Peter, Anne-Katrin Herbig, and Dmitry Khavinson. "Robert Jentzsch: Mathematician and Poet." *Mathematical Intelligencer* 30 (2008): 18–24.

Eddington, Arthur Stanley. *The Nature of the Physical World*. New York: Macmillan, 1928.

Eisenreich, Günther, and Ralf Sube, eds. *Dictionary of Mathematics*. New York: Elsevier, 1982.

Erfurt, Thomas of. *Grammatica speculativa*. Ed. and trans. G. L. Bursill-Hall. London: Longman, 1972.

———. *Grammaticae speculativae: nova editio* [attributed to Duns Scotus]. Ed. Mariano García. Florence: College of St. Bonaventure, 1902.

Erger, Johannes. *Der Kapp-Lüttwitz-Putsch: Ein Beitrag zur deutschen Innenpolitik, 1919–20*. Düsseldorf: Droste, 1967.

Fenves, Peter. *Arresting Language: From Leibniz to Benjamin*. Stanford, Calif.: Stanford University Press, 2001.

———. "Derrida and History: Some Questions Derrida Pursues in His Early Writings." In *Jacques Derrida and the Humanities: A Critical Reader*, ed. Tom Cohen, 271–95. Cambridge: Cambridge University Press, 2001.

———. "The Genesis of Judgment: Spatiality, Analogy, and Metaphor in Benjamin's 'On Language as Such and on Human Language.'" In *Walter Benjamin: Theoretical Questions*, ed. David Ferris, 75–93. Stanford, Calif.: Stanford University Press, 1996.

———. "Is There an Answer to the Aestheticizing of the Political?" In *Benjamin and Art*, ed. Andrew Benjamin, 60–72. New York: Continuum, 2005.

———. *Late Kant: Towards Another Law of the Earth*. London: Routledge, 2003.

———. "Marital, Martial, and Maritime Law: Toward Some Controversial Passages in Kant's *Doctrine of Right*." *Diacritics* 35 (Winter 2005): 101–20.

———. "Toward Another Teichology." In *Babel: für Werner Hamacher*, ed. Aris Fioritos, 142–50. Basel: Urs Engeler Editor, 2008.

Feuerbach, Ludwig. *Principles of the Philosophy of the Future*. Trans. Manfred Vogel. Indianapolis: Bobbs-Merrill, 1966.

———. *Werke*. Ed. Erich Thies. 6 vols. Frankfurt am Main: Suhrkamp, 1975.

Fichte, Johann Gottlieb. *Werke*. Ed. Immanuel Hermann Fichte. 11 vols. Berlin: Walter de Gruyter, 1971.

———. *The Science of Knowledge*. Trans. Peter Heath and Johns Lachs. Cambridge: Cambridge University Press, 1982.

Fineberg, Jonathan. *The Innocent Eye: Children's Art and the Modern Artist*. Princeton, N.J.: Princeton University Press, 1997.

Fink, Eugen. "Die phänomenologische Philosophie Edmund Husserls und die gegenwärtigen Kritik." *Kant-Studien* 38 (1933): 319–83.

Förster, Eckart. *The Final Synthesis: An Essay on the "Opus postumum."* Cambridge, Mass.: Harvard University Press, 2000.

Foucault, Michel. *The Order of Things: An Archaeology of the Human Sciences*. New York: Pantheon Books, 1971.

Frege, Gottlob. *Funktion, Begriff, Bedeutung: Fünf logische Studien*. Ed. Günther Patzig. Göttingen: Vandenhoeck and Ruprecht, 1986.

———. *Grundgesetze der Arithmetik*. 2 vols. Jena: Pohle, 1893–1903.

———. *Nachgelassene Schriften*. Ed. Hans Hermes, Friedrich Kambartel, and Friedrich Kaulbach. Hamburg: Meiner, 1969.

———. *Philosophical Writings*. Ed. P. T. Geach and Max Block. Oxford: Oxford University Press, 1966.

Freud, Sigmund. *Traumdeutung*. Ed. Otto Rank. 4th ed. Leipzig: Deutike, 1914.

Geiger, Moritz. *Die Bedeutung der Kunst: Zugänge zu einer materialen Wertästhetik*. Ed. Klaus Berger and Wolfhart Henckmann. Munich: Fink, 1976.

———. *Beiträge zur Phänomenologie des ästhetischen Genusses*. In *Jahrbuch für Philosophie und phänomenologische Forschung* 1 (1913): 567–684.

———. "Fragment über den Begriff des Unbewussten und die psychische Realität: Ein Beitrag zur Grundlegung des immanenten psychischen Realismus." *Jahrbuch für Philosophie und phänomenologische Forschung* 24 (1921): 1–137.

———. "Methodologische und experimentelle Beiträge zur Quantitätslehre." *Psychologische Untersuchungen* 25 (1907): 325–522.

———. "Phänomenologische Ästhetik." *Zeitschrift für Ästhetik und allgemeine Kunstwissenschaft* 19 (1924): 29–42.

———. *Die philosophische Bedeutung der Relativitätstheorie*. Halle: Niemeyer, 1921.

———. *Systematische Axiomatik der euklidischen Geometrie.* Augsburg: Filser, 1924.
———. "Zur Erinnerung an Ernst Meumann." *Zeitschrift für Ästhetik* 11 (1916): 189–93.
Geulen, Eva. "Toward a Genealogy of Gender in Walter Benjamin's Writing." *German Quarterly* 69 (Spring 1996): 161–80.
Grabmann, Martin. *Der Gegenwartswert der geschichtlichen Erforschung der mittelalterlichen Philosophie.* Vienna: Herder, 1913.
Grimm, Jacob, and Wilhelm Grimm. *Deutsches Wörterbuch.* Ed. Moriz Heyne, Rudolph Hildebrand, Matthias Lexer, and F. L. K. Weigand. Leipzig: Hirzel, 1965– .
———. *Kleinere Schriften.* Ed. Gustav Hinrichs. 4 vols. Berlin. Dümmler, 1881–87.
Goethe, Johann Wolfgang von. *Conversations of German Refugees / Wilhelm Meister's Journeyman Years.* Ed. Jane Brown. Trans. Krishna Winston. Princeton, N.J.: Princeton University Press, 1989.
———. *Farbenlehre.* Ed. Hans Wohlbold. Jena: Diederich, 1928.
———. *Goethe über seine Dichtungen: Versuch einer Sammlung aller Äusserungen des Dichters über seine poetischen Werke.* Ed. Hans Gerhard Gräf. Frankfurt am Main: Literarische Anstalt, 1901.
———. *Theory of Color.* Trans. Charles Lock Eastlake. Mineola, N.Y.: Dover, 2006.
———. *Werke.* Ed. under the commission of Archduchess Sophie von Sachsen. Weimar: Böhlau, 1890.
Goldschmidt, Viktor. *Farben in der Kunst: Eine Studie.* Heidelberg: Winter, 1919.
Grabmann, Martin. *Mittelalterliches Geistesleben: Abhandlung zur Geschichte der Scholastik und Mystic.* Munich: Heuber, 1926.
Guerlac, Suzanne. *Thinking in Time: An Introduction to Bergson.* Ithaca, N.Y.: Cornell University Press, 2006.
Gundolf, Friedrich. *Dichter und Helden.* Heidelberg: Weiss, 1921.
Hamacher, Werner. "Afformative, Strike." Trans. Dana Hollander. In *Walter Benjamin's Philosophy: Destruction and Experience,* ed. Andrew Benjamin and Peter Osborne, 110–38. London: Routledge, 1994.
———. "Intensive Sprache." In *Übersetzen: Walter Benjamin,* ed. Christiaan L. Hart Nibrrig, 174–235. Frankfurt am Main: Suhrkamp, 2001.
———. "Parusie, Mauer. Mittelbarkeit und Zeitlichkeit, später Hölderlin." *Hölderlin-Jahrbuch* 34 (2004–5): 93–142.
———. "Schuldgeschichte: Zu Benjamins Skizze 'Kapitalismus als Religion.'" In *"Jüdische und 'Christliche" Sprachfigurationen im 20. Jahrhundert,* ed. Ashraf Noor and Josef Wohlmuth, 215–42. Paderborn: Schöningh, 2002.
———. "Das Theologisch-Politische Fragment." In *Benjamin-Handbuch: Leben, Werk, Wirking,* ed. Burkhardt Lindner, 175–92. Stuttgart: Metzler, 2006.

———. "The Word *Wolke*—If It Is One." In *Benjamin's Ground*, ed. Rainer Nägele, trans. Peter Fenves, 147–76. Detroit: Wayne State University Press, 1988.
Hartmann, Eduard von. *Kategorienlehre*. Leipzig: Haacke, 1896.
Hausdorff, Felix, under the name of Paul Mongré. *Das Chaos in kosmischer Auslese: Ein erkenntniskritischer Versuch*. Leipzig: Naumann, 1898.
———. *Grundzüge der Mengenlehre*. Leipzig: Veit, 1914.
———, under the name of Paul Mongré. *Sant' Ilario: Gedanken aus der Landschaft Zarathustras*. Leipzig: Naumann, 1897.
Heidegger, Martin. *Becoming Heidegger: On the Trail of His Early Occasional Writings*. Ed. and trans. Theodore Kiesel and Thomas Sheehan. Evanston, Ill.: Northwestern University Press, 2007.
———. *Being and Time*. Trans. John Macquarrie and Edward Robinson. New York: Harper, 1962.
———. *Elucidations of Hölderlin's Poetry*. Trans. Keith Hoeller. Amherst, N.Y.: Humanity Books, 2000.
———. *Die Kategorien- und Bedeutungslehre des Duns Scotus*. Tübingen: Mohr, 1916.
———. *Erläuterungen zu Hölderlins Dichtung*. Frankfurt am Main: Klostermann, 1981.
———. *Frühe Schriften*. Ed. Friedrich-Wilhelm von Hermann. Frankfurt am Main: Klostermann, 1978.
———. *Kant and the Problem of Metaphysics*. Trans. Richard Taft. Bloomington, Ind.: Indiana University Press, 1990.
———. *Kant und das Problem der Metaphysik*. Frankfurt am Main: Klostermann, 1973.
———. *Sein und Zeit*. Tübingen: Niemeyer, 1979.
———. "Der Zeitbegriff in der Geschichtswissenschaft." *Zeitschrift für Philosophie und philosophische Kritik* 161 (1916): 173–88.
Heidegger, Martin, and Heinrich Rickert. *Briefe 1912 bis 1933 und andere Dokumente*. Ed. Alfred Denker. Frankfurt am Main: Klostermann, 2002.
Hegel, Georg Wilhelm Friedrich. *Lectures on Fine Arts*. Trans. T. M. Knox. 2 vols. Oxford: Oxford University Press, 1975.
———. *Phenomenology of Spirit*. Trans. Parvis Emad and Kenneth Maley. Bloomington, Ind.: Indiana University Press, 1988.
———. *Werke*. Ed. Eva Moldenhauer und Karl Markus Michel. 20 vols. Frankfurt am Main: Suhrkamp, 1971.
Heisenberg, Werner. *Physics and Beyond: Encounters and Conversations*. Trans. Arnold Pomerans. New York: Harper and Row, 1971.
Héring, Jean. "Bemerkungen über das Wesen, die Wesenheit und die Idee." *Jahrbuch für Philosophie und phänomenologische Forschung* 4 (1921): 495–543.

———. *Phénoménologie et philosophie religieuse: Étude sur la théorie de la connaissance religieuse*. Paris: Alcan, 1925.

———. "Sub specie aeterni: Réponse à une critique de la philosophie de Husserl." *Revue d'histoire et de philosophie religieuse* 7 (1927): 351–64.

Hessenberg, Gerhard. *Grundbegriffe der Mengenlehre*. Göttingen: Vandenhoeck and Ruprecht, 1906.

Hölderlin, Friedrich. *Essays and Letters on Theory*. Ed. and trans. Thomas Pfau. Albany: State University of New York Press, 1988.

———. *Poems and Fragments*. Trans. Michael Hamburger. Ann Arbor: University of Michigan Press, 1967.

———. *Sämtliche Werke*. Ed. Friedrich Beissner. 8 vols. Stuttgart: Cotta-Kohlhammer, 1948–85.

———. *Sämtliche Werke: Historisch-kritische Ausgabe*. Ed. Norbert von Hellingrath and Friedrich Seebaß. 6 vols. Munich: Müller, 1913–23.

———. "Untreue der Weisheit: Ungedruckte Handschrift aus den Sammlungen auf Stift Neuberg." *Das Reich* 1 (1916): 305–9.

Horkheimer, Max, and Theodor W. Adorno. *Zeugnisse, Theodor W. Adorno zum sechzigsten Geburtstag*. Frankfurt am Main: Europaische Verlagsanstalt, 1963.

Husserl, Edmund. *Briefwechsel*. Ed. Elisabeth Schumann and Karl Schuhmann. 10 vols. Dortrecht: Kluwer, 1994.

———. *Crisis of the European Sciences and Transcendental Philosophy*. Trans. David Carr. Evanston, Ill.: Northwestern University Press, 1970.

———. *Husserliana: Gesammelte Werke*. Ed. Husserl archive in Leuven, under the directorship of H. L. van Breda. 40 vols. to date. The Hague: Nijhoff, 1950– .

———. *Ideas Pertaining to a Pure Phenomenology and to a Phenomenological Philosophy*. Trans. Ted Klein and William Pohl. The Hague: Nijhoff, 1980.

———. *Ideen zu einer reinen Phänomenologie und phänomenologischen Philosophie*. In *Jahrbuch für Philosophie und phänomenologische Forschung* 1 (1913): 1–323.

———. *Logical Investigations*. Trans. J. N. Findlay. 2 vols. Amherst, N.Y.: Humanity Books, 2000.

———. *Logische Untersuchungen*. 2 vols. Halle: Niemeyer, 1900–1901.

———. "Philosophie als strenge Wissenschaft." *Logos* 1 (1911): 289–341.

———. *The Philosophy of Internal Time-Consciousness*. Ed. Martin Heidegger. Trans. James Churchill. Bloomington, Ind.: Indiana University Press, 1964.

———. "Vorlesungen zur Phänomenologie des inneren Zeitbewusstseins." Ed. Martin Heidegger. In *Jahrbuch für Philosophie und phänomenologische Forschung* 9 (1928): 367–498.

Imdahl, Max. *Farbe: Kunsttheoretische Reflexionen in Frankreich*. Munich: Fink, 1987.

Iso, Kern. *Husserl und Kant: Eine Untersuchung über Husserls Verhältnis zu Kant und zum Neukantianismus*. The Hague: Nijhoff, 1964.
Jacobs, Carol. *In the Language of Walter Benjamin*. Baltimore: Johns Hopkins University Press, 1999.
Jacobson, Eric. *Metaphysics of the Profane: The Political Theology of Walter Benjamin and Gershom Scholem*. New York: Columbia University Press, 2003.
Jean Paul. *Sämtliche Werke*. 29 vols. Berlin: Reimer, 1862.
Jhering, Rudolf von. *Der Zweck im Recht*. 2 vols. Leipzig: Breitkopf and Härtel, 1877–83.
Justinian. *The Digest of Justinian*. Ed. Theodor Mommsen and Paul Krueger. Trans. Alan Watson. Philadelphia: University of Pennsylvania Press, 1985.
Kaemmel, Thomas, and Philipp Sonntag. *Arthur Schoenflies: Mathematiker und Kristallforscher: eine Biographie mit Aufstieg und Zerstreuung einer jüdischen Familie*. Halle: Projekte-Verlag, 2006.
Kafka, Franz. *Briefe, April 1914–1917*. Ed. Hans-Gerd Koch. Frankfurt am Main: Fischer, 2005.
———. *Gesammelte Werke*. 12 vols. Ed. Hans-Gerd Koch. Frankfurt am Main: Fischer, 1994.
Kamke, Erich, and Karl Zeller. "Konrad Knopp." *Jahresbericht der Deutschen Mathematikervereinigung* 60 (1958): 44–49.
Kambas, Chryssoula. "Walter Benjamin liest Georges Sorel: *Réflexions sur la violence*." In *Aber ein Sturm weht vom Paradiese her: Texte zu Walter Benjamin*, ed. Michael Opitz and Erdmut Wizisla, 250–69. Leipzig: Reclam, 1992.
Kant, Immanuel. *Critique of the Power of Judgment*. Trans. Paul Guyer and Eric Matthews. Cambridge: Cambridge University Press, 2000.
———. *Critique of Pure Reason*. Trans. Paul Guyer and Alan Wood. Cambridge: Cambridge University Press, 1998.
———. *Gesammelte Schriften*. Ed. Königlich-Preußische [later, Deutsche] Akademie der Wissenschaften. 29 vols. to date. Berlin: Reimer; later, Walter de Gruyter, 1900– .
———. *Practical Philosophy*. Ed. and Trans. Mary Gregor. Cambridge: Cambridge University Press, 1996.
Kierkegaard, Søren. *The Concept of Anxiety*. Trans. and ed. Reider Thomte and Albert Anderson. Princeton, N.J.: Princeton University Press, 1980.
Klages, Ludwig. "Vom Traumbewußtsein." *Zeitschrift für Pathopsychologie* 3 (1914): 1–38.
Kleinberg, Ethan. *Generation Existential: Heidegger's Philosophy in France, 1927–1961*. Ithaca, N.Y.: Cornell University Press, 2005.
Knopp, Konrad. "Bemerkungen zur Struktur einer linearen perfekten nirgends dichten Punktmenge." *Mathematische Annalen* 77 (September 1916): 438–51.

———. "Ein einfaches Verfahren zur Bildung stetiger nirgends differenzierbarer Funktionen." *Mathematische Zeitschrift* 2 (1918): 1–26.

———. "Einheitliche Erzeugung und Darstellung der Kurven von Peano, Osgood und v. Koch." *Archiv der Mathematik und Physik* 26 (1917): 103–15.

———. *Funktionstheorie: Grundlagen der allgemeinen Theorie der analytischen Funktionen.* Berlin: Göschen, 1913.

———. *Theory of Functions.* Trans. Frederick Bagemihl. 2 vols. Mineola, N.Y.: Dover Publications, 1975.

Kobell, Luise von. *Farben und Feste: Kulturhistorische Studie.* Munich: Verlag der vereinigten Kunstanstalten, 1900.

Kraft, Werner. "Friedrich C. Heinle." *Akzente* 31 (1984): 9–21.

Kusch, Martin. *Psychologism: A Study in the Sociology of Knowledge.* London: Routledge, 1995.

Ladner, Gerhart. *Images and Ideas in the Middle Ages: Selected Studies in History and Art.* Rome: Edizioni di Storia e Letteratura, 1983.

Lask, Emil. *Die Lehre vom Urteil.* Tübingen: Mohr, 1912.

Leibniz, Gottfried Wilhelm. *Die philosophischen Schriften.* Ed. C. J. Gerhardt. 7 vols. Hildesheim: Olms, 1978.

———. *Philosophical Papers and Letters.* Ed. and trans. Leroy Loemker. Dordrecht: Reidel, 1969.

———. *Sämtliche Schriften und Briefe.* Ed. Preußische [later, Deutsche] Akademie der Wissenschaften under the directorship of Paul Ritter. 46 vols to date. Darmstadt: Reichl, 1923– .

Lévinas, Emmanuel. *Théorie de l'intuition dans la phénoménologie de Husserl.* Paris: Alcan, 1930.

———. *The Theory of Intuition in Husserl's Phenomenology.* Evanston, Ill.: Northwestern University Press, 1973.

Lindner, Burkhardt. "Derrida, Benjamin, Holocaust: Zur politischen Problematik der 'Kritik der Gewalt.'" *Zeitschrift für kritische Theorie* 3 (1997): 65–100.

Linke, Paul F. *Grundfragen der Wahrnehmungslehre: Untersuchungen über die Bedeutung der Gegenstandstheorie und Phänomenologie für die experimentelle Psychologie.* Munich: Reinhardt, 1918.

———. "Das Recht der Phänomenologie: Eine Auseinandersetzung mit Th. Elsenhans." *Kant-Studien* 21 (1916): 163–221.

Lipps, Theodor. *Ästhetik: Psychologie des Schönen und der Kunst.* 2 vols. Hamburg: Voss, 1906.

Locke, John. *Two Treatises of Government.* Ed. Peter Laslett. Cambridge: Cambridge University Press, 1960.

Lorentz, Hendrik, Hermann Minkowski, and Albert Einstein. *Das Relativitätsprinzip: Eine Sammlung von Abhandlungen.* Leipzig: Teubner, 1913.

Lotze, Hermann. *Logik: Drei Bücher vom Denken, vom Untersuchen und vom Erkennen*. Leipzig: Hirzel, 1874.
Lucka, Emil. *Die Phantasie: Eine psychologische Untersuchung*. Vienna: Braumüller, 1908.
Lukács, Georg. *Die Seele und die Formen*. Berlin: Fleischel, 1911.
———. *Soul and Form*. Trans. Anna Bostock. Cambridge, Mass.: MIT Press, 1974.
———. *Die Theorie des Romans*. Berlin: Verlag Cassirer, 1920.
———. *The Theory of the Novel*. Trans. Anna Bostock. Cambridge, Mass.: MIT Press, 1971.
Mac Lane, Saunders. "Mathematics at Göttingen." *Notices of the American Mathematical Society* 42 (1995): 1134–38.
Mandelbrot, Benoit. *The Fractal Geometry of Nature*. New York: W. H. Freeman, 1983.
———. *Les Objets fractals: Forme, hasard et dimension*. Paris: Flammarion, 1975.
Marty, Anton. *Die Frage nach der geschichtlichen Entwickelung des Farbensinnes: Nebst zwei Anhängen*. Vienna: Gerold, 1879.
Marx, Karl. *Capital: A Critique of Political Economy*. Trans. Ben Fowkes. 3 vols. New York: Vintage Books, 1977.
———. *Das Kapital: Kritik der politischen Ökonomie*. 3 vols. Berlin: Dietz, 1981.
McCole, John. *Walter Benjamin and the Antinomies of Tradition*. Ithaca, N.Y.: Cornell University Press, 1993.
Merleau-Ponty, Maurice. *The Phenomenology of Perception*. Trans. Colin Smith. London: Routledge, 1962.
Métraux, Alexandre. "Zur phänomenologischen Ästhetik Moritz Geigers." *Studia philosophica* (1969): 68–92.
Meyerson, Emile. *De l'explication dans les science*. 2 vols. Paris: Payot, 1921.
———. *Explanation in the Sciences*. Trans. Mary-Alice and David Sipfle. Boston: Kluwer, 1991.
Milgrom, Jacob. *Leviticus, 17–22: A New Translation with Introduction and Commentary*. New York: Doubleday, 2000.
Mohanty, Jitendranath. "Individual Fact and Essence in Husserl's Philosophy." *Philosophy and Phenomenological Research* 28 (1959): 220–30.
Moore, Gregory. *Zermelo's Axiom of Choice: Its Origins, Development, and Influence*. New York: Springer, 1982.
Moran, Dermot. *Introduction to Phenomenology*. London: Routledge, 2000.
Mosès, Stéphane. *The Angel of History: Rosenzweig, Benjamin, Scholem*. Trans. Barbara Hershav. Stanford, Calif.: Stanford University Press, 1993.
Müller-Freienfels, Richard. "Gefühlstöne der Farbenempfindungen." *Zeitschrift für Psychologie und Physiologie der Sinnesorgane* 46 (1907): 241–74.

Naas, Josef, and Hermann Ludwig Schmid, eds. *Mathematisches Wörterbuch*. Berlin: Akademie-Verlag, 1962.

Nägele, Rainer. *Literarische Vexierbilder: Drei Versuche zu einer Figur*. Eggingen: Isele, 2001.

Nancy, Jean-Luc. *The Experience of Freedom*. Trans. Bridgit McDonald. Stanford, Calif.: Stanford University Press, 1993.

Natanson, Maurice. *Edmund Husserl: Philosopher of Infinite Tasks*. Evanston, Ill.: Northwestern University Press, 1973.

Newton, Isaac. *Principia mathematica*. Trans. Bernard Cohen and Anne Whitman. Berkeley: University of California Press, 1999.

Nietzsche, Friedrich. *Beyond Good and Evil: Prelude to a Philosophy of the Future*. Trans. Walter Kaufmann. New York: Vintage Books, 1966.

———. *The Birth of Tragedy*. Trans. Douglas Smith. New York: Oxford University Press, 2000.

———. *Sämtliche Werke*. Ed. Giorgio Colli and Mazzino Montinari. 15 vols. Berlin: Walter de Gruyter, 1967–77.

Noeggerath, Felix. "Synthesis und Systembegriff in der Philosophie: Ein Beitrag zur Kritik des Antirationalismus." Diss., University of Erlangen, 1916.

Olesen, Søren. *Wissen und Phänomenon: Eine Untersuchung ontologischer Klärung der Wissenschaften bei Edmund Husserl, Alexandre Koyré und Gaston Bachalard*. Würzburg: Königshausen and Neumann, 1997.

Otero, Alfonso. "Die Eigenständigkeit der plenitudo potestatis in den spanischen Königreichen des Mittelalters." In *Epirrhosis: Festgabe für Carl Schmitt*, ed. Hans Barion, 597–616. Berlin: Duncker and Humblot, 1968.

Peltzer, Alfred. *Die ästhetische Bedeutung von Goethes "Farbenlehre."* Heidelberg: Winter, 1903.

Pfänder, Alexander. *Zur Psychologie der Gesinnungen*. In *Jahrbuch für Philosophie und phänomenologische Forschung* 1 (1913): 325–404.

Picker, Marion. *Der konservative Charakter: Walter Benjamin und die Politik der Dichter*. Bielefeld: Transcript, 2004.

Planck, Max. *Acht Vorlesungen über theoretische Physik, gehalten an der Columbia University in the City of New York im Frühjahr 1909*. Leipzig: Hirzil, 1910.

Plato. *Opera*. Ed. John Burnet. 5 vols. Oxford: Clarendon, 1905.

Pollitt, Jerome. "The Canon of Polykleitos and Other Canons." In *Polykleitos, the Doryphoros, and Tradition*, ed. Warren Moon, 19–25. Madison: University of Wisconsin Press, 1995.

Reich, Klaus. *The Completeness of Kant's Table of Categories*. Trans. Jane Kneller and Michael Losonsky. Stanford, Calif.: Stanford University Press, 1992.

———. *Die Vollständigkeit der kantischen Urteilstafel*. Berlin: Schutz, 1932.

Reichenbach, Hans. *Relativitätstheorie und Erkenntnis apriori*. Berlin: Springer, 1920.

Reicke. Rudolf. *Lose Blätter aus Kants Nachlass.* Königsberg: Beyer, 1889.

Reinach, Adolf. "The A Priori Foundations of Civil Law." Trans. John Crosby. *Aletheia* 3 (1983): 1–142.

———. *Die apriorischen Grundlagen des Bürgerlichen Rechtes.* In *Jahrbuch für Philosophie und phänomenologische Forschung* 1 (1913): 685–847.

Reitter, Paul. "Irrational Man: Gershom Scholem's Decisive Years." *Harper's Magazine* 316 (May 2008): 87–94.

Rickert, Heinrich. "Das Eine, die Einheit und die Eins: Bemerkungen zu Logik des Zahlenbegriffs." *Logos* 2 (1911–12): 26–78.

———. *Der Gegenstand der Erkenntnis: Einführung in die tranzendentale Philosophie.* Tübingen: Mohr, 1904.

———. "Vom Begriff der Philosophie." *Logos* 1 (1910–11): 1–34.

Ronell, Avital. *Stupidity.* Champaign, Ill.: University of Illinois Press, 2008.

Rosenstock-Huessy, ed. *Judaism Despite Christianity: The Letters on Christianity and Judaism Between Eugen Rosenstock-Huessy and Franz Rosenzweig.* University: Alabama University Press, 1969.

Rosenzweig, Franz. *Das älteste Systemprogramm des deutschen Idealismus: Ein handschriftlicher Fund.* Heidelberg: Winter, 1917.

———. *Briefe und Tagebücher.* Ed. Rachel Bat-Adam, Edith Rosenzweig-Scheinmann, and Bernhard Casper. The Hague: Nijhoff, 1979.

———. *Philosophical and Theological Writings.* Trans. and ed. Paul Franks and Michael Morgan. Indianapolis, Ind.: Hackett, 2000.

———. *The Star of Redemption.* Trans. William Hallo. Notre Dame, Ind.: Notre Dame University Press, 1985.

———. *Stern der Erlösung.* Frankfurt am Main: Kauffmann, 1921.

Ruskin, John. *The Elements of Drawing.* New York: Dover Publications, 1971.

Russell, Bertrand. *The Principles of Mathematics.* 2nd ed. New York: W. W. Norton, 1996.

Rüstow, Alexander. *Der Lügner: Theorie, Geschichte und Auflösung.* Leipzig: Teubner, 1910.

Sartre, Jean-Paul. *Being and Nothingness: A Phenomenological Essay on Ontology.* Trans. Hazel Estella Barnes. New York: Simon and Shuster, 1993.

———. *L'Être et le néant.* Paris: Gallimard, 1943.

Scheerbart, Paul. *Lesabéndio, ein asteroïden-Roman.* Munich: Müller, 1913.

Scheler, Max. *Der Formalismus in der Ethik und die materiale Wertethik.* In *Jahrbuch für Philosophie und phänomenologische Forschung* 1 (1913): 405–565.

———. *Formalism in Ethics and Non-Formal Ethics of Value: A New Attempt at the Foundation of Ethical Personalism.* Trans. Manfred Frings and Roger Funk. Evanston, Ill.: Northwestern University Press, 1973.

———. *Gesammelte Werke.* Ed. Maria Scheler and M. S. Frings. 14 vols. Bonn: Bouvier, 1986.

Schelling, Friedrich Wilhelm Joseph von. *Philosophical Investigations into the Essence of Human Freedom*. Trans. Jeff Love and Johannes Schmidt. Albany: State University of New York Press, 2006.

———. *Philosophische Untersuchungen über das Wesen der menschlichen Freiheit*. Frankfurt am Main: Suhrkamp, 1975.

Schestag, Thomas. "Interpolationen: Benjamins Philologie." In *Philo:zenia*. Ed. Thomas Schestag. Basel: Engeler, 2009.

Schmitt, Carl. *Politische Theologie: Vier Kapitel zur Lehre von der Souveränität*. Munich: Duncker and Humblot, 1922.

———. "Recht und Macht." *Summa* 1 (1917–18): 37–52.

Schoenflies, Arthur. *Entwicklung der Mengenlehre und ihrer Anwendungen*. Part 1, *Allgemeine Theorie der unendlichen Mengen und Theorie der Punktmenge*. Leipzig: Tuebner, 1913.

———. "Über die logischen Paradoxieen der Mengenlehre." *Jahresbericht der Deutschen Mathematiker-Vereinigung* 15 (1906): 19–25.

Scholem, Gershom. *Briefe an Werner Kraft*. Frankfurt am Main: Suhrkamp, 1986.

———. *From Berlin to Jerusalem: Memories of My Youth*. Trans. Harry Zohn. Schocken, 1980.

———. *Lamentations of Youth*. Ed. and trans. Anthony Skinner. Cambridge, Mass.: Harvard University Press, 2007.

———. *A Life in Letters, 1914–1982*. Ed. and trans. Anthony David Skinner. Cambridge, Mass.: Harvard University Press, 2002.

———. *Major Trends in Jewish Mysticism*. New York: Schocken, 1941.

———. *Tagebücher, nebst Aufsätzen und Entwürfen bis 1923*. Ed. Karlfried Gründer, Herbert Kopp-Oberstebrink, and Friedrich Niewöhner, in association with Karl Grözinger. 2 vols. Frankurt am Main: Jüdischer Verlag, 1995–2000.

———. *Von Berlin nach Jerusalem*. Frankfurt am Main: Suhrkamp, 1994.

———. *Walter Benjamin: Die Geschichte einer Freundschaft*. Frankfurt am Main: Suhrkamp, 1975.

———. *Walter Benjamin: The Story of a Friendship*. Trans. Harry Zohn. New York: Schocken, 1981.

———. "Walter Benjamin und Felix Noeggerath." In *Walter Benjamin und sein Engel: Vier Aufsätze und kleine Beiträge*, ed. Rolf Tiedemann, 78–127. Frankfurt am Main: Suhrkamp, 1983.

Schwartzman, Steven. *The Words of Mathematics: An Etymological Dictionary of Mathematical Terms Used in English*. Washington, D.C.: Mathematical Society of America, 1994.

Schwepenhäuser, Hermann. "Walter Benjamin über Gerechtigkeit." *Frankfurter Adorno Blätter* 4 (1996): 43–51.

Shell, Susan Meld. *Kant and the Limits of Autonomy*. Cambridge, Mass: Harvard University Press, 2009.

Shestov, Lev. *Potestas clavium*. Ed. and trans. Bernard Martin. Columbus: Ohio University Press, 1968.
Simmel, Georg. *Das Problem der historischen Zeit*. Berlin: Reuther and Reichard, 1916.
Smid, Reinhold. "'Münchener Phänomenologie': Zur Frühgeschichte des Begriffs." In *Pfänder-Studien*, ed. Eberhard Avé-Lallemant und Herbert Spiegelberg, 109–53. The Hague: Nijhoff, 1982.
Smith, Gary. "'Das Judentum versteht sich von selbst': Walter Benjamins frühe Auseinandersetzung mit dem Judentum." *Deutsche Vierteljahrsschrift für Literaturwissenschaft und Geistesgeschichte* 65 (1991): 318–34.
———. "Thinking Through Benjamin: An Introductory Essay." In *Benjamin: Philosophy, Aesthetics, History*, ed. Gary Smith, vii–xxxviii. Chicago: University of Chicago Press, 1983.
Spiegelberg, Herbert. *The Phenomenological Movement: A Historical Introduction*. 2nd ed. 2 vols. The Hague: Nijhoff, 1971.
Steiner, Uwe. *Die Geburt der Kritik aus dem Geiste der Kunst: Untersuchungen zum Begriff der Kritik in den frühen Schriften Walter Benjamins*. Würzberg: Königshausen and Neumann, 1989.
———. "Phänomenologie der Moderne. Benjamin und Husserl." In *Benjamin-Studien* I, ed. Daniel Weidner and Sigrid Weigel, 107–25. Munich: Fink, 2008.
———. "The True Politician: Walter Benjamin's Concept of the Political." *New German Critique* 83 (Spring–Summer 2001): 43–88.
Surzyn, Jacek. "Jean Héring." In *Phenomenology World-Wide: Foundations, Expanding Dynamisms, Life-Engagements; A Guide for Research and Study*, ed. Anna-Teresa Tymieniecka, 74–76. Berlin: Springer, 2002.
Taminiaux, Jacques. *The Metamorphoses of the Phenomenological Reduction*. Milwaukee: Marquette University Press, 2004.
Tanakh: A New Translation of the Holy Bible—The New JPS Translation According to the Traditional Hebrew Text. Philadelphia: Jewish Publication Society, 1985.
Trendelenburg, Friedrich. *Geschichte der Kategorienlehre*. Berlin: Berthge, 1846.
Trillhaas, Wolfgang. *Aufgehobene Vergangenheit: Aus meinem Leben*. Göttingen: Vandenhoeck and Ruprecht, 1976.
Van Buren, John. *The Young Heidegger: Rumor of the Hidden King*. Bloomington: Indiana University Press, 1994.
Vorwerk, Herbert. "Das Recht zur Gewaltanwendung." *Blätter für religiösen Sozialismus* 1 (1920): 14–16.
Wackernagel, Wilhelm, *Glossar zum altdeutschen Lesebuch*. Basel: Schweighauser, 1809.
Wassermann, Jakob. "Der Jude als Orientale." In *Vom Judentum: Ein Sammelbuch*, ed. Bar Kochba Association, 5–8. Leipzig: Wolff, 1913.

Watts, Pauline. "The Donation of Constantine, Cartography, and Papal 'Plenitudo Potestatis' in the Sixteenth Century." *MLN* 119 (2004): 88–107.
Weber, Max. *Politics as Vocation*. Trans. H. H. Gerth and C. Wright Mills. Philadelphia: Fortress Press, 1965.
———. *Politik als Beruf*. Munich: Duncker and Humblot, 1919.
Weber, Samuel. *Benjamin's -abilities*. Cambridge, Mass.: Harvard University Press, 2008.
———. "Der Brief an Buber vom 17.7.1916." In *Benjamin-Handbuch*, ed. Burkhardt Lindner, 603–9. Stuttgart: Metzler, 2006.
———. *Targets of Opportunity: On the Militarization of Thinking*. New York: Fordham University Press, 2005.
Weierstraß, Karl. "On Continuous Functions of a Real Argument That Do Not Have a Well-Defined Differential Quotient." In *Classics on Fractals*, ed. and trans. Gerald A. Edgar, 3–9. Reading, Mass.: Addison-Wesley, 1993.
Weigand, Rudolf. *Die Naturrechtslehre der Legisten und Dekretisten von Irnerius bis Accursius und von Gratian bis Johannes Teutonicus*. Munich: Hueber, 1967.
Weigel, Sigrid. *Body- and Image-Space: Re-reading Walter Benjamin*. Trans. Georgina Paul, Rachel McNicholl, and Jeremy Gaines. New York: Routledge, 1996.
Werner, Karl. *Die Sprachlogik des Johannes Duns Scotus*. Vienna: Gerold, 1877.
Wiesenthal, Liselotte. *Zur Wissenschaftstheorie Walter Benjamins*. Berlin: Athenäum, 1973.
Wittgenstein, Ludwig. *Remarks on Color*. Ed. G. E. M. Anscombe. Berkeley: University of California Press, 1978.
———. *Tractatus Logico-Philosophicus; Logische-philosophische Abhandlung*. Frankfurt am Main: Suhrkamp, 1999.
Wohlfarth, Irving. "Nihilistischer Messianismus: Zu Walter Benjamins 'Theologisch-Politischem Fragment." In *"Jüdisch" und "christliche" Sprachfigurationen im 20. Jahrhundert*, ed. Aschraf Noor and Josef Wohlmuthm, 141–214. Paderborn: Schöningh, 2002.
———. "Walter Benjamin and the Idea of Technological Eros: A Tentative Reading of 'Zum Planetarium.'" *Benjamin Studies/Studien* 1 (2002): 65–109.
Zeltner, Hermann. "Moritz Geiger im Gedächtnis." *Zeitschrift für philosophische Forschung* 14 (1960): 452–66.
Zilsel, Edgar. *Anwendungsproblem: Ein philosophischer Versuch über das Gesetz der großen Zahlen und die Induktion*. Leipzig: Barth, 1916.

Index

Adorno, Theodor, 1–2, 4, 188, 259n2, 288n4
Agamben, Giorgio, 98, 271n21, 285–86n18, 286–87n31, 287n35, 287–88n18
Allen, Grant, 60
Aristotle, 263n17, 279n22

Bahti, Timothy, 269n49
Balzano, Bernard, 114
Bambach, Charles, 276n35
Barthel, Ernst, 99, 139, 282n23
Baudelaire, Charles, 11, 149, 221, 251, 289–90n1
Bauch, Bruno, 127
Baumgarten, Alexander, 87, 270n12
Beck, Jakob Sigismund, 284n7
Benjamin, Walter, "Agesilaus Santander," 146; "Aphorisms on the Theme of Fantasy and Color," 85; *Arcades Project*, 49, 242, 282–83n25, 283n28; "Attempt at a Solution of Russell's Paradox," 127–28, 141, 144, 147; *Berlin Chronicle*, 18, 21; "Capitalism as Religion," 208, 269n48; "Color, Considered from the Perspective of the Child," 63–67, 76, 79, 80, 89; *Concept of Art Critique in German Romanticism*, 44, 45, 172–73, 263n1; "Destiny and Character," 70–71, 73–75; "Eidos and Concept," 50–55, 74, 133, 134, 144, 263–64n5; "For a Work on the Beauty of Colorful Pictures in Children's Books," 76–78, 79, 98, 102; *Germans*, 284–85n10; "Goethe's *Elective Affinities*," 17, 33, 171, 211, 269n4, 284–85n10; "The Infinite Task," 171–72; "Judgment of Designation," 128–30; "Life and Violence," 208; "The Meaning of Time in the Moral World," 241; "Metaphysics of Youth," 267–68n35; "Methodical Modes of History," 232–34; "Moral Instruction," 215; "Notes Toward a Work on the Category of Justice," 13, 107, 188–89, 192, 196, 197–202, 204–7, 208, 214, 218, 222, 273n8, 257–58; "On Language as Such and on Human Language," 15, 52, 117–18, 137, 141–49, 173, 202–4, 224, 228; "On Painting or Sign and Mark," 99–101, 223; "On Perception," 159–61, 173, 178; "On Shame," 67–73, 76, 266–67n27, 268n39, 40; "On the Concept of History," 243–34; "On the Program of the Coming Philosophy," 12, 31–32, 48, 152–86, 215; "On Transcendental Method," 174; *One-Way Street*, 19, 222; *Origin of the German Mourning*

Play, 1–3, 6, 8, 13, 18, 19, 54, 61, 75–76, 122, 123, 128–29, 182–86; "Outlook into Children's Books," 265–66n25, 269n50; "Politics," 208–9, 219, 225, 230, 231, 235; "Psychology," 228–29; "The Rainbow: Dialogue on Fantasy," 9, 45, 67, 79–91, 94–97, 102, 247–55, 269n1, 270n15; "The Rainbow; or, The Art of Paradise," 9, 79–80, 91–98, 101, 102, 254–55; "Reddening in Rage and Shame," 66; "Task of the Translator," 11, 57, 149–51; "Theological-Political Fragment," 3, 230; "Theory of Art Critique," 172; "Theses on the Problem of Identity," 54–55, 169–71, 281n15; "Tragedy and *Trauerspiel*," 123; "Two Poems of Friedrich Hölderlin," 15, 19–43, 44, 45, 60, 65, 91, 102, 107, 116, 212; "Toward the Critique of Violence," 208–26, 231, 234, 285–86n18; "Word and Concept," 57, 60; "World and Time," 224–25, 230, 244
Berger, Klaus, 260n9
Bergson, Henri, 6, 15, 58, 118, 240; *Essay on the Immediate Data of Consciousness*, 30–32; *Matter and Memory*, 262n9
Berliner, Henoch, 113, 275n27
Biale, David, 271n1, 287n34, 288n7
Bible. *See* individual books of the
Bloch, Ernst, 208, 230
Block, Richard, 270n15
Blumenthal-Belmore, Herbert, 98, 271n21
Braque, Georges, 98
Brentano, Franz, 60, 227
Brod, Max, 189
Brüggemann, Heinz, 266n26
Buber, Martin, 39, 68, 104, 105–6, 117, 156, 157, 272n4; *Daniel*, 105, 269n1; *On Judaism*, 263n14, 271–72n1, 280n5; "The Password," 272n5

Cantor, Georg, 43, 113, 114, 140, 169, 171, 182, 278n20, 278–89n21, 279n22, 283n29, 274n21
Cassirer, Ernst, 5, 32; *Concept of Substance and Concept of Function*, 26, 27, 28, 52–53, 229, 259n4, 263–64n5, 282n23; *Problem of Knowledge*, 261–62n6
Caygill, Howard, 281n17
Cesàro, Ernesto, 243
Chagall, Marc, 98
Clairvaux, Bernard of, 220
Cohen, Hermann, 5, 6, 12, 20, 25–26, 32, 43, 92, 164, 167, 168, 172, 173, 174, 179, 180, 210–12, 215, 227, 234, 275n29, 283n26; *Aesthetics of Pure Feeling*, 7, 8, 82–86, 93, 269n4; *Ethics of Pure Will*, 187, 210, 286n19, 286n21; *Kant's Theory of Experience*, 161, 184–85, 263n17; *Logic of Pure Knowledge*, 12, 26, 158–61, 182, 183, 185; *Principle of the Infinitesimal Method*, 5, 259n4, 263n17; *Religion of Reason*, 227–28, 281n18
Cohen, Paul J., 278–79n21
Corngold, Stanley, 273n12
Crossley, Martin, 274n20
Cusa, Nicholas of, 286–87n31

Da Vinci, Leonardo, 61
Derrida, Jacques, 10, 97, 222, 261n17, 278n16, 279n29, 288n15
Deuber-Mankowski, Astrid, 261n5, 269n4
Diamond, Cora, 277n6
Dilthey, Wilhelm, 156, 157, 166, 280n4
Dummett, Michael, 277n10
Duns Scotus, John, 55–56, 57–58

Eddington, Arthur Stanley, 282–83n25
Edgar, Gerald, A., 274n21
Eedel, Kurt, 259n4
Einstein, Albert, 114, 157, 177–78, 260n10, 282n24, 282–83n25
Empedocles, 162
Euclid, 114

Index 309

Feuerbach, Ludwig Andreas, 152–53, 154, 155, 167
Fichte, Johann Gottlieb, 30, 173, 181
Fineberg, Jonathan, 266n26
Fink, Eugen, 51, 265n14
Flaubert, Gustav, 31
Förster, Eckart, 281n20
Foucault, Michel, 266n26
Frege, Gottlob, 125–28, 130, 157, 158, 167, 168–69, 277n6, 277n10
Freud, Sigmund, 81–82
Frobenius, Ferdinand, 109, 112

Garve, Christian, 175
Geiger, Abraham, 7, 235
Geiger, Ludwig, 7, 60
Geiger, Moritz, 6–9, 11, 46, 47, 82–83, 85, 105, 113, 235, 260n9, 10, 261n15, 264–65n11, 272n2, 282n23; *Axiomatics of Euclidean Geometry*, 260n10; *Contributions to the Phenomenology of Aesthetic Enjoyment*, 7, 8, 47, 95, 270n17; "Phenomenological Aesthetics," 272n2; *Philosophical Significance of the Theory of Relativity*, 260n10, 282n23
George, Stefan, 33
Genesis, Book of, 11, 143–48, 202–4, 228
Geulen, Eva, 268
Gödel, Kurt, 278–89n21
Goethe, Johann Wolfgang von, 22, 33, 43, 84, 267n30, 268n37; *Elective Affinities*, 231, 266–67n27; *Theory of Color*, 61–70, 266–67n27; *West-Eastern Divan*, 267n31; *Wilhelm Meister's Journeyman Years* (including "The New Melusine"), 231–32, 234–378, 242; *Wilhelm Meister's Years of Apprenticeship*, 63
Goldschmidt, Viktor, 60
Gombrich, Ernst, 266n26
Grabmann, Martin, 55, 57, 265n19
Gräf, Hans-Gerhard, 61
Grimm, Jacob, 287n33
Grünewald, Matthias, 252

Guerlac, Suzanne, 262n9
Gundolf, Friedrich, 33, 262n12
Gutkind, Erich, 178

Hallmann, Johann Christian, 186
Hamacher, Werner, 222, 262n11, 269n48, 270n14, 279n27, 283n26, 288n4, 288–89n18
Hausdorff, Felix, 111, 113; *Chaos from Cosmic Selection*, 108–10, 238, 273n13; *Sant' Ilario*, 108
Hartmann, Eduard von, 57
Heidegger, Martin, 2, 3, 6, 15, 9, 151, 177, 186, 196, 215, 235; *Being and Time*, 19, 70–73, 265n24, 268n42, 268–89n44, 276n34; "Concept of Time in Historical Scholarship," 15, 118–21, 123, 206, 208; *Doctrine of Judgment in Psychologism*, 265n24; *Duns Scotus's Doctrine of Categories and Meaning*, 57–59, 118; *Elucidations of Hölderlin's Poetry*, 261n2; *Kant and the Problem of Metaphysics*, 196, 284n9; "New Research in Logic," 58, 265n20
Heinle, Friedrich, 17, 21, 37, 67, 113, 251, 269n50, 270n11, 275n24, 290n2
Heisenberg, Werner, 282–83n25
Hegel, Georg Friedrich, 87, 106, 158, 175, 184, 272n6
Héring, Jean, 1, 49–50, 54, 161, 184, 264n8
Herzl, Theodor, 275n22
Hessenberg, Gerhard, 126–27, 277n8
Heym, Georg, 275n21
Hilbert, David, 260n10
Hirsch, Immanuel, 206
Hölder, Otto, 243
Hölderlin, Friedrich, 5, 15, 19–43, 49, 84, 143, 150, 173, 182, 240, 263n13, 263n17, 283n30, 288–89n18; "Blödigkeit" (Infirmity), 27, 28–31, 34–42, 91, 262n7; "Brot und Wein" (Bread and wine), 33; "Dichtermuth" (Poetic courage), 27, 28–30, 36–38,

91; "Hälfte des Lebens" (Middle of life), 44–45; "Herbst" (Autumn), 42; "Patmos," 17; "Urteil und Sein" (Judgment and being), 262n8, 281n15; "Wie wenn am Feiertage" ("As when on a holiday"), 33
Holländer, Käthe, 113, 235
Humboldt, Wilhelm von, 84
Husserl, Edmund, 1–3, 4, 10–11, 12, 20, 32, 73, 105, 111, 148, 167, 172, 184, 223, 232, 264n8; *Crisis of the European Sciences*, 264n7; *Ideas Pertaining to a Pure Phenomenology*, 2–3, 7, 9, 10, 46, 47–48, 49, 50–51, 53, 57, 70, 71, 93, 94, 100, 163, 165, 264n7, 264n10, 270n16, 278n18; *Logical Investigations*, 1, 5, 6, 8, 11, 24, 46, 47–48, 50, 52, 57, 58, 67–70, 72, 74, 133, 135, 151, 163, 227–28, 263n6, 264n10, 268–69n44, 278n16; "Philosophy as Rigorous Science," 8, 23, 24, 161–63, 166

Imdahl, Max, 266n26
Innocent III, Pope, 286n31
Iso, Kern, 280n9

Jacobs, Carol, 279n26
Jacobson, Eric, 271n1, 284n4, 285n14
Jentzsch, Robert, 113, 275n24
Jhering, Rudolf von, 211
Justinian, 284n6

Kafka, Franz, 68, 189, 285n12; "In the Penal Colony," 204, 207, "Report for an Academy," 268n38, 272n4
Kambas, Chryssoula, 285–86n18
Kandinski, Wassily, 98
Kant, Immanuel, 1, 4, 5, 35, 47, 110, 152–57, 163, 165–67, 170, 171, 172, 173–74, 175–76, 177, 178, 182, 183–85, 187–215, 218, 219, 222, 224, 234, 241, 270–71n18, 280n10, 281n20, 286n20; *Critique of Judgment*, 7, 8, 9, 11–12, 48, 82–85, 92, 97, 198, 209, 210; *Critique of Practical Reason*, 153–54, 190, 196, 197, 210, 211, 215; *Critique of Pure Reason*, 11, 25–26, 30, 60, 122–23, 137–39, 142–43, 153–54, 155, 159, 161, 168, 169, 174, 180, 209, 210, 238; *Metaphysics of Morals* (including *The Doctrine of Right*), 13–14, 187, 187–97, 199–200, 201, 202, 211, 213, 284–85n10; *Nova dilucidatio*, 168; *Opus postumum*, 175, 180; *Prolegomena to any Future Metaphysics*, 153, 154, 155, 157; *Toward Eternal Peace*, 190–91, 195, 209; *True Estimation of Living Forces*, 139
Kellner, Leon, 275n22
Kierkegaard, Søren, 103–4, 144; *Concept of Anxiety*, 73, 147; *Stages on Life's Way*, 103
Klages, Ludwig, 270n15
Knopp, Konrad, 110–11, 112, 114, 115, 242–43, 274n16, 274n21
Kokoschka, Oskar, 98
Kobell, Luise von, 60–61
Kraft, Werner, 261n18
Kush, Martin, 261n4

Lask, Emil, 57
Leibniz, Gottfried, 176–67, 180, 184, 243, 281–82n21
Lévinas, Emmanuel, 264n6, 264n8
Leviticus, Book of, 225–26, 287n34, 287n35
Lindner, Burkhardt, 286n18, 269n2, 285–86n18
Linke, Paul, 50–54, 184, 264–65n11; *Basic Questions of Perception*, 51; "The Law of Phenomenology," 46, 50–51, 164n5, 265n14
Lipps, Theodor, 260n10, 270n17
Locke, John, 195
Lotze, Hermann, 57, 127
Lucka, Emil, 60
Lukács, Georg, 31, 262n10, 269n1
Lyser, Johann Peter, 269n50

Mac Lane, Saunders, 261n15
Mach, Ernst, 106
Mandelbrot, Benoit, 111, 274n21, 288n17
Mann, Thomas, 156, 279n3
Marty, Anton, 60
Marx, Karl, 213, 270–71n18
Matthew, Gospel of, 200–201
McCole, John, 270n15
Mennicke, Carl, 215–56
Merleau-Ponty, Maurice, 266n26
Meyerson, Emile, 282–83n25
Milgrom, Jacob, 287n34
Mohanty, Jitendranath, 264n10
Moore, Gregory, 278n21
Mongré, Paul. *See* Hausdorff, Felix
Moran, Dermot, 260n11
Mosès, Stéphane, 272ß3n7
Müller-Freienfels, Richard, 267n31

Nägele, Rainer, 271n19
Nancy, Jean-Luc, 280n7
Newton, Issac, 61, 62, 114, 177, 178, 180
Ng, Julia, 273n15, 274n20, 276n30
Nietzsche, Friedrich, 20, 32–33, 167; *Beyond Good and Evil*, 152–53, 154, 155; *Birth of Tragedy*, 20, 39, 81; *Thus Spoke Zarathustra*, 103, 154, 268n40
Noeggerath, Felix, 47, 112–13, 275n23
Novalis (pseudonym of Georg Philipp Friedrich Freiherr von Hardenberg), 22, 173

Olesen, Søren, 288n3
Otero, Alfonso, 287n31

Peltzer, Alfred, 61
Pfänder, Alexander, 7–8
Picasso, Pablo, 98–99
Picker, Marion, 261n5, 272n5
Planck, Max, 177, 282n23
Plato, 1, 12, 76–77, 115, 162, 178–89, 183–84, 270n15; *Phaedrus*, 4; *Symposium*, 76; *Timaeus*, 185
Pollack, Dora, 106, 272n6, 275n22
Pollitt, Jerome, 270n8

Pollack, Max, 106
Polycleitus, 83

Radt, Grete, 84, 91, 247
Radt, Friedrich, 6, 46–47, 84–85, 259n5
Reich, Klaus, 168
Reichenbach, Hans, 282n23
Reicke, Rudolf, 284–85n10
Reinach, Adolph, 7, 260–61n12, 264n8
Reitter, Paul, 271–72n1, 276n30
Rickert, Heinrich, 6, 15, 57, 58, 59, 65, 118, 119, 259–60n8, 265n21, 276n35; "On the Concept of Philosophy," 164, 265n24, 280n10; *Object of Knowledge*, 259n6; "Unit, Unity, and One," 281n16
Ronell, Avital, 262n7
Rosenstock-Heussy, Eugen, 272–73n7
Rosenzweig, Franz, 262n8, 272–73n7, 280–81n14
Ruskin, John, 266n26
Russell, Bertrand, 10, 120, 125, 157, 158, 167; "Russell's Paradox," 125–30, 134, 141, 144, 147, 169
Rüstow, Alexander, 277n5

Sartre, Jean-Paul, 268n41
Scheerbart, Paul, 208–9, 230–31
Scheler, Max, 7, 268n41
Schelling, Friedrich Wilhelm Joseph von, 30, 175, 280–81n14
Schestag, Thomas, 288n5
Schiller, Friedrich, 20, 33, 36
Schlegel, Friedrich, 45, 173
Schmitt, Carl, 285–86n18, 286n22
Schocken, Zalman, 238–39
Schoenflies, Arthur, 113, 125–26, 127, 140, 275n25, 277n5
Scholem, Gershom, 1, 4, 12, 13, 14–16, 40, 43, 48, 50, 51–53, 57, 73, 76, 103–24, 127, 132, 136, 145, 146, 149–50, 154, 161, 169, 177–78, 197, 204, 205–8, 231–32, 234–43, 263n17, 269n48, 271–72n1, 273n13, 275n22, 275n28, 29, 276n31, 283n26; *From*

Berlin to Jerusalem, 264–65n11; *Major Trends in Jewish Mysticism*, 238; "On and Against Cubism," 98–100, 103; "Potpourri Regarding a Mechanistic World-Image," 114; *Walter Benjamin: The Story of a Friendship*, 107, 110, 111, 112, 116, 188–89, 190, 200, 263n16, 269n50, 271n21, 272n6, 272–73n7
Schopenhauer, Arthur, 20
Schweppenhäuser, Hermann, 116–17, 266–67n27, 277n11, 278n15, 284n2
Shakespeare, William, 286n21
Shell, Susan Meld, 286n18
Shestov, Lev, 162
Simmel, Georg, 14, 116, 117, 118, 122
Skinner, Anthony, 271–72n1
Smid, Reinhold, 264–65n11
Smith, Gary, 259n4, 272n3
Socrates, 4
Sombert, Werner, 208
Sorel, Georges, 219
Spiegelberg, Herbert, 260n11
Steiner, Uwe, 259n3, 285n18
Steiner, Rudolf, 62, 267n29
Strauss, Ludwig, 105, 268n40
Strindberg, August, 101

Taminiaux, Jacques, 268–89n44
Tennyson, Alfred Lord, 243
Thomas of Erfurt, 55, 58, 265n22
Tiedemann, Rolf, 116–17, 266–67n27, 277n11, 278n15

Tillich, Paul, 215–16
Trendelenburg, Adolf, 57
Trillhaas, Wolfgang, 260n9

Van Buren, John, 276n34
Vorwerk, Herbert, 215–17, 220–21, 286n28

Wackernagel, Wilhelm, 287n36
Walden, Herwarth, 98
Wassermann, Jakob, 263n14
Watts, Pauline, 286n31
Weber, Max, 208, 216, 286n28
Weber, Samuel, 261n1, 263n15, 265n23, 272n5, 278n14
Weierstraß, Karl, 111, 113, 274n19, 274n21
Weigand, Rudolf, 285n11
Weigel, Sigrid, 268n35
Werner, Karl, 57
Wiesenthal, Lieselotte, 277n6
Wilmans, Friedrich, 263n13
Wittgenstein, Ludwig, 270–71n18, 277n6
Wohlfarth, Irving, 270n15, 288n4
Wohlhold, Hans, 267n29
Wölfflin, Heinrich, 6, 46–47
Wyneken, Gustav, 17

Zeltner, Hermann, 260n9
Zilsel, Edgar, 273n1

MERIDIAN

Crossing Aesthetics

Giorgio Agamben, *Nudities*

Hans Blumenberg, *Care Crosses the River*

Bernard Stiegler, *Taking Care of Youth and the Generations*

Ruth Stein, *For Love of the Father: A Psychoanalytic Study of Religious Terrorism*

Giorgio Agamben, *"What is an Apparatus?" and Other Essays*

Rodolphe Gasché, *Europe, or the Infinite Task: A Study of a Philosophical Concept*

Bernard Stiegler, *Technics and Time, 2: Disorientation*

Bernard Stiegler, *Acting Out*

Susan Bernstein, *Housing Problems: Writing and Architecture in Goethe, Walpole, Freud, and Heidegger*

Martin Hägglund, *Radical Atheism: Derrida and the Time of Life*

Cornelia Vismann, *Files: Law and Media Technology*

Jean-Luc Nancy, *Discourse of the Syncope: Logodaedalus*

Carol Jacobs, *Skirting the Ethical: Sophocles, Plato, Hamann, Sebald, Campion*

Cornelius Castoriadis, *Figures of the Thinkable*

Jacques Derrida, *Psyche: Inventions of the Other*, 2 volumes, edited by Peggy Kamuf and Elizabeth Rottenberg

Mark Sanders, *Ambiguities of Witnessing: Literature and Law in the Time of a Truth Commission*

Sarah Kofman, *Selected Writings*, edited by Thomas Albrecht, with Georgia Albert and Elizabeth Rottenberg

Arendt, Hannah, *Reflections on Literature and Culture*, edited by Susannah Young-ah Gottlieb

Alan Bass, *Interpretation and Difference: The Strangeness of Care*

Jacques Derrida, *H.C. for Life, That Is to Say...*

Ernst Bloch, *Traces*

Elizabeth Rottenberg, *Inheriting the Future: Legacies of Kant, Freud, and Flaubert*

David Michael Kleinberg-Levin, *Gestures of Ethical Life*

Jacques Derrida, *On Touching--Jean-Luc Nancy*

Jacques Derrida, *Rogues: Two Essays on Reason*

Peggy Kamuf, *Book of Addresses*

Giorgio Agamben, *The Time That Remains: A Commentary on the Letter to the Romans*

Jean-Luc Nancy, *Multiple Arts: The Muses II*

Alain Badiou, *Handbook of Inaesthetics*

Jacques Derrida, *Eyes of the University: Right to Philosophy 2*

Maurice Blanchot, *Lautréamont and Sade*

Giorgio Agamben, *The Open: Man and Animal*

Jean Genet, *The Declared Enemy*

Shoshana Felman, *Writing and Madness: (Literature/Philosophy/Psychoanalysis)*

Jean Genet, *Fragments of the Artwork*

Shoshana Felman, *The Scandal of the Speaking Body: Don Juan with J. L. Austin, or Seduction in Two Languages*

Peter Szondi, *Celan Studies*

Neil Hertz, *George Eliot's Pulse*

Maurice Blanchot, *The Book to Come*

Susannah Young-ah Gottlieb, *Regions of Sorrow: Anxiety and Messianism in Hannah Arendt and W. H. Auden*

Jacques Derrida, *Without Alibi*, edited by Peggy Kamuf

Cornelius Castoriadis, *On Plato's 'Statesman'*

Jacques Derrida, *Who's Afraid of Philosophy? Right to Philosophy 1*

Peter Szondi, *An Essay on the Tragic*

Peter Fenves, *Arresting Language: From Leibniz to Benjamin*

Jill Robbins, ed. *Is It Righteous to Be? Interviews with Emmanuel Levinas*

Louis Marin, *Of Representation*

J. Hillis Miller, *Speech Acts in Literature*

Maurice Blanchot, *Faux pas*

Jean-Luc Nancy, *Being Singular Plural*

Maurice Blanchot / Jacques Derrida, *The Instant of My Death / Demeure: Fiction and Testimony*

Niklas Luhmann, *Art as a Social System*

Emmanual Levinas, *God, Death, and Time*

Ernst Bloch, *The Spirit of Utopia*

Giorgio Agamben, *Potentialities: Collected Essays in Philosophy*

Ellen S. Burt, *Poetry's Appeal: French Nineteenth-Century Lyric and the Political Space*

Jacques Derrida, *Adieu to Emmanuel Levinas*

Werner Hamacher, *Premises: Essays on Philosophy and Literature from Kant to Celan*

Aris Fioretos, *The Gray Book*

Deborah Esch, *In the Event: Reading Journalism, Reading Theory*

Winfried Menninghaus, *In Praise of Nonsense: Kant and Bluebeard*

Giorgio Agamben, *The Man Without Content*

Giorgio Agamben, *The End of the Poem: Studies in Poetics*

Theodor W. Adorno, *Sound Figures*

Louis Marin, *Sublime Poussin*

Philippe Lacoue-Labarthe, *Poetry as Experience*

Ernst Bloch, *Literary Essays*

Jacques Derrida, *Resistances of Psychoanalysis*

Marc Froment-Meurice, *That Is to Say: Heidegger's Poetics*

Francis Ponge, *Soap*

Philippe Lacoue-Labarthe, *Typography: Mimesis, Philosophy, Politics*

Giorgio Agamben, *Homo Sacer: Sovereign Power and Bare Life*

Emmanuel Levinas, *Of God Who Comes To Mind*

Bernard Stiegler, *Technics and Time, 1: The Fault of Epimetheus*

Werner Hamacher, *pleroma--Reading in Hegel*

Serge Leclaire, *Psychoanalyzing: On the Order of the Unconscious and the Practice of the Letter*

Serge Leclaire, *A Child Is Being Killed: On Primary Narcissism and the Death Drive*

Sigmund Freud, *Writings on Art and Literature*

Cornelius Castoriadis, *World in Fragments: Writings on Politics, Society, Psychoanalysis, and the Imagination*

Thomas Keenan, *Fables of Responsibility: Aberrations and Predicaments in Ethics and Politics*

Emmanuel Levinas, *Proper Names*

Alexander García Düttmann, *At Odds with AIDS: Thinking and Talking About a Virus*

Maurice Blanchot, *Friendship*

Jean-Luc Nancy, *The Muses*

Massimo Cacciari, *Posthumous People: Vienna at the Turning Point*

David E. Wellbery, *The Specular Moment: Goethe's Early Lyric and the Beginnings of Romanticism*

Edmond Jabès, *The Little Book of Unsuspected Subversion*

Hans-Jost Frey, *Studies in Poetic Discourse: Mallarmé, Baudelaire, Rimbaud, Hölderlin*

Pierre Bourdieu, *The Rules of Art: Genesis and Structure of the Literary Field*

Nicolas Abraham, *Rhythms: On the Work, Translation, and Psychoanalysis*

Jacques Derrida, *On the Name*

David Wills, *Prosthesis*

Maurice Blanchot, *The Work of Fire*

Jacques Derrida, *Points . . . : Interviews, 1974-1994*

J. Hillis Miller, *Topographies*

Philippe Lacoue-Labarthe, *Musica Ficta (Figures of Wagner)*

Jacques Derrida, *Aporias*

Emmanuel Levinas, *Outside the Subject*

Jean-François Lyotard, *Lessons on the Analytic of the Sublime*

Peter Fenves, *"Chatter": Language and History in Kierkegaard*

Jean-Luc Nancy, *The Experience of Freedom*

Jean-Joseph Goux, *Oedipus, Philosopher*

Haun Saussy, *The Problem of a Chinese Aesthetic*

Jean-Luc Nancy, *The Birth to Presence*

CPSIA information can be obtained
at www.ICGtesting.com
Printed in the USA
JSHW041418160522
25977JS00001B/5